Yale University Press books may be purchased in quantity for educational, business, or promotional use. For information, please e-mail sales.press@yale.edu (U.S. office) or sales@yaleup.co.uk (U.K. office).

Set in Sabon type by Keystone Typesetting, Inc.
Printed in the United States of America.

Library of Congress Cataloging-in-Publication Data
Tal, Alon, 1960–
All the trees of the forest : Israel's woodlands from the Bible to the present / Alon Tal.
p. cm. — (Yale agrarian studies series)
Includes bibliographical references and index.
ISBN 978-0-300-18950-6 (hardcover : alk. paper)
1. Forests and forestry — Israel — History. 2. Afforestation — Israel — History.
I. Title. II. Title: Israel's woodlands from the Bible to the present.
III. Series: Yale agrarian studies.
SD345.I75T35 2013
634.9 — dc23
2013012773

A catalogue record for this book is available from the British Library.

This paper meets the requirements of ANSI/NISO Z39.48-1992 (Permanence of Paper).

10 9 8 7 6 5 4 3 2 1

publication of this book is enabled by a grant from
Jewish Federation of Greater Hartford

To my siblings:
Gabriella Tal, who always adored trees, and came to live among them
Aliza Stark, who steadily nourishes our family with her strength, love, and wisdom
Oren Rosenthal, whose creative spirit always enriches and inspires

Contents

Preface

My aim in this book is to document the story of Israel's woodlands in a compelling and accurate fashion. The reader can approach the book as an academic undertaking in Israel's environmental history, with all of the attendant documentation. It can also serve as an international case study, presenting the unique trial-and-error history of the country's afforestation efforts, past and present. There are many lessons for land managers, foresters, and environmentalists in search of insights about ongoing global efforts to keep drylands from becoming deserts and degraded landscapes from remaining wastelands. Like most reliable nonfiction, it seeks validation, perhaps too compulsively, in citations and references.

Although filled with stories—some of them pretty good—*All the Trees of Forest's* primary objective is to inform rather than entertain. Yet, it will soon be clear to readers that despite modest efforts to cloak the writing in third-person, academic pronouns, this is a deeply personal work. Israel's forests mean a lot to me, and I am involved in their "story."

As a child growing up in North Carolina, I saw the woods not just as a shortcut to elementary school but as a doorway into a dominion where imagination could run free among the soft foliage, piney odors, and sweet sounds of the earth. Here, a magical world was beckoning and the real world could wait. Years later I would learn that psychological research shows children's mental acuity is enhanced by visits to the forests. In surveys all over the globe, respondents report that their spiritual well-being is enriched by spending time under the trees' canopy. As a youngster, long before I would marvel at the mechanical wonders of xylem transporting water and minerals from the roots to the leaves, I too was aware that trees were special.

Except for the prevalence of conifers, Israel's forests look very different from those of my childhood. They enjoy less rainfall; people planted them; they have a political meaning. But once I grew accustomed to these trees, they had a charm of their own. It is not coincidental that my Hebrew name, *Alon,* means "oak." These woodlands to me symbolize the potential that humans have for leaving a healthier ecological footprint, reversing the ongoing destruction to our planet.

To ensure objectivity, I have avoided personal versions of the more recent episodes in the history of Israeli forestry. But full disclosure requires that I mention that for more than two decades I have been watching events closely. To some extent *All Trees of the Forest* is an extension of my first book, *Pollution in a Promised Land* and the corresponding Hebrew volume, which reviewed Israel's environmental history. Chapter 3 of this earlier book focuses on the history of the Jewish National Fund (JNF; in Hebrew the Keren Kayemeth L'Yisrael, or KKL), the corporation that oversees Israel's forestry work.

When I wrote that chapter, I was still chairman of a public interest, environmental law organization, the Israel Union for Environmental Defense, or as it is known in Hebrew, Adam Teva V'din. At the time, the organization was suing the JNF for what we believed to be ill-advised planting practices. (The full account of the case can be found in chapter 8 of the present volume.) At some point during the writing process, I came to realize that in the race to litigation I had caricaturized the "respondents," as lawyers in adversarial systems are often wont to do. It became clear to me that Israeli forestry was unlike other environmental matters that had engaged me as an external critic and public interest advocate. The international board that oversees the JNF is democratically elected. I became convinced that if I wanted to influence the management of woodlands and open spaces of Israel, it would be well to roll up my sleeves and take a place at the table.

A fortuitous series of circumstances and modest political maneuvering have allowed me to serve as an elected member of the board of the JNF since 2002. The organization's chairman at that time, Yehiel Leket, was unhappy with my election, as it appeared to him to be a hostile takeover attempt by an old antagonist, so he publicly announced his intention to appeal the election results. A common friend intervened and got us to sit down and talk a little about forests. Eventually I ended up quoting the wisdom of the late American president Lyndon Johnson: "Better to have him on the inside pissing out, than on the outside pissing in." Leket, in turn, decided to let bygones be bygones and took a leap of faith by appointing me chairman of a new organizational subcommittee for sustainable development, charged with drafting new policy documents about forestry practices. We soon became friends. Four years later, I was appointed chairman of the JNF's Committee for Land Development, which oversees implementation of these forestry policies.

From my time working with Israel's forestry program I have learned a great deal. Mostly I learned to appreciate the intelligence and integrity of Israel's foresters. But after ten years on the board, thinking about forestry in Israel, I also realized that many pieces of the puzzle I had managed to put together were still missing. There were important questions for which life-long forest professionals did not yet have answers. While many of these are technical, others fall squarely into the realm of environmental ethics and national priorities. It seemed well for me to take some time off to research, reflect, and write about matters a little more systematically.

All the Trees of the Forest is my attempt to consider the past history of Israel's forests, their current condition, and where they should be in the future. I would like to think that this book will be instructive for those interested in the subject of forestry in general, for those who are concerned about Israel's environment, or simply for people who care about trees. In one sense, this book offered an opportunity to "zoom in" relative to my earlier writing on forests as well as to update. I hope that the presentation is enhanced by the greater access to information I have had as an insider about these issues. Being on sabbatical from my political work has also offered me the space and distance to ponder what the Israeli experience might mean for a planet that needs to do a far better job of conserving its existing forests and planting new ones. In that sense, the book also tries to "zoom out," to see where Israel's experience might be relevant in a global context.

There is always the danger that being too close to a topic can soften or blur a writer's judgment. Aware of such pitfalls, I have tried not to be too nostalgic or excessively forgiving. Even without the pretext of academic analysis, I believe in constructive criticism and "tough love." At the same time, offering a fair account also means giving credit where credit is due. Any failings born of familiarity are compensated, I hope, by the richer nuances that emerge from up-close observations.

Returning to the forests I described in *Pollution in a Promised Land,* it seems to me that, overall, the picture I presented there remains largely intact and accurate. But a great deal has happened in Israel and in the world over the fifteen years since I offered that review. There may be other narratives about trees in Israel, but I believe that this book offers the fullest account of how Israel's inimitable forests actually came to be and what needs to be done if they are to become fully sustainable. There is considerable disinformation in circulation about the history and present ecological dynamics of Israel's forests. I hope that this book offers an accessible and factually precise account of this remarkable story, dispels misconceptions, and contributes to a more fruitful discussion about their future.

While writing this book, I was clearly affected by the research conducted by my hosts at the Stanford Center for Conservation Biology. I came to see my own thinking about forests as narrow and one-dimensional. Forests provide a lot more than just picturesque scenery, pleasant shade for picnics, and an abiding sense of satisfaction at restoring a dusty, degraded landscape to arboreal green. They deliver many of the crucial services that sustain life.

It is also clear to me now that Israel's forests remain a grand experiment—perhaps more precisely, a series of experiments. The methods used to establish and maintain them have been unorthodox and frequently unsystematic. In many cases our foresters have been lucky or blessed with good intuition. Sometimes they were unlucky and the results were unhappy. But ultimately, the success of the ongoing venture will depend on the kinds of questions we continue to ask and our openness to answers and creative solutions. Thus far, few natural resource management initiatives in the world have been as nimble and open to suggestions as Israel's forestry program.

Human evolutionary history started in the trees and it is impossible to think of a future without them. The world's original forestlands, while not as degraded as those in Palestine a century ago, are not in good shape today. It will take a concerted international intervention to stop deforestation and restore the most renewable of our natural resources. But to change present trends and for the woodlands to remain healthy, afforestation will need to be based on science, intelligence, and humility. It is my hope that this story of Israel's forests may provide others with images and ideas on how to bring this about.

Acknowledgments

Many people helped me in the preparation of this book, and I am grateful to them all.

Since joining the board of directors a decade ago, I have learned much from the staff and leadership at the Keren Kayemeth L'Yisrael (JNF) in Israel and am grateful for their support. David Brand, director of forestry, was particularly obliging and positive throughout the duration of this project. Itzik Moshe, over the years, has been a patient and erudite teacher in the field. Omri Bonneh remains an encyclopedia of knowledge and insight. Gershon Avni is as dedicated to the rehabilitation of Israel's forests as anyone in the world and has explained a great deal to me in our work together. Orr Karassin is a close friend and colleague with whom I have been fortunate to work for many years now and from whom I continue to learn. The present and past JNF chairs, Yehiel Leket and Efi Stenzler, have been allies in moving the organization toward sustainability.

An author cannot help but be influenced by his immediate environment. Accordingly, in the United States, the faculty and staff associated with the Stanford University Center for Conservation Biology surely influenced my thinking and the present book. Collectively, they were unhesitatingly generous in their hospitality and assistance. Professor Gretchen Daily, my hostess at Stanford, was unfailingly encouraging. For almost two decades she has been in the front lines of global efforts to better characterize ecosystem services, showing how compatible scholarship and action can be. The weekly meetings of her "Daily lab" and discussion with the dedicated and competent staff of the Natural Capital project helped to sharpen my analysis about many issues that arose while writing this book. Many inquiries found answers during intermittent discussions with Professor Daily, the center's founder Paul Ehrlich, Hal Mooney, along with the excellent associated faculty and graduate students. And I cannot imagine anyone more friendly and helpful than the center's administrator, Janet Elder.

The Stanford University library staff in its sundry branches was consistently accommodating in identifying relevant materials and extending due dates. The Star-Lacks of Palo Alto were kin from the second we landed in Palo Alto; to share a year with Josh and his remarkable family was a blessing that I shall always treasure.

To the team at Yale University Press: Jean Thomson Black, who retains her healthy bias for trees as a Yale School of Forestry graduate, shepherded the book along masterfully; Sara Hoover and Jeff Schier solved every problem expeditiously; Andrew Frisardi, meticulous and smart—a combination not all copy editors enjoy.

Review and editorial suggestions were received from many colleagues. These include: Shmuel Arbel, Gershon Avni, Zvika Avni, David Brand, Omri Bonneh, Greg Bratman, Gretchen C. Daily, Paul Ginsburg, Dan Goor, Rachelle Gould, Pinhas Kahana, Moti Kaplan, Gail Kimbell, Zvi Mendel, Itzik Moshe, Gidi Ne'eman, Yossi Riov, Alon Rothschild, Uriel Safriel, Gabriel Schiller, David Schorr, Bill Slott, Elaine Solowey, and Robyn Tal. Paul Ginsburg offered particularly detailed and useful comments and proofreading in real time when the chapters were in their infancy. Rachel Aharoni, who is working with me on the Hebrew version of this book, contributed greatly to general precision and accuracy in the references. And of course any and all of the remaining errors that they did not manage to flag are surely my own.

My mother, Dr. Yonina Rosenthal, remains by her own admission "my biggest fan and most merciless critic." As in all of my previous books, she found the time to bring her considerable writing skills to early drafts, offering suggestions for innumerable corrections and thoughtful revisions. For this, I am profoundly grateful.

Finally, I would express my gratitude to my family: Mika and Hadas, who valiantly "held the fort" in Israel while we were away; Zoe, who enthusiastically adapted to her new role as California girl, ever cheerful as we came to know the new neighborhood and local sporting events; and most of all Robyn, who keeps our literal and figurative gardens blossoming wherever we are and who is, now as always the most marvelous partner.

All the Trees of the Forest

1

Degradation and Restoration

For there is hope that a tree, once cut down, will sprout again. And that the tender branch will not come to an end. Though its roots may grow old in the soil, and its trunk seems to die in the ground; with the very scent of water, it will bud, and bring forth boughs like a plant.

—Job 14:7–9

Israel is unlike other countries. Not yet seventy years of age, it is at once a developing and a postindustrial nation. The state is a multicultural mishmash. Its citizens have immigrated from ninety nations, and thirty-three languages and dialects are spoken on its streets each day. Israel has the highest per capita concentration of daily newspapers, theater subscriptions, high-tech startups, drip irrigation systems, yeshivas, Facebook profiles, and armored tanks in the world. Relative to its population size, it leads the world in Nobel prizes, but also pays its teachers less and has more severe air pollution than most Western countries. The atypical is typical. Not surprisingly, its forests are also unique.

Israel's woodlands tell an extraordinary story. They carry the scars of relentless military invaders and past conquests. Most of the trees, however, are contemporary—an expression of recent national zeal to restore the woodlands of the Bible and make a harsh climate more hospitable. Many stands still show signs of ecological blunders. These rows of crowded, spindly, sticklike trees serve as reminders of earlier, misguided attempts to create timber plantations on barren, dry soils. Other forests thrive, living proof that human resourcefulness can craft natural successional processes that produce a fetching, vegetative mosaic. Here, the shade from contrasting tree types presides over a rich and blossoming understory that is more garden than thicket.

The foliage also reflects the country's exceptional location at the crossroads of three continents. Trees provide the botanical foundation for the myriad ecosystems that combine to make the country an internationally recognized biodiversity "hotspot." Israeli forests constitute an ongoing experiment in transforming desertified drylands and degraded soils into prosperous parks, rangelands, and ecosystems. They are the product of a forestry agency that has no government status, but is, in actual fact, a public corporation—the Jewish National Fund (JNF; or in Hebrew, Keren Kayemeth L'Yisrael)—which, technically, is owned by the Jewish people around the world. With so many committed stock holders, Israel's modest woodlands constitute an international concern. To make matters more complicated, the trees have been recruited by the region's competing political narratives, becoming part of the larger national conflict that has festered for over a century.

It might seem that Israel's forests are so small as to be irrelevant in the international discourse about global forestry programs. Today, the country's woodlands fill over 8 percent of the country's territory—but then, Israel is a tiny country. All told, its forests amount to little more than one hundred thousand hectares—about 1/60,000, of the wooded area on the planet—seemingly inconsequential.

Yet these trees and their history are worth considering. The chronicle of Israeli afforestation may be highly instructive to a world that seeks to restore its timberlands. Because it is such an extreme case, it offers a distinctive model, for better or for worse. To understand why Israeli woodlands and forestry experience matter, a few words about the global situation are in order.

No natural resource on the planet has been more affected by human activities than the earth's forests. Some eight thousand years ago, at the end of the pre-agricultural age, when there were just five million people on the globe, forests covered 6.2 billion hectares, or 47 percent of the land on the planet.[1] Since then, a great deal has been lost.

Estimates of the actual dimensions of deforestation vary dramatically: *Na-*

tional Geographic quotes one analysis that counts three-quarters of the original forests as missing. Other experts say that no more than 20 percent is lost.[2] The 2005 UN–World Bank Millennium Ecosystem Assessment, with over a thousand participating scientists, is probably the most ambitious and reliable inventory of the planet's resources to date. It calculated that humans have destroyed some 40 percent of the earth's original forests.[3] Modern technology, along with massive population growth during the twentieth century, accelerated the phenomenon dramatically — with half of historic deforestation taking place during the past hundred years.[4] The precise figure ultimately depends on definitions of woodlands, baseline dates, and other arcane classifications. But even accepting the most positive projections, trees have paid a high price for the spectacular proliferation of the human species and the ongoing quest for human prosperity.

Despite the massive devastation, forests still cover 31 percent of the planet's land surface and the rate of destruction appears to be slowing.[5] The UN Food and Agricultural Organization (FAO) estimates that between the years 2000 and 2010 some 5.2 million hectares of forests were cut or burned. This is a staggeringly high figure, but it is 38 percent lower than the 8.3 million hectares lost during the previous decade.[6] Such numbers are almost too enormous for mere mortals to grasp. Suffice it to say that, today, annual deforestation destroys a forested area the size of Costa Rica, while a decade ago there was a Poland-size annual loss.

If there is progress and reason for optimism, it is because many countries have finally come to understand that trees constitute a renewable resource. This has translated into new public policies and the launching of large-scale afforestation efforts.[7] Associated gains, rather than any significant drop in old growth deforestation, are what offer a glimmer of hope.[8] As the world begins to restore at least a portion of its devastated woodlands, questions of quality emerge. Ecological integrity is not only a function of quantity; rather, questions such as the following are germane: "How much land should be planted?" "What we should plant?" "Where we should plant?" "How we should plant?" "What ecosystem services can the new stands provide?" "How should we treat the surviving ecological systems on deforested or degraded lands?" These are not merely technical matters for foresters in the field, natural resource management wonks, economists, and tree huggers: they raise fascinating issues and ethical dilemmas.

Forests provide habitat for much of the world's wildlife. When considering the harmonious relationship between animals and trees, people may imagine deer munching on leaves, birds nesting on boughs, and possums hanging upside-down from branches (even as the last image is actually a myth). In fact, the primary contribution made by trees to wildlife may be indirect, through their nurturing of other smaller plants.

Forests are ecosystems.[9] Diverse natural *homes* are created around the great organic pillars, which set the boundaries and establish the ecosystem character. (As children are often taught in grade school, *eco* actually comes from the Greek *oikos,* or "household.") Plants of every variety grow along the forest floor and even directly on trunks and branches of trees. They typically need less space to thrive than the bulkier trees do. It may take several hectares of land to ensure sufficient space for a sustainable population of trees in the wild, while a plant community might flourish in tiny openings no more than a square meter. This may explain why there are five times more identified plants than tree species.[10] Accordingly, if planned correctly, trees can produce an entire world beyond their immediate bark and leaves.

There is no shortage of explanations for why people are so enamored of trees. There may well be primordial, biological, or even evolutionary connections. Even as humans have come down to earth, most primates still live in them. It is well to remember that if trees had not been there to deliver early *Homo sapiens* their shelter, construction materials, fruits, nuts, boats, and tools, civilization would not have happened.[11]

Although modern humans have the ability to manipulate countless other materials, wood still provides much of the raw materials for buildings, furniture, glues, dyes, and of course the paper on which this book is printed. Half of the commonly prescribed drugs in America come from compounds taken from the trees in the world's forests.[12] Add to this list of products essential ecological services such as pollination, erosion prevention, water purification, pest control, shade, and of course oxygen.[13] There are also intangible benefits.

Nature offers humans a rich menu of landscapes for spiritual and emotional rejuvenation, contemplation, recreation, and reclaiming a sense of wonder. While beaches surely have their admirers, no landscape appears to be as *universally* compelling as forests. Even the odd individual, for whom forests carry no emotional force, can recognize that a comprehensive, rational appraisal should place an inestimably high value on them.

Forests' significance has gained even greater recognition in recent years due to their role in regulating and maintaining the planet's climate. The UN Intergovernmental Panel on Climate Change has emerged as the definitive scientific voice in sorting out the range of controversies regarding global warming. (Its contribution was important enough to earn it the 2007 Nobel Peace Prize.) While its calculations vary, the panel's consensus report estimates that soil and vegetation sequester 2.6 gigatons of carbon dioxide annually. Forests absorb the lion's share of this CO_2.[14] As old growth forests disappear without being replaced, carbon escapes to the atmosphere and terrestrial sequestration drops. Land use changes presently contribute 17 percent to annual global greenhouse gas emissions, with calculations dominated by deforestation in the tropics.

The literature, both popular and academic, associated with global forest dynamics is discouraging. It focuses on land degradation, fragmentation, logging, flora and fauna extinctions, as well as the direct and indirect drivers behind the alarming arboreal losses. But present trends suggest that humanity might be turning a corner and the aggregate increase in forest lands could soon exceed that which is harvested. As the world launches its collective reforestation campaign, it is important to get it right.

The story of Israel's forestry therefore is relevant for land managers, environmental advocates, and policy makers who seek a sustainable strategy for reforestation and afforestation efforts. It is also of interest for ordinary people who care about the future of the planet. During its sixty-five-year history, Israel proved that trend is not destiny by expanding its forested areas more than eightfold. In 1948, the planted stands and remnants of natural woodlands occupied less than 2 percent of the area of the State.[15] By 2005 that figure had increased to some 8.5 percent,[16] and should easily cross the 10 percent mark before stabilizing in a couple of decades. A land that was synonymous with erosion, desertification, and human neglect, is enjoying an environmental makeover.

This exercise in ecological rehabilitation occurred in a country where 97 percent of the ground is classified as "drylands," making it particularly relevant for half of the planet where water will always be scarce. The very alacrity with which Israel's foresters and the politicians who backed them embraced the country's forestry challenges ensured that mistakes would be made. And indeed they were — with economic, political and ecological consequences. But, in retrospect, lessons were learned and new approaches and policies for managing woodlands evolved.

Compared to its neighbors, Israel appears to care a great deal about trees. The UN Food and Agriculture Organization (FAO) reports that 1.5 percent (or 9,000 hectares of the 620,000 total area) of present-day Palestine is forested. The good news there is that the situation is stable: numbers have neither increased nor decreased during the past twenty years, despite huge population surges. Conditions in neighboring Jordan are similar. Forests comprise 1.1 percent or 86,000 hectares of the countryside.[17] About half was planted (largely during the days of the British Mandate), while the other half is defined as naturally regenerated forest.[18] Since 1962, protective regulations have been enacted and logging has been prohibited for decades, leaving the small sample

The face of degradation: Palestine after the Ottoman empire. Less than 2 percent of original woodlands remained, as reflected in this 1939 photograph of the Jerusalem hills. (Zoltan Kluger, KKL-JNF Photo Archives)

of surviving forestlands intact.[19] While falling short of a controlled investigation, the stability in Palestine and Jordan allows them to serve as a quasi-experimental control group. The comparison with Israeli woodlands is instructive. There have been a few negative side effects, but most environmental indicators in Israel confirm that conditions among the "treatment group," with its hundreds of millions of trees, are favorable.

All the Trees of the Forest tells the story of this intervention and offers an account of Israel's forests. The first half of the account falls squarely under the discipline of environmental history. To understand the present policy dilemmas, it is important to grasp the physical and biological characteristics of Israel's newly established woodlands, along with the personalities, institutions, and political forces that produced them.

The second half of the book examines policy dilemmas that arise when engaging in aggressive afforestation. Some of them, like the tensions forestry policies created with Israel's Arab minority, might appear to be idiosyncratic or place specific. In fact, similar postcolonial dynamics exist in many countries around the world where herders' livelihoods depend on finding vegetation wherever they can.[20] Finding a sustainable place in wooded rangelands for

The face of ecological restoration: the Jerusalem forest today. (Yossi Zamir, KKL-JNF Photo Archives)

nomadic shepherds may not be a pressing challenge for forestry services in Europe, but it can make the difference between nourishment and famine for millions of people in the African Sahel. Afforestation in semi-arid regions raises hopes for land restoration across huge areas of Asia, Africa, and the Americas. It also raises complex issues: Should we be planting fast-growing trees on lands that were probably never forested just because we can? What sort of ecosystem service benefits would justify a departure from biological authenticity and diversity? How important is it for forests to be able to regenerate naturally? To what extent should we intervene to prevent forest fires and how? What role should the public have in planning forests and how accessible should they be to visitors? Is there a role for a timber industry in the drylands? And everyone needs to be thinking about carbon sequestration and the role that trees might play in addressing climate change, perhaps the world's paramount ecological challenge.

This book is an attempt to extend Israel's unique approach to some of the universal issues that foresters face. Because of its small size and the frenetic nature of its afforestation program, Israel can serve as a time machine. Readers can "fast-forward" ahead to see where other, hitherto untested dryland forestry programs might be in a few decades. While ever cognizant of the bigger picture, it is well not to lose sight of the details—the actual biological mechanisms and the sociological dynamics, so often overlooked, that make a local forestry program successful or unsuccessful. In trying to "see the forest through the trees," the site-specific lessons and the way ahead for Israel's woodlands remain important. Thanks to the Bible and the Holy Land's status for four religions, much of humanity has been inspired by ancient stories of the land of Israel. As the world is increasingly aware of its own ecological follies and global vulnerability, Israel's forests today can offer valuable lessons. They may even offer hope.

2

Israel's Forests

From the Bible to the British

I have brought low the high trees and raised the shrubs, I have dried out the evergreens and made the dry vegetation flourish.

—Ezekiel 17:24

Characterizing Natural Israeli Woodlands

The first reports of organized political life in Canaan from the Bible include an uncompromising policy directive to conduct massive deforestation operations. Surprisingly, the instructions came from Moses' successor, Joshua Ben Nun.

According to the book of Joshua, upon arriving in the Promised Land, the tribes of Ephraim and Manassah (the two sons of Joseph) were allowed a glimpse of the territories that had been allocated to them. These took the form of undulating, craggy woodlands in the Mediterranean heartland, containing what today is Haifa and its southern environs. The tribes were appalled at the meager amount of arable acreage available for farming, given their sizable populations. But Joshua told them to worry not: "If you are great people, then get up to the wooded country, and cut it down for yourselves—there in the land of the Perizzites and of the giants. If Mount Ephraim is too narrow for you . . . the mountain shall be yours; for it is a forest, and you will cut it down: and its products will be yours" (Joshua 17:14–18). Happy to reestablish productive homesteads in their ancestral homeland, presumably, the Israelites didn't think twice.

Israel's afforestation work fundamentally is an exercise in ecological rehabilitation.[1] To assess the effectiveness of the afforestation program and plan for the future, operational objectives need to be defined: Just what sort of ecology and landscape do we seek to restore? But this preliminary question cannot be answered, without responding to another, ostensibly simple but actually complicated query: what did the land of Israel look like in days of old and what sort of botanical profile did it have?

The answer is complicated because there is probably no place on the planet whose landscape has undergone as many transformations as the land of Israel. Recently found archeological artifacts show prehistoric human cave dwellers in the Qesem caves east of Tel Aviv, as far back as four hundred thousand years ago.[2] Continuous human settlement persevered for many thousands of years. So it is not even clear what the historic basis for comparison should be. No resource reflected the constant vicissitudes in the landscape more than Israel's trees.

The dispute over a valid vegetative baseline is hardly a theoretical, academic debate. In 2012, the Society for Protection of Nature in Israel published a formal critique of present forestry program in the drylands, asserting that Israel's desert landscape remained relatively unchanged during the past *two* thousand years and should not be altered.[3] In its position paper responding, the JNF Forestry Department explained that today's afforestation program seeks to restore the authentic landscape of the Negev as it was *ten* thousand years ago.[4]

In the Bible's description of patriarchal shepherds tending their flocks there were clear environmental implications associated with this intensive human interaction with the land and the woodlands. Scientists postulate that six thousand years ago, goat and sheep husbandry was already sufficiently widespread to affect the local vegetation.[5] Indeed, a recent analysis by agricultural expert No'am Seligman argues that the lion's share of land degradation in Israel due to human mismanagement took place in ancient times: "Clearing the perennial vegetation cover to prepare land for cultivation at the beginning of agriculture conceivably accelerated the rate of erosion and it is possible that soil on the hills was lost already thousands of years ago."[6]

Many ecologists believe that when the pastoral system was in balance, the domesticated herbivores and the flora of the land exhibited a symbiotic relationship. Grazing animals created a healthy "disturbance" that enriched the biodiversity of the land by opening up dense maquis woodlands to sunlight. They also suppressed some species that would otherwise be dominant and overwhelm delicate wildflowers and geophytes.[7] Clearly, there were excesses and the ancients understood the perils of overgrazing and had intuitive notions of "carrying capacity." In Genesis 13:5 we learn that Lot, who was moving about with Abram, also had flocks and herds and tents. But the land could not support them while they stayed together. And so in the first documented case of sustainable stocking limits, they divided up the rangelands.

The ecosystems that emerged from these prolonged interactions were anything but natural. Thus, by the time Joshua, Judges, and Kings ruled the ancient land Israel, the word *wilderness* was long since a misnomer. For forests in the land of Israel, "pristine" as a state synonymous with imperceptible human influence was already a romantic illusion that belonged to the distant past.

Defining a vegetative baseline is further complicated, because like so much in the Middle East, politics tends to sully objective observations and scientific debate. Modern Palestinian nationalistic narratives do not particularly see the erosion and deforestation as notable characteristics of historic Palestine. Rather they focus on the sustainability of Palestinian *fellahin* (peasant) agriculture—and the dignity and beauty broadcast by those olive trees and prickly pear (*sabra*) cactuses that endured centuries of hardship as symbols of the Palestinian people's bonds to their homeland:[8]

> Palestinians draw connections between their ancient presence in Palestine and that of the ancient olive tree rooted in the land of Palestine. The traces of the olive tree in Palestine date back to 8000 B.C. Olives are grown on qattayen ["terraces"], which are hand-built terraces of stacked stones that encircle the mountains and hills. The hills then look like one large staircase. The terraces require constant care and maintenance to continue to hold the soil and prevent land erosion. This setting of the olive tree requires the persistence of traditional methods of farming and maintenance.[9]

Almost from the outset in the early twentieth century, many Arabs came to see afforestation efforts in Palestine as a ruse for real estate appropriation, either by uninvited colonial British rulers or by the Zionist enemy. To acknowledge the extent of the massive deforestation would play into the traditional Zionist perception of the land's vegetative emptiness, with its parallel demographic assumption of a land empty of indigenous people. The pervasive absence of trees reinforced the invisibility of local communities.[10] This has affected the way many Palestinians choose to paint environmental history. For example, there are "learned" Palestinian Web sites dedicated to explaining why reports of nineteenth-century ecological desolation in Palestine are overstated.[11]

And then there are other preconceptions, fueled by internal Israeli environmental politics that make an impartial analysis difficult. After opposing the "pine monocultures" historically favored by Israeli foresters, some environmental advocates today seek "affirmative action" on behalf of other, broadleaf, nonconiferous species that for many years were left out of Israeli forests.[12] When naturalists start with an agenda that favors certain tree species over others to compensate for "historic discrimination," it is hard to imagine a dispassionate biological discourse. Notwithstanding political agendas, natural bias or legitimate opinions about how Israel's forest should look today, there are certain objective facts that should inform any discussion of the subject.

The Biblical Nursery

There is no doubt that the ancient land of Israel was dotted with woodlands. To be sure, Mediterranean climates produce natural stands that are sparser and shorter than the forests of temperate and tropical lands. These "maquis" scrublands are typical biomes in Mediterranean regions. Forests that receive modest precipitation look different than the towering, green "cathedrals" found around the world where rainfall is more abundant. Yet, for the locals these sprawling, shady groves dominated by four meter high oak trees, are no less meaningful and magical.

By the third day trees find their place in the Genesis creation story. Despite the semi-arid conditions, they soon filled up the land of Israel and spread beyond. The Bible has four words for forests—presumably less than an Inuit vocabulary for snow or Arabic nomenclature for camels[13]—but still an important noun: *yaar,* which is the modern Hebrew for "forest," appears more than forty times in the Old Testament and constitutes the predominant term; while *choresh,* the Hebrew for "woodlands," is mentioned only five times. Other terms include "choreshah" (a small grove) or "pardes"—today commonly translated as "orchard."[14] In addition, geographic designations are used synonymously with certain kinds of woodlands. Sharon, the region between the Taninim and the Yarkon stream, also refers to thick oak dominated stands (e.g., 1 Chronicles 5:16 or 27:29);[15] "Lebanon" denotes forests with more towering trees (2 Kings 19:23; Isaiah 37:24 or 60:13); and "Carmel" indicates more of a maquis shrublands (Isaiah 29:17 or 32:15).[16] Moreover, there are dozens of tree species that the Bible specifically mentions.[17]

As biblical travelers did not enjoy access to GPS coordinates or roadside attractions, trees often provided important landmarks, playing a key role in the national mythology of the Israelites: the patriarch Abraham's path in life is defined by the trees he encounters and plants: Genesis 12:6 describes him setting up camp the first day he arrives in Canaan alongside a noteworthy: "oak of Moreh" near Nablus. And it was near another grove of these trees in Mamre that he receives the happy news that his ninety-year-old wife Sarah will make him a father at age one hundred (Genesis 18:1–13). In Genesis 21:33 Abraham commemorates his truce with the rival king Avimelech by planting a tamarisk near Be'er Sheva.[18] Besides providing landmarks, trees of sundry types are cited in lieu of formal gravestones.[19] In perhaps the first recorded real estate transaction in history, Abraham's deal with Efron the Hittite for a familial burial site, includes all the trees surrounding the Machpelah cave.

There are memorable interactions associated with trees throughout the Bible. Moses instructs his spies to size up the land of Canaan and ascertain: "Are there trees therein or not?" (Numbers 13:20). (And of course there were!) Deborah the Judge holds court by her palm tree (Judges 4:5). Second Samuel 18:9 tells how the rebellious Absalom meets his unhappy end, fleeing from his father, King David—trapped "between heaven and earth," when his hair is caught in an impassable thicket of terebinth trees. And everything became complicated in the Garden of Eden because of that irresistible Tree of Knowledge. The conspicuous mention of so many individual trees does create the impression that they were probably lonely forms, on an empty landscape.

The deification of trees was not limited to the Druids of ancient Britain. Apparently, it was common to many early Near Eastern cultures as well, making it totally taboo for the children of Israel. For the ancient Israelites, trees provided an outstanding example of God's handiwork. Yet, they were certainly not to be worshiped. This explains why in Deuteronomy 16:21 Moses forbids the planting of trees anywhere in the vicinity of the altar of the Lord. The Carmel and the Tabor, evidently, were hot spots for pagan rites involving trees. The prophet Amos 9:3 warns that idolaters could not hide among the brush, that the Lord would seek them out. Isaiah 44:14–16 rails against the worship of trees by idolaters. And the legends of Abraham destroying the idols in his father's shop speak of his burning the merchandise, implying that they were wooden rather than stone images.[20]

At the same time, the Israelites constantly attributed human characteristics to trees. The psalmist repeatedly refers to trees' idiosyncratic, humanlike qualities: from the prosperity of species planted alongside a perennial stream (Ps. 1:3), to the righteousness of those who will blossom and flourish like the palm or the cedar (Ps. 92:13), or to trees' propensity to rejoice in jubilant gratitude to their creator (Ps. 96:12).

Much can be learned about biblical perception of tree personalities from the memorable passage from the book of Judges when the trees themselves argue over who would be an appropriate botanical monarch. After all his other siblings were slain by his ambitious brother, Avimelech, Yotam presents the proceedings of deliberations by local trees as a metaphor for his own challenge to his brother's rule:

> The trees went forth to anoint a king over them; and they said to the olive-tree: You should reign over us. But the olive-tree said to them: Should I leave my fatness, seeing that by me they honor God and man, and go to hold sway over the trees? And the trees said to the fig-tree: Come and reign over us. But the fig-tree said to them: Should I leave my sweetness, and my good fruitage, and go to hold sway over the trees? And the trees said unto the grape vine: Come and reign over us. And the grape vine said to them: Should I leave my wine, which cheers God and man, and go to hold sway over the trees? Then said all the trees unto the bramble [*atad,* technicallyboxthorn lycium]: Come and reign over us. And the bramble said to the trees: If in truth you anoint me king over you, then come and take refuge in my shadow; and if not, let fire come out of the bramble, and devour the cedars of Lebanon. (Judges 9:8–15)

Presumably, Yotam is implying that his brother is no different than a bramble bush that offers none of the potential benefits found in the goodly fruit trees. The bramble also constantly threatens to influence life for the worse as combustible brush that accelerates forest fires.

The personality of Israel's trees has continued to engage Scripture enthusiasts throughout the ages, as they integrated horticultural, psychological, and physical sciences in their attempt to better understand the Bible. Based on a series of clever word associations from the Hebrew texts, Old Testament scholar Irving Koller writes: "Trees, like people, can not only be divided according to their trunks, boles, roots and branches, but also by their different dimensions, and lifestyles. You can find both 'righteous and evil species' each with a taste and a smell of its own. . . . The carob is honest and innocent, weak and thin, whose spirit suffers constantly, as it gets by frugally from Sabbath to Sabbath. . . . Or there are righteous trees, whose eyes are turned towards the heavens like the willow or the dates that seek to ascend upward."[21]

Rabbis continued to wax hyperbolic about the astonishing qualities of Israel's trees. This passage from the Talmud, compiled two thousand years ago, is particularly evocative: "Rabbi Hanina said: 'When I came up (to the Holy Land) from exile, I tried to measure the girth of a carob trunk by tying together my own and my father's belt to the stomach of my animal, but they did not suffice.' Rabbi Hanina also said: 'When I came up from exile, I cut down a carob tree and my whole hand was filled with honey.'"[22] "Rabbi Yohanan said: 'I remember [the time] when a child would break open a carob pod and a line of honey would run down over both his arms.'"[23]

Trees were to be treasured as a source of life *and* a source of wonder. Harvesting timber was allowed when required. But fruit trees enjoyed a special status; their wood could only be hewn if it held a higher market value than its produce. Indeed, the same Rabbi Hanina acknowledges in the Talmud that his son Shivahat passed away "for the sole reason that he cut down a fig tree before its time."[24] Eliezer Ben Hercanus, a noted first-century rabbi with an academy based in Lod wrote that, like people, "when a tree producing fruit is cut down, its voice goes out from one end the world to the other, but is not heard."[25]

The Biblical Tree Profile

That the land of Israel sported a rich variety of fruit and forest trees in ancient times is indisputable. The range and composition of these woodlands, however, remains an open debate. Of course there are the usual taxonomic arguments—which start by asking, "What is a tree?" For instance, technically, date palms—common trees in Israel—are not even trees but monocots. They contain no wood per se and grow from the center out, like blades of grass.[26] Disputes over such classifications are only part of the problem. What of the biological authenticity of the names we ascribe to biblical tree?

As the ancients left no dendrology or taxonomic textbooks behind, figuring out just which of today's trees fit the biblical botanical designation is a mighty challenge indeed. Dendroarcheologist and tree maven Professor Nili Lipschitz points out that there are at least thirteen names of trees mentioned in the Bible that we can't correlate with a modern species.[27] And there is no shortage of other trees with regard to which the conventional classification is challenged. For instance, biblical flora maven Nogah HaReuveni was convinced that the *atad* plant had been incorrectly translated as the lowly boxthorn shrub. Trying to make sense of Yotam's previously cited prophesy of resistance against his powerful brother Avimelech, HaReuveni suggests that the *atad,* ultimately crowned as king of the trees, must be a powerful, towering variety, suggesting the jujube (*Ziziphus spina-christi,* or *shizaf* in Hebrew). This is the same tree that became immortalized when it contributed the briars for Jesus' crown of thorns. A relatively new species to the region that appears to have migrated up from Africa's valleys ages ago, the jujube is hardly a prickly shrub: with proper watering it can reach a height of twenty meters.[28]

Another long-standing debate involves the assertion that the Aleppo or

Jerusalem pine is in fact the Tree of Oil (mentioned in the Bible six times) as originally argued by pioneering Israeli botanist, Michael Zohary.[29] Others claim that it is actually a wild olive species.[30] As the biblical narratives span many centuries, it is also possible that plant titles were forgotten or changed, or that new nicknames evolved.

The father of modern Zionism, Theodore Herzl, suffered the embarrassing fate of tree misclassification. On November 28, 1898, during his one trip to Palestine, he planted a symbolic sapling in Motza outside Jerusalem. Herzl was told that he was planting a cedar and cheerfully recorded the event in his diary. In fact it was a cypress. (The tree was soon uprooted by vandals, typically assumed to be Orthodox Jewish political opponents, and not by indignant plant taxonomists.)[31]

Trees evolve and today may be genetically very different than their ancestors of the past. For instance, carbon-14 tests have confirmed that several pits from date palms discovered in caves near Masada had been lying dormant since the Jewish revolt against the Romans in the year 66 A.D.[32] Dr. Elaine Solowey, of the Arava Institute for Environmental Studies, is an accomplished "tree doctor" and researcher, having developed innovative germination techniques. In the botanical equivalent of Jurassic Park, she managed to germinate the world's oldest seed in 2005 — and a male date palm seedling soon sprung up. The leaves are considerably longer than any of today's date species and are genetically distinct. The tree was affectionately dubbed "Methuselah," after the indestructible protagonist from Genesis who lived to a ripe old age of 969. Unfortunately, male date trees do not bear fruit. So Solowey had to coax additional ancient pits back to life, in the hope of finding a female, to discover how dates tasted in the time of the Romans.

Characterizing Historic Forest Distribution

Even assuming that the ancient genetic authenticity of Israel's present tree assemblage could be confirmed with confidence, delineating past distributions and the extent of local forests long gone is a practically impossible task. To begin with, an operational definition for '"forests" is hardly self-evident. One study identified more than eight hundred different definitions, in countries around the world, for forests and wooded areas.[33] Even definitions codified by the United Nations can be inconsistent. On the one hand, in 2008, the UN Food and Agricultural Organization (FAO) defined forests as woodlands that are a minimum of 0.5 hectares in size (one and a quarter acres) or five meters high, with 10 percent crown cover.[34] By contrast, the UN criteria for "forests" eligible as carbon sequestering units under the Climate Change Convention are entirely different. Here, stands must cover no less than one hectare (two and a half acres) with tree crown cover of more than 10–30 percent and a minimum height potential of two to five meters at maturity.[35] In the drylands, a copse will typically be quite different than in the rich temperate and tropical forests that people living in nonarid regions (or those writing regulations pursuant to the Kyoto Protocol) know. Frequently, natural woodlands in semi-arid regions, such as, scrub oak "maquis" areas, neither meet technical UN standards as forests, nor the subjective expectations of Western visitors.

The first systematic effort to consider the historic botanical boundaries that characterized ancient Israel was made by the British officer and surveyor Claude Reignier Conder. He explained that while there had surely been many forested regions, it should not be assumed that this essentially dryland countryside ever contained anything approaching full forest coverage: "The watershed of the country forms the limit of the thickets; the western slopes exposed to the fresh sea-breeze, are covered with shrubs, the eastern are bare and desert; this natural phenomenon is no doubt unchangeable, and a minute examination of the country tends to show that the eastern districts which are now without wood, were also treeless in Bible times."[36]

The most serious effort to reconstruct the boundaries of the historic vegetative cover throughout Israel was undertaken over fifty years ago by geographer Hanah Margalit, as part of her doctoral dissertation. In 1955, Margalit prepared detailed maps (with her Hebrew University colleague Asher Schick) for the first atlas of Israel.[37] Integrating the many travel reports that had survived through the ages, ancient maps, and British aerial photography from the end of World War I, they sketched the landscape that existed in Ottoman Palestine before the modern British-Zionist era. Margalit not only demarcated the few remaining and decimated state forests, but also reconstructed the agriculture and crop profile that existed in centuries past. The nineteenth-century picture that emerges has most human settlements located in the mountains (presumably for security reasons) with the remaining coastal plain and valley areas largely unsettled and denuded of trees.

Yosef Weitz, who for much of the twentieth century oversaw forestry in Israel, spent at least fifty years mulling over the range of the country's historic forest stands. Given his personal vision and zealous mission of Jewish reforestation, Weitz had a tendency to interpret the little existing historical data optimistically. His projections of long disappeared indigenous woodlands were undoubtedly extended to maximal dimensions. In his seminal *Forests and Afforestation in Israel* Weitz claims that the tribe of Ephraim alone occupied an area of, 150,000 hectares (or some 375,000 acres) with tree cover. In the Jerusalem region, he estimates that terebinths naturally covered 40 percent of the hillsides.[38]

Trees evolve in a natural succession, so no "snapshot" can fully represent distribution of tree types over time or faithfully portray the way Israel's woodlands looked at any given moment. Yet, forests ultimately reach equilibrium or a so-called climax state where species composition remains essentially unchanged as long as the trees are undisturbed. For considering the authentic composition of Israel's forests, there is a rich literature about biblical plants based on a variety of historical sources that offer many clues.[39] Weitz relied heavily on botany professor Michael Zohary. Based on the literature and his impressions from the field, Zohary conjectured that there were four identifiable forest types evenly distributed throughout the country: (1) pine forests, (2) deciduous Tabor oak forest, (3) evergreen oak maquis and forests, (4) carob-dominated stands.[40] Because so little was left intact, it would take a more sophisticated combination of palaeoecological, archeological, and historical research to produce a "best guess" of actual historic arboreal distribution. These subsequent assessments show that forests in the past looked very different than most Israeli forests today.

Tel Aviv University professor of palynology and Quaternary geology Aharon Horowitz was among the first to utilize sophisticated chemical and biological laboratory tests on several thousand-year-old pollen samples from the north of Israel. Horowitz identified a mixture of oak, pine, olive, and pistachio trees, but concluded that precise retrospective projections of their distribution over time was practically impossible. Forests had waxed and waned with different climatic conditions and human activities: "These periods are separated by intervals in which arboreal share is considerably lower. The bottoms of these 'lows' occur around 2250 B.C. and again about 950 B.C. after which the arboreal pollen curve remains low until the present day."[41]

For three decades, Tel Aviv University archeologist Nili Lipschitz has conducted research that generates a somewhat higher-resolution picture. Based on her analyses of pollen residues and wood remnants found in excavations, she and her colleagues conclude that oaks were the dominant tree species throughout ancient Israel, during the period spanning 8000 B.C. to 1500 A.D..[42]

In a book that summarizes over eighty of her publications, Lipschitz convincingly refutes Zohary's more evenly distributed projections: "Dendroarchaeological research has proven that the *Quercus calliprinos* (Oak)—*Pistacia palaestina* (Terebinth) association was the only dominant association all over the Mediterranean region since the Pre Pottery Neolithic A period. One of the native elements of this association was *Olea europea* (Olive) but this species was present in small percentages only."[43] The oak's slow but perceptible replacement of abandoned olive orchards in the Galilee and Judea—as well as their domination in closed nature reserves—today seem to support this scenario of local botanical succession.

Based on his observations of their present efficacious regeneration, ecologist Avi Perevolotsky believes that pine trees were also a significant part of the climax assemblage.[44] But his is a minority position.

While these controversies might appear to be the petty hairsplitting of academics, the arguments are highly relevant and resonate loudly in the rough-and-tumble, real-world conflicts over Israeli afforestation policy and conservation.

The Pathology of Deforestation

The land of Canaan was settled extensively long before the Israelite return from Egypt and the rule of the Judges, when archeological evidence of their presence begins. A single formula, therefore, cannot adequately characterize the composition of the "original" tree stands. Human intervention constantly influenced the scenery, and usually not for the better. Warfare was a particularly salient driver of deforestation, and the trees of Palestine paid heavily for a litany of conquerors' rapacity. Accounts of battles throughout the ages suggest that modern abuse of nature—such as the application of Agent Orange by American troops in Vietnam—is only the most recent link of a merciless chain. Vegetative destruction was certainly part of conventional combat strategy in the ancient Near East.

This is borne out by the biblical commandment in Deuteronomy 20:19 that enjoins the Israelites from existing practices of destroying fruit trees in time of war. This precept would eventually be transformed into a broad principle of Jewish law, Bal Tashchit ("do not destroy"), which forbids waste and upon which an entire theology of sustainability is based.[45] But the Bible's attempt to leave the natural world out of the wars of Canaan appears to be the exception to the prevalent violence against the countryside.

The Romans were particularly ruthless in their wartime practices. Before the Roman emperor Titus ultimately succeeded his father to the throne, one of his stepping stones was to squelch the "increasingly annoying" Jewish rebellion. Flavius Josephus, a Jewish officer in the war, became a Roman historian following his capture. Josephus recorded many examples of how deforestation was used as a military tactic. The following account describing Titus's siege of Jerusalem in the year 70 A.D. is typical:

> And now the Romans, although they were greatly distressed in getting together their materials, raised their banks in one and twenty days, after they had cut down all the trees that were in the country that adjoined to the city, and that for ninety furlongs round about, as I have already related. And truly the very view itself of the country was a melancholy thing; for those places which were before adorned with trees and pleasant gardens had now become a desolate country every way, and its trees were all cut down: nor could any foreigner that had formerly seen Judea and the most beautiful suburbs of the city, and now saw it as a desert, but lament and mourn sadly at so great a change: for the war had laid all the signs of beauty quite a change.[46]

Titus also gave orders to "to 'set the surrounding areas on fire' and when all the trees round about the city had been felled in making the original mounds, the soldiers transported fresh timber from places ninety stadia [13.5 kilometers] away."[47]

Trees regenerate, however, and there were clearly periods of renewal. In 670, some four hundred years before the Crusader period, Arculf, a bishop from France, visited the Holy Land and recorded forests in the Hebron region that provided timber to the residents of Jerusalem via camel.[48] Based on his report, it appears that no other woodlands survived in the forty-odd kilometers separating the two cities. Mount Tabor was another of the rare sites where there was a sighting of "thick woods."[49]

When the Crusaders arrived in the eleventh century they found similar woodlands intact. Nineteenth-century German historian Hans Prutz summarizes the countryside described in Crusader chronicles:

> Though Palestine is for the most part a hilly country, the Franks esteemed it highly on account of the fertility of its soil. Today the forests of Syria and Palestine appear sparse in comparison with former times, when most of the hills were wooded. The shores of the Sea of Galilee were covered with trees, the slopes of Mount Tabor were under forest, and the Judean Hills as far as Jerusalem and the Hebron hills, which today are almost bare, then supplied the inhabitants of Jerusalem with all the wood they needed. Around Beirut there were forests of stone pine, of which a relic survives today, the hills around the Banyas were thickly wooded and in general, the slopes of Lebanon were covered with pine forests in which the cedar was more common than it is today.[50]

Pine trees are mentioned only once in the Bible, and in the past there was some argument about whether any species of these conifers was among the natural local mix of trees. Eventually, pollen samples would resolve the debate, showing that pine trees had always been part of the endemic assemblage, long before the Crusaders arrived. Support for the indigenousness of pine trees in general, and for their presence in medieval Palestine, can be found in the writings of al-Muqaddasi, a tenth-century geographer based in Jerusalem. Al-Muqaddasi claimed that "in the province of Palestine there are thirty-six things which together are found only there," and ranked pine nuts (*snoberim*) at the top of this list.[51]

For the most part, travel logs and the artistic impressions from Palestine that survive the first millennium report how few trees there are, rather than how many. Reports of vegetative diminution and environmental desolation became even more pronounced during the so-called Mamluk period, between 1250 and 1516, when a Muslim regime based in Egypt controlled Palestine. Subsequently, deforestation and land degradation continued during the early Ottoman period, from 1516, until the Napoleonic invasion at the end of the eighteenth century. There is a consensus that the Mamluk conquerors, who put an end to Crusader rule, brought a particularly harsh colonial presence to the land of Israel and its hapless inhabitants. Mamluks burned and sacked towns and villages, uprooted orchards, and destroyed wells. Public health problems were rampant. For instance, in 1351, the Black Death was reported in Palestine and by 1500 the estimated population had declined to its lowest post-biblical levels: a mere two hundred thousand people.

A few first-hand reports from the period describe acute ecological decline. For example, there are the letters written by noted rabbi Moses ben Nahman ("the Ramban," or Nahmanides), who spent his last three years, 1267–70, in the region.[52] Probably the best overall geographic description from this period comes from a German visitor, Bernhard von Breydenbach, whose detailed testimonial of his journey to the Holy Land in 1483–84 contains a treasure chest of geographic descriptions and illustrations.[53] Another example is Helen Gertrude Preston's 1903 dissertation, which concludes that initial deforestation was even more accelerated after Crusader rule.[54]

Located in the "triangle" in Israel's heartland on the hills overlooking Wadi Ara, Umm-el-Fahem is one of the oldest Arab communities in Israel, first documented in writing in 1262. The village's name, "the mother of charcoal," reflected the local economic specialization: preparing charcoal from the surrounding forests and groves. Today, as one passes through the offices of the city's art museum, one sees a large image celebrating the local heritage: an archival photograph of trees piled high, waiting to be burned internally to prepare charcoal. Yet, this was hardly a sustainable enterprise. In a Lorax-like myopia, the town quickly extirpated the local woodlands and had to bring wood from greater and greater distances. By the twentieth century, most of the residents had become farmers.

After the Ottoman empire defeated the Mamluk Egyptian regime for control of Palestine in 1516 the situation did not improve appreciably. Grim arboreal reports became more common during the four hundred years that the Turks ruled over what they saw as an essentially peripheral province. The fact that its forests were disappearing probably went unnoticed by decision makers in Constantinople. The British cleric Henry Maundrell provides a discouraging portrait of late seventeenth-century Palestine in his *Journey from Aleppo to Jerusalem,* published in 1697. This work contains several relevant passages bemoaning a disfigured landscape. Reports from the 1783 visit of French traveler Constantin François Volney (who years later, when visiting America, was accused and extradited by the John Adams administration for being a spy) describes the oak forest in Caesarea as the "only existing forest in *Syria.*"[55] The first modern map of the area, prepared by another Frenchman, Pierre Jacotin, drafted during Napoleon's brief invasion. It confirms the existence of a single twenty-by-four-kilometer strip of forest.[56] Little else remained.

There were many drivers behind the massive deforestation during this period, These included unrestricted grazing, the lack of alternative fuel sources, and the abject poverty of local farm communities that were incapable of organizing reforestation efforts.[57] It is hard to identify a clear administrative approach to forestry during the long years of Ottoman rule, except perhaps tax policies that led to the unintended consequence of providing incentives for deforestation. Several accounts from the nineteenth century relate that five hundred trees in Hebron were cut down to evade demands from Ottoman tax collectors. Then there is the case of the lovely apricot trees that adorned the top of the Mount of Olives. Presumably to avoid the associated taxes, the grove was not replanted after locusts leveled it in 1864.[58]

Not all of the ecological loss during this period, however, can be blamed on the Turks. Some scholars argue that Napoleon's 1799 campaign initiated the series of events that led to eradication of the rich woodlands in the Sharon.[59] Seemingly to its credit, a token effort to address the deforestation problem was made in 1860, when the Ottoman administration invited French experts to help them design forestry legislation and an associated government agency. But in practice, little came of the initiative.[60] In 1903 a Forest Law was finally enacted which contained measures to protect existing forests and to create a framework for afforestation. But implementation was negligible and Ottoman officials reportedly were happy to grant cutting permits for a price.[61]

Throughout the nineteenth century, Western visitors were deeply disappointed when they compared the actual state of Palestine to the verdant and even luxuriant imagined images which centuries of European paintings and illustrations had fostered.[62] Because of his inimitable, jocular style and long-term popularity, Samuel Clemens — otherwise known as author Mark Twain — did more than any visitor to engrave into public consciousness an indelible picture of a desolate Holy Land during the twilight of the Ottoman rule. Sent to the region by a San Francisco newspaper, Twain serialized the travel logs from his 1867 tour of Palestine. His generally gloomy descriptions of a neglected landscape were then immortalized in the best-selling book *The Innocents Abroad*. Twain's descriptions of a desolate Holy Land were consistent with the Zionist political narrative which perceived Palestine as unkempt and forgotten and are still frequently recounted in Israeli histories. Twain describes a land with "no timber of any consequence — none at all to waste upon fires" — and he offers vivid descriptions of almost total deforestation as he crisscrosses the country. After boarding the boat home in the Jaffa Port he summarizes his impressions of Palestine:

> Of all the lands there are for dismal scenery, I think Palestine must be the prince. The hills are barren, they are dull of color, they are unpicturesque in shape. The valleys are unsightly deserts fringed with a feeble vegetation that has an expression about it of being sorrowful and despondent. The Dead Sea and the Sea of Galilee sleep in the midst of a vast stretch of hill and plain wherein the eye rests upon no pleasant tint, no striking object, no soft picture dreaming in a purple haze or mottled with the shadows of the clouds. . . . It is a hopeless, dreary, heart-broken land. . . . Palestine is desolate and unlovely. And why should it be otherwise?[63]

George Marsh, although largely forgotten today, was much more famous than Mark Twain when he visited Palestine in 1864. At age sixty-three Marsh was a Dartmouth graduate, a certified attorney, a former U.S. congressman, a past American ambassador to the Ottoman empire and Greece, and an accomplished linguist. He was also passionate about reducing human impact on the earth and highly erudite in his presentation. Frequently called "the first environmentalist" because of these positions, Marsh argued convincingly in favor of conservation and was among the original geographers to identify the process of desertification. After his visit, Marsh wrote a natural history book, *Man and Nature*. Among other matters in this volume, he considers Palestine's condition and whether the poor fertility of the soils was a result of climatic changes or of land mismanagement by the local peasant communities.

> So long as the cisterns were in good order, and the terraces kept up, the fertility of Palestine was unsurpassed, but when misgovernment and foreign and internecine war occasioned the neglect or destruction of these works — traces of which still meet the traveler's eye at every step — when the reservoirs were broken and the terrace walls had fallen down, there was no longer water for irrigation in summer, the rains of winter soon washed away most of the thin layer of earth upon the rocks, and Palestine was reduced almost to the condition of a desert.[64]

Some argue that Marsh's and Twain's glimpses of Palestine's degraded landscape were heavily influenced by the seasonal extremes in Israel.[65] By August or September, after six months without rain, the countryside invariably looks desiccated and forlorn. But there was nothing seasonable about the conspicuous absence of trees. Even though their descriptions of eradicated forests in the 1860s surely reflect the prevailing arboreal reality at the start of the twentieth century, there were some areas in the country where trees already fared better. Nascent efforts at afforestation began in the 1880s starting with the arrival of a small group of Templars from Germany, who soon set up a tidy European neighborhood in Haifa. Members of this peripheral sect of German Protestants were unceremoniously excommunicated by a Lutheran church that did not take kindly to their excessive enthusiasm for the Holy Land and its ancient prophets. Some two thousand Templars would reach Palestine as pilgrims and were soon dispersed in a network of six colonies and neighborhoods. There they remained in somewhat isolated German enclaves until their ultimate expulsion by the British Mandate as hostile aliens in World War II. (Pro-Nazi sentiment was not unknown, even among second- and third-generation Palestinian Templars.) From the start this group proved that, with great patience and determination, it is possible to make trees flourish again in the land of Israel. They began by planting cypress and pine trees along the main streets of their Haifa and Jerusalem colonies.[66] After failed attempts to gain land concessions from Ottoman officials, the Templars purchased a few privately owned plots from real estate owners on the Carmel and set about establishing several groves of Italian stone pines on its mountains.[67]

The first trees of the Zionist pioneers soon followed. This group came from Europe at the end of the nineteenth century with green aspirations and an ideology committed to returning forests to the desolate homeland. In 1897, the great Jewish philanthropist Edmond de Rothschild had several hundred thousand eucalyptus saplings delivered to the new settlers in Hadera, primarily to assist with their swamp-draining activities. (Certain species of eucalyptus trees are still used for drying marshy areas with high water tables by consuming copious quantities of water, which then evaporates through the leaves.)

Ongoing assistance by Rothschild's agents in establishing new Jewish agricultural villages typically also included support for planting ornamental trees and groves, along with fruit plantations.[68] Subsequent forestry work in the small Jewish community of Ottoman Palestine was primarily initiated by the Jewish National Fund (JNF), whose first forests were characterized by considerable trial and even more error. These efforts, along with a few isolated tree-planting initiatives by the Ottoman regime in Be'er Sheva and Jaffa,[69] did not amount to much, and most of the trees in the JNF pilot plots were abandoned and died during World War I. In short, afforestation efforts during the Ottoman empire were imperceptible in terms of the overall vegetative profile and in retrospect had little symbolic significance.

Aaron Aaronsohn, a brilliant agronomist, established a research station in Palestine during the final days of Ottoman rule and evaluated the success of sundry international tree species under local conditions. It is interesting to note his lack of enthusiasm for conifers, primarily due to concerns about their flammability. In 1913 Aaronsohn prepared a strategy for "renovating" the forests that he submitted to the governor of the Beirut district with a series of proposed policy reforms. Among his major concerns was the scourge of overgrazing. Aaronsohn cited American research showing that a hundred goats could wipe out seven and a half hectares of vegetation within a year. He called for an end to nomadic grazing with stock limits placed on local herds as well as levies on livestock and tax-free status for afforested lands. He also recommended a cessation of all charcoal exports to Egypt and a ban on harvesting trees for heat during the flowering season. Although apparently commissioned, there are no signs that Aaronsohn's report was ever seriously considered.[70]

Nevertheless, at the end of the nineteenth century several natural forest stands remained. Between 1871 and 1877 two British military engineers, Lieutenants Claude Reignier Conder and Horatio Herbert Kitchener, traversed Ottoman Palestine on behalf of the Palestine Exploration Fund. Their observations were compiled in a four-volume, annotated compendium that describes the local geography and archeology in fastidious detail. The report confirms the sparse vegetation, with a few notable exceptions: "The whole of Carmel is now wild and uncultivated except, around two villages, 'Esfia and el Dalieh.' The mountain consists of hard grey limestone, covered more or less thickly with brush wood. On the watershed is a thin layer of chalk and stunted pines are found on the very top of the mountain. The red soil of the slopes appears to be very rich as the wild growth, which is remarkably luxuriant, indicates."[71]

The German geographer Leo Anderlind's description of his visit to the Carmel from 1884 also contains a rich account of terebinth, arbutus, carob, laurel, and of course, oak trees. "In the center and higher parts of the mountain, the forest is tall containing a considerable quantity of Aleppo pine over sizeable areas. In certain places, about three kilometers south east of the Carmelite monastery, the dominant trees are 10–12 year old pines. . . . In a deep ravine six kilometers southeast of the Carmelites I counted out one hundred pines growing in natural scrub. These were between twenty and thirty-five years old, standing seven meters high and measuring about thirty-five cm in girth."[72]

Such appraisals are clearly supported by Conder's detailed maps, which offer a reliable confirmation of extremely modest, nineteenth-century tree cover. His report expounds on this point: "It must not, however, be supposed that Palestine is entirely devoid of woods. A thick forest of oak extends between Carmel and Nazareth with underwood below the trees in parts. An open woodland occurs on the low hills south of Carmel and in the northern part of the plain of Sharon remains of the mighty wood of Strabo and the Forest of Assur of the twelfth century; from these oaks according to some schools, Sharon takes its name and they form some of the prettiest scenery of the Holy Land."[73]

Even these bedraggled arboreal survivors could not withstand the exigencies of the Great War and the hasty construction of a railway to the Turkish army's southern front. Responding to exhortations from their German allies, in 1915, the Ottoman command sought to seize the offensive and capture the strategic Suez Canal. In order to transport the necessary troops, a railway infrastructure was critical. An entire line, stretching from Lebanon to the southern border town of Be'er Sheva, was built in an astonishingly brief nine month period, completed in May 1916. To run the supply lines as well as provide fuel to the massive machinations of modern warfare, the Ottoman regime indiscriminately helped itself to any local trees it could reasonably harvest, with little regard for property rights or ecological uniqueness. According to the retroactive British forestry reports, 60 percent of the country's economically viable olive trees were used for rail fuel.[74] A recent history of Palestine summarizes the results: "Nothing was spared. Oaks, cedars and olive trees disappeared, not to mention the beautiful variety offered by the natural forests. This was an act of annihilation that undermined the livelihoods of many Palestinians. The destruction was in vain. Jamal's armies were defeated. And the railway lines were used two years later by the British expeditionary force coming from Egypt under the command of General Allenby."[75]

By 1918, except for miniscule patches of woods in the inaccessible Carmel and Mount Meron, precious little was left of the forests of Israel. Zionist settlers in the Sejira village (in the Lower Galilee on the way to Mount Tabor) planted hundreds of hectares of forest at the start of the century. Its fate was typical of Ottoman Palestine's few new woodlands: The Jewish settlers painstakingly maintained and managed the forest between 1901 and 1913. When firewood became a valuable commodity during World War I, neighboring Arab villagers began to indulge in illicit logging. Years later, Lova Schechter, a Sejira farmer, described a traumatic incident. Her husband, who was on guard duty, chanced upon dozens of Arabs, mostly women from the neighboring villages, in the thick of the forest. In a well-planned operation, they had come with their axes and saws to fell the trees and take them home on the backs of their waiting mules and donkeys. Her husband charged the villagers on his horse and they left in a panic. Subsequently, security around the forest was expanded. But it was a losing battle. The forest was doomed by its proximity to the Afula train station, a major stop on the Damascus-Cairo line. By order of Turkish authorities it was clear cut to provide fuel for the military trains.[76]

In a recent academic review of the environmental record of the Ottoman era, Hebrew University geographers Ruth Kark and Noam Levin summarize the steady deforestation that took place across the Ottoman empire:

> Human population increase and the development of agricultural and built-up areas, usually resulted in the loss of forest cover. Vast forest areas once covered the Middle East, the majority of which disappeared as a result of damage by man and animals. (Lebanon, Jordan and Iraq are good examples of the disappearance of forests, due to human activity.) A similar situation prevailed in Palestine until the beginning of the modern era. The main reasons for the destruction of forests in Palestine were Bedouin invasions and uncontrolled grazing, anthropogenic fires, and the cutting of trees for wood coal, timber and lime kilns.[77]

Kark and Levin find an analysis by Zionist agronomist Arthur Rupin from 1918 reliable enough to quote: "The destruction of forests has been going on steadily, especially in the coastal regions, the vicinity of cities, and wherever good roads or railways permit the transportation of lumber . . . but the worst enemy of forests are the herds of sheep and goats."

Arboreal Collapse

Fluctuations in tree composition and distribution were the norm in the land of Israel, whose woodlands for thousands of years reflected the exigencies and myopia of its rulers and residents. Even so, ancient Israel never boasted the abundant, green forests of northern Europe or South America. With an oak and terebinth climax baseline, human activity quickly came to influence the ancient landscape of Israel, producing a dynamic patchwork quilt of forests, fruit orchards, fields, and rangelands. Regrettably, as time went on, any semblance of a healthy balance was lost. Trees were forfeited, apparently without much regret, to intermittent violent military campaigns, chronic overgrazing, and the farms of the local populations.[78]

In 2006, UCLA ecologist Jared Diamond published the best seller *Collapse*, in which he deciphers the conditions that led to the collapse of societies throughout history. Rather than citing acute cataclysmic events such as smallpox or wars, in case after case he blames chronic environmental disequilibrium for the human disasters. Based on diverse cases from Easter Island to the Anazi Indians of the American southwest, from the Norse colonies in Greenland to the Mayans, Diamond lists eight environmental factors—including unsustainable hunting and fishing, soil erosion, and overpopulation—that led to the sudden disappearances of entire civilizations. But the primary phenomenon that set societies on the road to ruin was deforestation.

Considering this factor in the sudden extinction of the human settlement on Easter Island, Diamond writes: "The reason for Easter's unusually severe degree of deforestation isn't that those seemingly nice people really were unusually bad or improvident. . . . They had the misfortune to be living in one of the most fragile environments, at the highest risk for deforestation, of any Pacific people."[79]

The land of Israel is primarily comprised of drylands that are not particularly robust and ultimately are highly vulnerable to human abuse. Consistent with the well-documented desertification syndrome, it follows that the tremendous poverty, erosion, depopulation, poor agricultural yields, and low life expectancy that characterized Palestine during the Ottoman reign were also among the misfortunes produced by deforestation.

The new technological capabilities of the twentieth century, coupled with the stormy upheaval of a world at war, left the remaining forest stands of Palestine decimated. Decimation actually connotes 10 percent losses, as the word comes from the practice in the Roman army where officers would remove or "decimate" one-tenth of mutinous soldiers. To refer to the decimation of the forests of Palestine, therefore, is a gross understatement. By the end of Ottoman rule, a liberal estimate suggested that some thirty-seven thousand hectares of natural woodlands remained—less than 2 percent of the land in Ottoman Palestine.[80] Even if one removes the deserts, where forests could never have existed from the calculation, over 98 percent of the original, natural woodlands had vanished.

In our own time, the Internet offers history enthusiasts easy access to remarkable photographic collections from the early twentieth century in Palestine. The black-and-white images of triumphant World War I Allied soldiers, mysterious tombs of prophets, or classic Jerusalem vistas are invariably charming, but they all present the same bleak picture of a land without trees.[81] The series of aerial photographs taken by British planes prior to their invasion of Palestine in 1917 confirm this ailing diagnosis more systematically.

The land of Israel was poorer without the many ecosystem services that its forests had once provided local residents at no cost. The loss of tree cover left the country susceptible to soil erosion and fertility loss, degrading habitats and scarring landscapes that had once inspired prophets and pilgrims.

During the thirty years following World War I, these trends would change. The sustained government afforestation programs of the British Mandate and Zionist settlers would redefine the way subsequent generations perceive the countryside. They also established a system of forestry that still aspires to restore the woodlands of the Bible, whatever that may have been.

3

A Mandate for Trees

The king gave silver like pebbles, and made the cedars as plentiful as the sycamores of the lowlands.

— 1 Kings 10:27

The Unfulfilled Promise of British Saplings

"When General Allenby's army swept over Palestine, in a campaign as brilliant and decisive as any recorded in history, it occupied a country exhausted by war. The population had been depleted; the people of the towns were in severe distress; much cultivated land was left untilled; the stocks of cattle and horses had fallen to a low ebb; the woodlands, always scanty, had almost disappeared."[1] Thus opens Herbert Samuel's 1921 inaugural report as high commissioner and commander-in-chief of Palestine, only a year after assuming his position as the "first Jewish leader of Palestine in 2,000 years." It would take another year for the League of Nations to formally approve the British Mandate for Palestine that remained in force until 1948.

As World War I wound down, France and England haggled in the small

Italian town of San Remo over the geographical spoils of the victors. From the conquered Ottoman lands, the French took Syria, while the British empire negotiated Iraq for itself along with "southern Syria" as a separate entity: Palestine. Under Article 22, the League of Nations recognized the legality of mandatories for territories "not yet able to stand by themselves under the strenuous conditions of the modern world . . . until such time as they are able to stand alone."[2] Great Britain divided its Palestinian holdings into two separate legal realms: an autonomous kingdom of Trans-Jordan, run by the Hashemite royal family to the east of the Jordan River and a British-run mandate from the Jordan to the Mediterranean. The British government had already announced in the 1917 "Balfour declaration" that it "viewed with favour the establishment in Palestine of a national home for the Jewish people." For the Liberal Party–led coalition, Samuel was a logical choice to take the helm.

Born to Jewish aristocracy in Liverpool, Samuel was an Oxford graduate who at age thirty-two would serve in Parliament. Subsequently, as a member of the Liberal Party, he held three separate government ministerial portfolios before receiving the high commissioner post. Herbert Samuel was also an open and unapologetic Zionist, making the appointment controversial.[3] He arrived in Palestine with a clear agenda that included economic development and ecological restoration for an abandoned Holy Land. Trees played a prominent role in this vision for the future. In his first report as commissioner he explains:

> It is obvious to every passing traveler, and well-known to every European resident, that the country was before the War, and is now, undeveloped and under-populated. The methods of agriculture are, for the most part, primitive; the area of land now cultivated could yield a far greater product. The summits and slopes of the hills are admirably suited to the growth of trees, but there are no forests. Miles of sand dunes that could be redeemed are untouched, a danger by their encroachment to the neighbouring tillage.[4]

Such proclamations were more than mere lip service. Afforestation was one of the Mandate's top priorities. Among the very first pieces of legislation passed by the incoming civil administration was a 1920 Ordinance for the Regulation of Forest Lands and the Protection of Trees.[5] The first operational provision of the ordinance declared that all forests in Palestine, except those on private property, were under the protection, control, and management of the Mandatory government. Section 6 allowed the high commissioner, if he deemed it advisable in the public interest, to authorize "forest officers" to enter into possession of, protect, control, and manage forest lands on private property.[6] Armed with the requisite legal powers, Samuel proceeded apace with his plan. Within a year of arriving he could proudly report: "Forest areas are being

Mandate connections circa 1934: Herbert Samuel, Britain's first high commissioner in Palestine (second from right, with mustache); with Menahem Ussishkin, chair of JNF (with beard, second from left); on far right is Abraham Granot, who succeeded him, and on the far left is Dr. Akiva Ettinger, an agronomist who founded the JNF Forestry Department. The men convene on the rooftop of JNF offices in Jerusalem. (Yosef Shweig, KKL-JNF Photo Archives)

demarcated and a staff of forest rangers and guards has been appointed. The destructive felling of the few remaining trees in the country has been stopped; forest nurseries have been established, and some hundreds of thousands of trees have been planted by the Government or by private landowners. Such are the first beginnings of a process which should add largely to the productiveness of Palestine."[7]

It was a truly energetic start. For a new political authority still consolidating military victory, forestry could easily have remained a low-priority program. The official zeal is particularly noteworthy as, unlike most other colonial forestry programs of the time, economic exploitation for timber exports was not a salient motivation.[8] The British Mandate's initial alacrity to confront historic deforestation trends created expectations for a "green revolution" in Palestine. Political support was uncompromising for the small but smart cadre of professional British foresters who shared a solemn ideological commitment for the task and enjoyed virtually unlimited legal authority to move forward.

The three decades of Mandate rule should have been the best of times for Israel's forest regeneration. In fact the results were minimal.

The area afforested by the British Mandate government during its thirty-year regime reached 5,400 hectares. This was roughly 0.2 percent of the lands in Palestine or 0.5 percent of its nondesert regions. (Even these modest numbers dwarf the 1,288 hectares planted by the parallel Zionist efforts at the time by the JNF and some 6,000 hectares on lands owned by the Rothschild family's Palestine Jewish Colonization Association.)[9] By way of comparison, in 1951, during its second year of independence, a penniless Israeli government initiated planting on 5,640 hectares, more than the area planted during the entire Mandate period.[10] Nevertheless, sixty years later, the Mandate forestry program still resonates deeply among researchers, ecologists, and local Israeli foresters. Apparently, it is not just the quantity of trees but the quality of the stands that matters—even if in retrospect, it was a symbolic effort.

Any evaluation of forestry policy under the British Mandate, therefore, must start with an explanation for the disappointing performance. Mentioned time and again in the government's own internal reports of the period, the obstacles that encumbered successful policy implementation are straightforward and easily summarized. It was not just steep, rocky, and eroded lands that resisted British forestry efforts. The local Arab population also did not welcome the trees. This hostile reception translated into chronic disregard for grazing restraints, intermittent vandalism, arson, and even violence against forestry officials. Moreover, the perplexing land registration dynamics, the political ambivalence that soon reigned among senior decision makers in London and Jerusalem, World War II, along with inadequate program funding and professional personnel, constituted a recipe for failure. To some extent, the disappointing results can also be attributed to the lofty aspirations that the British foresters set for themselves from the outset. Even under ideal conditions, their ambitious afforestation objectives would have been hard to attain.

But there remains "a half-cup" that the Mandate's forest program undeniably filled. An extensive system of forest reserves was established initially by the agricultural ministry's Forestry Service. After 1936, its successor, an independent Forests Department, created sanctuaries which, for the very first time, provided woodlands with a modicum of protection for regeneration, free of human insults and harmful interventions. The importance of this precedent is often overlooked. Notwithstanding conflicting orientations and personalities, the influence of the British Mandate on Israel's subsequent afforestation methods is considerable. The British forestry program and the knowledge it produced left its mark on the foresters who inherited the reserves.

This chapter describes the efforts made by British foresters: their aspirations, their policies, and ultimately their frustrations. A great deal about their forestry program is known today, as the British were meticulous in their documentation and left behind a rich variety of reports.[11] Several excellent retrospective reviews of the Mandate forestry legacy have been published;[12] they provide far more detail than is necessary to understand the effect that the period has had on modern forestry. Had the Mandate forestry program enjoyed greater political stability, its impact on the land of Israel would undoubtedly have been far more profound. Nonetheless, millions of the trees in today's forests were planted during the Mandate. So too were the seeds of Israel's ambitious afforestation initiative that would ultimately transform the local landscape.

British Afforestation Ideology

For the British, afforestation was first and foremost an economic development initiative. To be sure, like High Commissioner Samuel, many Mandate officials were imbued with a sense of higher purpose in restoring the Promised Land to its former botanical glory. Chief among them were Mandate officials who oversaw the government bureaucracy: the chief forest officer, F. J. Tear, and the subsequent "conservator(s) of the forest," Gilbert Sale and Amihud Goor. Professionals with considerable understanding of the technical details of planting, they all shared an unspoken, ideological impulse or metaphysical reverence for the sacred nature of their commission. Coming from a fecund island of gardens and horticulture, many British bureaucrats had an affinity for green, shaded landscapes. A belief that forests had the power to make harsh, unforgiving, Middle Eastern climates more accommodating was prevalent,[13] and had been for some time.[14] From an official policy perspective, trees primarily were perceived as having instrumental value in attaining three essential objectives highlighted in the new ordinance: *soil conservation, wood supply, and pasturage.*[15]

These goals were ultimately codified in 1946 as a formal mission statement, "the Forest Policy of the Government": (a) "to re-afforest the hilly and waste non-agricultural land in order to conserve water supplies, to prevent soil erosion and denudation and to afford shelter and protection to adjoining crops and orchards; (b) to curtail the encroachment of sand dunes; and (c) to bring into economic use land not suited to agriculture and horticulture by the production of timber, fuel and other forest produce."[16]

In 1920 it seemed that, for the foreseeable future, Palestine would remain an agrarian economy and that it was hemorrhaging soil. The data were not particularly precise, but according to internal British statistics the equivalent of two thousand hectares of agricultural soil were washed off Palestine's lands each year into the Mediterranean. The eastern side of Judea's watershed contributed another million tons that flowed toward the Jordan Valley.[17] The severity of the problem was confirmed by visiting experts who documented blue waters of the Mediterranean turning brown during rain storms. One external estimate projected that the soil of Palestine was a full meter poorer than it was in days of old, when terraces were well maintained and tillage sustainable.[18]

Housed in the Department of Agriculture,[19] and infused with its soil conservation perspective, the new Forestry Service believed that there was "no economic factor in a predominantly agricultural country which is of greater importance than the productivity of the soil." Yields of Palestinian fellaheen's subsistence farms were already notoriously low.[20] If soil loss continued, the long—term implications for agricultural productivity appeared disastrous.

The inundation of local settlements and farms by the coastal sand dunes was one of the rare environmental problems of sufficient severity to engage the previous Ottoman administration. One assessment estimated that between 1870 and 1930, sand dunes along the coast moved inland at a rate of four to six meters per year.[21] The Ottoman administration even paid for Jewish afforestation projects to slow it.[22] The British saw the sand drift phenomenon as just another piece of the deforestation pathology, enacting a Sand Drift Ordinance in 1922, which relied almost entirely on tree planting to prevent the chronic economic damages. The fundamental policy orientation was one that today would be classified as regenerating "ecosystem services."

Erosion not only contributed to poverty among the Arab farmers; it also created safety hazards such as seasonal landslides that accompanied winter storms on the steep hills around Tiberias.[23] In 1933, the resulting destruction and bedlam caused the death of thirty-six people, with massive damage to homes, citrus groves, and fields.[24] Deposition of gravel and dirt clogged stream beds and made roads impassable. Forestry experience at the time suggested that if the erosive hillsides were covered with woodlands (or other vegetation), the flow of discharge could be slowed in times of flooding and the soil retained. Vegetative cover provided the additional benefit of capturing rainwater, which could then percolate into underlying aquifers. The British experts claimed that terracing combined with trees would create new natural streams and springs.[25]

Sand dune stabilization was deemed even more critical to maintaining basic safety and supporting agricultural production infrastructure. Between 1917 and 1925 the government had to hire scores of workers to remove the million tons of sand that covered the tracks on the coastal railway line to Gaza.[26] The government realized that the right kind of trees could thrive on sand dunes and restrain them, protecting critical transportation arteries; Australian acacia, it turned out, were extremely adept at doing just that.[27]

The conventional wisdom among foresters held that if a 15 percent tree coverage were established, it would transform the livelihoods of the Palestinian peasant community, especially in erosive hilly regions.[28] While the 15 percent figure was written into enough reports to become a benchmark, not all senior administrative officials in the Colonial Office in London embraced such ambitious afforestation efforts. By the Mandate's end, however, the goal had actually increased, with the head of the department calling for at least 20 percent of the country to be set aside for afforestation as appropriate for a Mediterranean region. (Italy had 18 percent of its land under forest at the time and Spain 14 percent.)[29] Regardless of final planting targets, the Mandate brought with it a consensus that planting trees was central to Palestine's new lease on life and future prosperity.

The Mandate Plants

No sooner had the British conquered the southern districts of Palestine, than its occupying military government issued a ban on cutting or uprooting carob and pistachio trees. Protection for olive trees was soon to follow.[30] The cannons were still warm from the war, but the military government was already putting people to work: between 1918 and 1919 some 370,000 trees were planted. The following year the number swelled to two million. While most never reached maturity, the mix of trees selected remains of great interest today.

Gershon Avni is among Israel's most senior foresters. He spent forty years at the JNF, rising through the ranks to eventually oversee forestry and land development in Israel. One of the first "academics" to be hired in the JNF Forestry Department, he brought the critical thinking and analytical rigor of an agronomist to the job when he started his forestry career in 1973. As he grew intimately familiar with the forests in Israel's heartland, Avni found himself comparing the old stands planted by the British Mandate with the more recent Israeli forests and was consistently impressed with the British results.

> I looked at several areas, and saw a rich variety of species. In the Sharon, for example, I'd find stone pine, tamarisk, terebinth, oaks of every type. The maps were in my hand and told me that they had been planted by the British. I asked around if anyone was left from that period and they told me about an

old timer named Amihud Goor—who still lived near the Habimah Theater in Tel Aviv. It turned out that Goor had actually been in charge of the British Mandate's Forests Department. I tracked down his books, where he wrote about the afforestation work he had done in Africa and said to myself: I have to meet this guy.[31]

Avni had already found himself in something of a professional disagreement with his bosses and even the scientific staff in the JNF Forestry Department. During his first day at the organization, as part of his orientation, he accompanied the senior in-house forestry expert, Dr. Moshe Kolar in his rounds. Kolar was a devout advocate of dense pine plantings—and Avni had the audacity to ask him why:

"We plant them densely so they'll grow up straight and won't have side branches; this prevents the formation of knots that lowers the quality of the timber," Kolar explained.

"But what do you actually do with the wood," asked Avni ingenuously.

"Well as a matter of fact, they're used for production of particle board so they are broken down," he acknowledged.

"So why does it matter whether the wood has branches or eyes or not?" retorted the young upstart.

"From then on—I was on his black list," Avni recalls with a smile.[32]

Meeting Goor was an encounter with an entirely different kind of scientist. Like so many men of the period, the dramatic global events affecting the Jewish people during the twentieth century contributed to a fascinating biography. Born in 1898 as Amihud Grasovsky, he was immediately immersed in the mechanics of biology, growing up in Mikveh Israel, Palestine's first agricultural high school, where his father Yehudah was a teacher. (Fifty years later he would shorten and hebraize his name.) Goor served as a sergeant in the Jewish Brigade in World War I, after which he pursued scientific training at the world's best universities, receiving an undergraduate diploma at Oxford and a master's degree at the University of California, Berkeley, before going on to the Yale University School of Forestry. His doctorate was ultimately published by the university in a 1929 monograph entitled *Some Aspects of Life in the Forest.* In this work, Goor presents findings from experiments conducted on white pine, red pine, hemlock, red oak, and chestnut oak under varying light intensities. Surprisingly, this proved *not* to be a dominant factor in forest reproduction.[33]

By then he was already drawing praise as a forest officer in Palestine's still incipient Forestry Service, where he would work for twenty years, frequently representing the service at international conferences of foresters. With a sister married to Herbert Samuel's son Edwin, the Grasovksy clan was one of the

Chief forest conservator for the Mandate and then head of Israel's Forestry Department, Amihud Goor, with seedlings in the Ilanot research station. (Photo courtesy Alon Nathan [Goor's grandson])

few local Jewish families that mingled freely with the Mandate's colonial elite. Despite this social status, Goor's promotion in 1946 from assistant conservator of the forest to the head of the agency was highly irregular, given the pervasive distrust among British officials of "the natives" and their obsessive desire to display neutrality between the feuding Jewish and Arab communities. Only when his predecessor, the indefatigable Gilbert Sale, was caught in a highly publicized sex scandal, and there was no time to find a replacement, did Goor receive the promotion as "acting conservator."[34]

By then his professional acumen and competence were beyond reproach. Nonetheless, Goor's political allegiances were clear: he is credited with providing aid to the Jewish paramilitary Haganah, which frequently conducted its clandestine maneuvers under the cover of the forest.[35] Although distrustful of politics and politicians, Goor served briefly and gallantly as a military governor of Jerusalem during the 1948 Arab siege. For his interventions to save a convent from starvation in this capacity, he was later awarded a medal of honor by the leading international forestry organization.[36]

Thirty years afterward, the young forester Gershon Avni knocked on his door:

> Goor must have been well into his eighties then, as he lived to a ripe old age. He was already in a wheel chair, and a bit imperious and I remember being a bit surprised at the way he gave out orders to the maid. (The maid, clearly fond of him, calmed me down because she saw I was uncomfortable.) We soon got to talking and he was simply generations ahead of us in his thinking. Goor kept emphasizing that the first thing we had to do was to make sure that tree species were compatible with their habitat. I had been thinking the same thing for some time, and he completely reinforced the ideas that I had been

> afraid to raise. He was such an expert, and had done so much that his opinions truly gave me confidence.[37]

Goor had more than twenty years to see the results of his theories in practice. Of all the Mandate's senior forestry officials, he was by far the best-trained scientifically and the most able to develop and articulate the scientific assumptions implicit in the Mandate's approach to afforestation. While still heading Israel's government Forestry Department, Goor periodically served as an expert consultant to the Food and Agriculture Organization of the United Nations, which eventually commissioned him to write a book. Printed in 1955, *Tree Planting Practices for Arid Areas* was for many years considered the definitive work in the field. The manual can actually be seen as the collected wisdom of the British Mandate Forests Department. It prescribes in precise detail the different stages of a successful dryland forestry program. From designing nurseries ("Half a million potted plants or about 1,000,000 plants in tin cans should be grown per hectare, but only about 250,000 naked-rooted plants in rows can be raised in the same area") to specifying work days per hectare required for cleaning land, preparation of soil, planting, hoeing and weeding, and creating firelines.[38] In the opening section Goor expresses the theoretical approach to indigenousness and species selection which was at the heart of British planting policies:

> Restoration to original forest types may, therefore have to pass through intermediate stages, using pioneer or nurse crops, often of xerophelious bush species before high forest species can be successfully planted. Yet even under these conditions, some of these denuded areas can be reforested; but to do so, certain fundamental phytogeographical facts must be ascertained It should be possible from the remaining vestiges of vegetation to reconstruct a picture of the original forests, and this could serve as a guide in choosing species to be planted. Indigenous trees, planted properly, are the best assurance of success for afforestation.[39]

Recognizing the imminent end of British rule, as the last forest curator of the British Mandate, Goor submitted an "expanded" final report in 1947 that well reflects this perspective. In addition to presenting countless data tables enumerating nursery saplings raised and dunams of land planted, he includes a section called "Management of Natural Forest" that reflects a healthy balance between preservation and afforestation. Here Goor describes the "dense maquis" that survive on five or six government forest reserves containing a variety of nonconiferous species such as oak and terebinth.[40] But Jerusalem pine (*Pinus halepensis*) was also planted where the surface rock is a soft chalky limestone. Goor explains: "Most of the areas of natural forest are so damaged that any form of management other than simple protection is out of the question." Yet, he describes a few areas where natural growth was "filled up by planting or sowing conifers."

Anything but an ecological purist, Goor in his arboreal gospel sees a role for nonindigenous trees. He draws on international examples: "With proper judgment, the introduction of exotics should be tried. Monterey pine (*Pinus radiata*) does better in New Zealand than in its native habitat, California and several of the eucalypts grow faster in California than in their natural habitat on the continent of Australia. . . . The likelihood of success under such different conditions can, however, be estimated for different species."[41]

In the second appendix to his final report, a "List of Trees and Shrubs Native to Palestine," Goor lists some eighty species that attained "a height of over one meter under natural conditions."[42] Most of these were actually planted at some point in the new forests. At the same time, they did not think twice about introducing new "exotic" species, some of which caused considerable ecological mischief.[43] For example, the report includes extensive photographs of saplings planted to arrest sand dune drift in Gaza, revealing a variety of tamarix, pines, and acacia. The latter, an Australian import called the blue wattle (*Acacia saligna*) was a favorite of British foresters for stabilizing dunes. Unfortunately, this tree has emerged as a pernicious invasive species, dominating coastal landscapes and upsetting the natural ecological balance to this day.

The annual reports submitted dutifully by the Forest Service and then by the Forestry Department testify to the consistent commitment to tree diversity. Take, for example, the 1934 description of the seeds and plants used at the fourteen active government nurseries: "Local collection has been developed considerably in comparison with previous years and amounted to 404 kilogrammes of some 39 species. There were not yet enough developed pine trees for collection, and so pine seeds were purchased from Syria for direct sowing."[44]

The planting combinations in each Forest Reserve are documented—each with a unique mix of species, reflecting the geographical diversity of the land. The 1935 catalog of planting is representative: At the thirty-two declared Forest Reserves, hundreds of thousands of saplings were planted, with distribution evenly divided between such local and exotic species as cypress, carobs, Chinese sumac, eucalyptus, and Australian acacias. Aleppo (Jerusalem) pine trees were planted abundantly, but rarely comprise more than 30 percent of seedlings at a given site. In some reserves they were not planted at all.[45]

Haim Blass was one of the few full-time foresters working for the JNF during the Mandate period, and he eventually came to manage its central region after Israeli independence. The JNF worked closely with the Mandate Forests Department in those days, enjoying its assistance with logistics, and its

A new science of dryland forestry emerges from Palestine. Albert Einstein plants a tree in the Galilee (1923). (KKL-JNF Photo Archives)

provision of saplings and superior professional training. But the JNF did not always adopt the official way of doing things. Blass explains one of the fundamental differences in style: "The British had a perspective that sought to preserve the landscape. They opposed clearing off the land and planting everything from scratch. We figured that if you cleared off all the natural trees (except maybe for carobs) you'd have a much easier time planting the pines and the cypress after that. The British left much of the existing vegetation — bushes and shrubs and so the new seedlings couldn't get all the water they needed to survive. Mandate foresters might have to go back and replant three or four times. And it also meant that you had trees of varying ages in the same stands. There were definite disadvantages to our approach, but we felt that ultimately it was more efficient."[46]

In absolute terms, the annual budget of the Forestry Service between 1922 and 1927 was paltry — averaging seventeen thousand pounds annually.[47] (This was only 5 percent of the annual *surplus* that the Herbert Samuel government was proud to send back to the British treasury — based solely on internal Palestinian revenues.)[48] Over time and with inflation the budget increased, but never exceeded thirty thousand pounds in actual expenditures, which amounted to 0.45 percent of the Mandate government's budget.[49]

In relative terms, it can be argued that the Forest Service budget was not trivial. The entire budget of the Department of Agriculture in 1930 was only £77,054; tree protection and planting received more than 25 percent of its funds.[50] Nonetheless, Mandate foresters were quick to compare their meager resources with those of neighbors in Cyprus, whose forestry service had been around since 1889. Even though the island's land area was only a third the size of Palestine's, it had more than three times the forested area and its forestry budget was roughly double that of the Mandate's. Cyprus also employed twice the field staff, with many more trained managers.[51] One report acknowledged that there were no personnel in Palestine trained in entomology and so problems associated with pests and insects were rarely noticed or treated by forest rangers.[52] But even with limited resources, Palestine's Forest Service managed to conduct a full scientific program of field experiments and to develop a sophisticated and environmentally sensitive approach to its "re-afforestation" program.

The Forest Reserve Paradox

The British forestry strategy relied on two primary initiatives for attaining its ambitious afforestation objectives: planting forests and protecting them. The former was more costly and with its limited budget, afforestation proceeded at a modest pace—ultimately averaging only two hundred hectares of successfully forested land a year. Planting needed to be supplemented by a strong regulatory infrastructure of conservation. Without protection, young saplings would be eaten and trampled by herds of goats and cattle in a classic tragedy-of-the-commons dynamic.

The British foresters felt that there was almost a total absence of local conservation culture or monitoring capacity for silviculture. In 1920, justifying their recommended takeover of village forests, Forest Service officials wrote: "Much of this oak forest is found on common village land, and though one of two villages has appointed ghaffirs (junior policemen) to look after their trees, this system has not worked very well."[53] They recommended that the Mandate's forestry laws not only empower the government to protect trees on public lands, but also take private lands under its protection and pay owners the profits from associated government sales of timber and charcoal—after deducting management expenses.[54] The high commissioner was amenable and a law was hurriedly prepared and enacted.

Like the Cypriot legislation on which it was modeled, the Mandate's 1920 Woods and Forest Ordinance based its afforestation work on establishing forest reserves on uncultivated lands.[55] The Mandate's forest reserves were never intended to be the untouched refuges for flora and fauna that they are envisioned today. In actual fact, during the 1920s there were precious few trees left to preserve in Palestine. Nor were the reserves seen as idyllic habitats where evolution could proceed undisturbed by human intervention. Forest reserves, first and foremost, were places for people to plant trees without interference.[56]

Once declared, new woodlands were *not* public commons, although they were still open to the public. The lands did not confer *usufruct* or "utilization of property rights" except for pruned branches or dead and dry firewood that could be freely removed.[57] Unless a permit was acquired, an entire litany of activities was prohibited on forest reserve lands:

- removing forest products (including logs, charcoal, sap or resin, tree oils, weeds, vines, thatch, leaves, fruit, seeds, roots, bark . . . plants; turf, soil, or minerals);
- animal grazing or even passing through the forest;
- uprooting or burning trees, removing bark, or damaging them in any other way;
- building or living in the forest; and
- burning grass or starting fires without taking measures to prevent their spread.[58]

Even previous users of the forest needed to attain a timber license and permits for grazing.[59] Later regulations would go as far as prohibiting cigarette smoking from mid-March until November during the dry season.

The motivation of the Mandate Forest Service was to allow for robust forests to take hold—forests that would ultimately provide grazing rangelands and wood supplies to local populations. Planting trees was expensive and made no sense at all if grazing could not be controlled during the formative years when saplings need to establish themselves. But it was hard to convey this long-term perspective to the fellaheen and Bedouin beneficiaries in the surrounding communities, many of whom relied on the woodlands for heating, lumber, and livestock fodder.

A 1930 census by the veterinary service showed that there were 965,877 grazing animals in Palestine, almost half of which were goats.[60] The Forest Service was authorized to accommodate the local pastoralists, but never licensed more than 70,000 animals,[61] less than 8 percent of this number. This set the stage for confrontation. Ironically, the measures designed to protect the young saplings often spawned the hostility that led to their destruction.

In practice, only public lands that were completely unsuited for future agricultural cultivation due to their slope or soil type were selected and declared in the government's *Palestine Gazette* as lands designated to become forest reserves. This, of course, was easier said than done. Despite the fanfare, the 1920 law proved flawed and required revisions. Many large tracts were "gazetted" and proposed as forest reserves; in practice, very few were formally established.

For a forest reserve declaration to be finalized, a slow process of demarcation was required. Land surveyors had to set the formal boundaries for the reserves. With only two on staff, there was a perennial shortage of surveyors and the annual reports gripe about their being commandeered by other agricultural projects.[62] When the measurements were finally made, declaration became an impossibly protracted procedure, as local Arab residents were quick to claim large swaths of these undeveloped lands. "Encroachments," usually took place at night and were difficult to monitor.[63] These took the form of building huts, planting trees, or just leaving signs of cultivation to create prima facie evidence of previous presence and right to title. Section 11 of the ordinance required that oral objections submitted by the public to reserve declarations be heard by the district court. The state was left with the burden of proof to show that contested lands in fact had *not* been utilized.

The associated measurements and litigation required of the Forest Service paralyzed afforestation efforts. The British foresters became extremely frustrated at what they saw as the "multitude of claims and groundless assertions of ownership."[64] They eventually realized that the process of forest declaration would have to be part of a comprehensive "cadastral" system of land registration, in which land settlement officers determined property rights, encouraged development, and established borders between village and public lands. But this was time-consuming and proved overwhelming. Six years into the process, the foresters were losing most of the bureaucratic battles their own ordinance had created. The orderly procedures of the British legal system were ineffective when they came up against the pervasive lawlessness of the Levant. It can surely be argued that Arab pastoralists were simply operating under an alternative normative system and that the forces at play in Palestine were no different than trespassing dynamics in public lands around the world. In either case, the results for afforestation were the same. Years later, the Mandate government described the implications of the initial policy failures: "By 1926, it became clear that if no protective measures were taken, the vegetation would all disappear long before the arrival of the Settlement Officers in the hills. So a new system of forest reserves was established."[65]

The amended Forest Ordinance of 1926 was designed to remove the obstacles to declaring reserves. Many provisions from the initial 1920 ordinance were left unchanged. The amendments were designed to expedite public ownership of forests and encourage private landowners to both embrace conservation and afforestation as individual land use strategies. This posed an interesting philosophical dilemma. Tel Aviv University legal scholar David Schorr explains that despite the prevailing commitment to free markets and the wisdom of the "invisible hand" that guided capitalist economies of the empire at the time, colonial foresters were well aware of market failures and externalities which interfered with optimal social outcomes and led to the ruin of the commons. "Colonial officials generally believed in the superiority and necessity of state control over forest lands as a condition for their successful management on a sustainable basis, whether for preventing desiccation or for commercial timber harbor."[66] Zionists also applauded a process which would better define property rights across Palestine, creating opportunities to purchase land and establish new settlements.

Yet there were many colonial officials in London who for practical and ideological reasons were not enamored of the government getting into the timber business.[67] The underlings of the secretary of state for the colonies already understood the perils of "top-down" policies, asserting that "excessive State interference destroys initiative and sense of responsibilities, causes a reaction and may produce results opposite of what was intended."[68] They also feared that continued government intervention would alienate already hostile Arab communities. After taking the concerns of their London handlers into consideration, the local foresters in Palestine successfully lobbied for the upgraded legislation.

The new 1926 Forest Ordinance clipped the preexisting public standing granted to antagonistic land owners that was so easily used to stall and scuttle declarations on government-owned lands. It also allowed the taking of private property for protected forest reserves when the high commissioner felt it justified by the public interest (i.e., to ensure the supply of produce to contiguous villages, prevent interruption of water supply or damage to neighboring agricultural lands).[69] Moreover, when he felt that forested lands were especially vulnerable, the high commissioner could declare reserves "closed" and withhold all licenses for grazing or felling.[70] With the new streamlined procedures for reserve declarations in place, the Mandatory Forest Service had the authority to carry out its grand design. Between 1926 and 1928, 6,440 hectares of reserves were "gazetted."[71]

When the League of Nations met in the summer of 1929, reports from Mandatory powers were on its agenda. One of the questions put to Palestine's high commissioner, John Chancellor, involved afforestation progress. Chancellor admitted that the problem of land ownership remained vexing, explaining that the government was in a "difficult position as it had no clear title to any considerable area of land suitable for afforestation. . . . That being so, it had adopted the policy of declaring certain lands to which it had some title to be forest reserves." The high commissioner announced his intention to "shortly declare as a forest reserve about one million dunams which would ultimately be planted or regenerated." He added optimistically that "the bare slopes of the Judean Hills were responding to conservation measures in the most extraordinary way; trees planted on rocky slopes grew freely, especially the Aleppo pine, which was indigenous."[72]

So, planting trees and forest reserve declaration continued hand in hand. Foresters insisted that "the methods used in demarcating Forest Reserves were peculiarly mild and flexible and Forest Officers had obtained the approval of the villagers to the boundaries of the New Reserves."[73] That, at least, was how the Mandate saw it. The following excerpts from the 1935 Forest Service report describes the activities associated with just two of the six reserves declared that year, Meiroun and Jabal Sasha:

> Meiroun, Serial No. 235, area L 1,964 dunams, situated in the Safad subdistrict, east of Megido villages. Although not precipitous, much of the soil has been denuded, leaving perhaps fifty percent or more of rock outcrop. The land has been continuously overgrazed and very little woody vegetation remains. The intention is to plant soil patches and pockets and to restore natural vegetation by protection. The high rainfall in this locality should render possible the growing of timber. . . .
>
> Jabal Sasha Serial No. 237, area: 1,907 dunams, situated on the Nazareth hills on steep rocky land with scant natural vegetation. This has been taken up for reafforestation for the production of timber, firewood, charcoal and other produce and also to arrest soil erosion and to conserve rainfall.[74]

Afforestation advocates enjoyed a temporary sense of progress. Nobody was more energetic about expanding the reserves than Gilbert Sale, who in 1936 succeeded Tear as curator of the forests. A professional forester from England and a devout Christian, he was also an avid Zionist. Until an inappropriate romantic liaison with a Jewish woman led to his transfer to Trinidad, he was unfaltering in lobbying on behalf of his department's interests.[75]

Even so, the gap between declared protected lands and the situation in the field remained enormous. Due to rampant noncompliance with grazing restrictions, officials acknowledged frankly that vegetation inside open forest reserves seldom was noticeably better than outside, in the "wastelands" over which the department had no control. Top officials had always complained of insufficient and unskilled manpower. (To manage closed reserves effectively, "one woodman" was needed for each fifty hectares of closed forest, a level of staffing that proved unrealistic given budgetary constraints.)[76] Ultimately, personnel was not the reason why British forestry was failing. The real problem was that British forestry policies were singularly unpopular among Arab denizens. As Jewish land acquisitions increased, the issue of displaced Arab fellaheen became politically charged.[77] Actual Zionist land holdings were still modest and invariably exaggerated. Typically, they were limited to "marginal" wetlands or steep rocky areas that had never been cultivated. Even as late as 1948, Jewish lands never exceeded 6 percent of the territory in the Palestine mandate. The problem of fellaheen unemployment was real enough, however, largely driven by the exponential growth in the Arab population, which went from 375,000 in 1922 to 734,000 in 1944. To meet growing needs, subdivision led to subdivision among rural families, and arable land was in increasingly short supply.[78]

Many Arabs looked to the surrounding open spaces for a solution. The 1931 census indicated that only 3 percent of Palestinian Arabs made their living solely from herding.[79] Still, fellaheen families often relied on livestock as a supplementary source of protein. For pastoralists seeking pasture for their flocks, these rangelands, albeit highly eroded, offered the only available biomass. Once afforestation began, reserves could be out of commission for as much as ten years. Naturally, shepherds chose to circumvent or simply ignore the new prohibitions.

There were also speculators who simply saw an opportunity to establish land claims. An official report from the Forestry Service succinctly described the phenomenon: "The local attitude towards land tenure is such that a cultivator is considered to have a very good claim to the land on which he works, unless other persons or the state can conclusively prove that they have a prior and better title. This being so, the rapid clearing and ploughing of suitable plots within the Forest Reserves has become a common practice, the offenders then pleading that they have cultivated there for generations and thereafter being very hard to dislodge."[80]

From the local peasantry's perspective, it was not clear how a program designed to nurture economic development could usurp herders' historic rangelands. Forest reserve declarations might make sense in the rational British world of animal husbandry, with its optimization schedules and long-term time horizons. (Indeed, today such policies are called "sustainable development.") But for many Arabs, it posed an existential threat to their livelihoods.

From the perspective of the Mandate, the Forest Ordinance offered a generous deal for local communities. The government was committed to massive tree planting, the ultimate beneficiaries being Palestine's rural populations. Private lands with fifty trees or more received an exemption from property tax.[81] The rights of fellaheen to forest products near their homes—including firewood, timber, and pasture—were recognized even as they were regulated.

There was also an expectation of quid pro quo. Local village *mukhtar*s (community leaders) were required by law to cooperate in sundry activities, ranging from processing villagers' requests for timber to helping map forest boundaries.[82] In the coastal area they were responsible for supplying the labor of all males over fourteen years old, for a maximum of six days, to address the problem of sand dune reclamation work. In the event of forest fires, villagers within a five-kilometer radius were expected to join fire-fighting efforts in reserves. Unlicensed goats, sheep, and cattle were to be distanced from newly planted forest reserves or owners would face fines. Additionally, a growing list of tree species were protected, in and outside of reserves, including oak, olive, carob, and terebinth.

The local Arab population remained unconvinced. After a decade of efforts, the British grew resigned to Arab indifference to its forestry agenda. After praising Jewish afforestation, a 1931 official report concluded: "The Arabs on the other hand have no interest in the forest and they are only concerned with olives and other fruit trees."[83]

Failing at persuasion, the British turned to coercion. By its own admission, Forest Service enforcement of grazing, squatting, and felling proscriptions in reserves was not effective, but neither was it nonexistent. The 1935 Forest Service report happily announced: "The number of persons prosecuted is the highest recorded since the Forest Service was formed and has risen from 1,460 last year to 1,832. Fines amount to L.P. 896,060. The increase may be ascribed to financial stringency, poor harvest and grazing due to deficient rainfall, land encroachments following the rise in land values and doubled in part to the increased vigilance of the forest wardens."[84] For the hapless shepherd, whose goats and sheep had nothing else to eat, government forests became the object of hostility. Such confrontations were hardly unique at the time, nor have they ceased to characterize the dynamics facing conservation agencies in developing countries.[85]

It is clear from reports that Forest Service officials had no illusions about their popularity and realized they were failing in the arena of Arab public opinion.[86] In retrospect, their most important challenge was neither psychological nor horticultural.[87] Rather, they needed to create meaningful *economic* benefits and market these opportunities to Palestine's Arab community. The establishment of a timber industry, charcoal production, collection of pruned kindling for heating and produce of fruit and olive trees became a priority. Palestine produced almost none of its own wood, and lumber absorbed considerable foreign currency. In 1936, timber imports came to 1,344,000 Palestine pounds, money that presumably might have been earned by domestic enterprises and local loggers. In its annual reports, the service dutifully documented its efforts to increase timber yields and promote miscellaneous forest products. By 1934, a list of thirty-two different wood products had been compiled including: fruit cases, veneer, charcoal, and even tanning materials.[88] But Palestine was ultimately a dryland country, where towering hardwoods would never thrive. Most of the new trees were still too young for harvesting and profits were modest.

Planting trees did serve as a public works program that primarily benefited Palestine's Arab population. There were some cases, such as the planting of the Sha'ar HaGai (Bab el-Wad) forest in the late 1920s where unemployed Jews received temporary work with the Forest Service.[89] Laborers received the modest remuneration of 175 mil per dunam.[90] (The British were also careful to have the rival ethnic groups work in separate teams; Arab workers planted trees at the same time at nearby at Wadi Kof.)[91] The number of job openings, however, was too small to make a dent in the ever burgeoning numbers of frustrated, unemployed peasants.

Money was found for public relations campaigns, replete with brochures containing expensive photographs. A nonprofit organization, Men of the Trees in Palestine, was established by Richard St. Barbe Baker to galvanize support. As early as 1920, a national Arbor Day was declared to commemorate trees and their many blessings (set on Tu Bishvat, the traditional Jewish birthday for trees).[92] But the Arab residents of Palestine weren't celebrating.

Thus, the paradox of the British Mandate and its forest reserves: Without protection from humans and their livestock, afforestation didn't stand a chance. Yet, because of economic and political dynamics, reserves, once declared, were deeply resented by Palestine's majority.

A key institutional upgrade took place in April 1936, when years of lobbying by Gilbert Sale finally paid off. The Forest Service was transformed from its marginal position in the Agriculture Department into an autonomous Department of Forests, with independent authorities and budget, reporting directly to the high commissioner. The move was accompanied by an increase in funds for ongoing activities, including research. Sale became department head and the first "Conservator of the Forest."[93] He is remembered fondly by Israeli foresters as a dedicated public servant who was sympathetic to Jewish national and silvicultural aspirations. Years after the Mandate, JNF forester during this period, Yerahmiel Kaplan, coincidentally studied at Oxford with Sale's son, who confirmed the family's warm memories of Palestine.[94]

Sale brought his missionary zeal to the tasks ahead, quickly dividing the country into a northern and southern division, with nine ranges. By then his staff had grown to three hundred forest officers. He was happy to supplement his team with local expertise, utilizing entomologists from Hebrew University in an attempt to address pest infestations.[95] For long-suffering forest advocates, it was a promising moment when some institutional "muscle" could finally apply fifteen years of field experience on a more meaningful scale. But this was not to be.

April 1936 was also the month when years of discontent and simmering nationalistic hostilities reached a boiling point. An angry Arab revolt spilled out of the Mandate's troubled ethnic melting pot. In retrospect, this signaled the beginning of the end for the Mandate. It was also a significant setback for afforestation plans and Palestine's fragile forests, shattering any illusions that British restoration project could reach meaningful dimensions.

Revolt

In April 1936, Arabs terrorists forced themselves onto a bus and robbed the passengers in order to "fund their revolution," killing the two Jewish drivers. Jewish retribution came swiftly: the next night, two Arabs were murdered. A match was lit and the explosion was immediate. Usually divided by tribal loyalties, an Arab High Command came together, dominated by "hard-liners," most notably by the mufti of Jerusalem, Haj Amin al-Hussein. Ostensibly, there was a nonviolent component to "the 1936 revolt" that called for a general strike by the Arab population against Jewish businesses. It lasted until October. Arab demands reflected a consensus regarding their national struggle: a ban on Jewish immigration to Palestine, a ban on transfer of Arab-owned land to Jews, and a new, more representative government.[96]

The revolt is remembered for the viciousness of its violence: The initial Arab rampage included an all-out attack on Jewish property, uprooting of crops, and the murder of eighty Jewish civilians. Later, thousands of Arabs in irregular militias waged a guerrilla war that targeted Jewish and British assets. By the time the conflict subsided in 1939, 415 Jewish civilians had been killed. Initially, the British authorities were caught by surprise, unprepared for the ferocity of the uprising. They ineffectually attempted to impose curfews in cities and to restore order.[97] Eventually, they realized that they had a full-fledged revolution on their hand and brought in the necessary reinforcements to squelch it. When two divisions (twenty-five thousand British troops) took off the proverbial kid gloves the results were not pretty.[98] By British estimates three thousand Arabs lost their lives — a third due to internal score settling, with two thousand dying in combat or hanged by the Mandate government. Arab figures speak of 19,792 casualties, over 5,000 dead, and 14,760 wounded during the three-year conflict.

The forests of the Mandate were not spared the mayhem; trees, as always, offered a relatively easy target. As anarchy raged throughout Palestine, the understaffed Mandate foresters were helpless to stop the vandalism, sabotage, and arson. In some cases, battles between British forces and rebels took place inside forests and produced uncontrolled conflagrations.[99] Although most of the forest officers were Arab, they were affiliated with the Mandate. Along with their family members, they became targets of armed attacks.[100] The forestry *Annual Reports* were discontinued until 1940, when the conflict subsided. Military censors were still heavy-handed, and the composite 1936–39 report understates the magnitude of the trauma: "As the conditions in the country grew worse, control in the rural areas was loosened. . . . Various stations were destroyed by armed bands and serious damage to the Growing Stock was done in the neighbouring plantations. Besides wanton damage to buildings and vegetation, there were widespread thefts of poles and fuel, and illicit grazing throughout the Closed Forest Reserves. Only in a few areas in specially favourable situations, was effective protection possible."[101]

During the first year of unrest, official accounts report openly that arson had destroyed 100,000 trees in government forests along with 55,000 trees on JNF lands. That figure was ultimately updated to 275,000 trees burned on three hundred hectares.[102] Actual losses were far greater, with many forest stands never really recovering from what were reported as only partial losses. The British Mandate's official 1938 account to the League of Nations gives a few specific examples of just two months of damage: "In the Northern District disorders in the towns decreased and activity tended to concentrate in the countryside. The disturbances took three principal forms: (i) Malicious destruction.—During the months of May and June, 30,000 trees were destroyed by forest fires. In the Haifa rural district, damage amounting to £P.4,000 was done to fruit orchards . . ."[103]

For Sale and his newly appointed Forests Department it was a particularly disheartening time, helplessly watching much of their painstaking efforts go up in smoke. Sale called for an enforcement wing within the Forestry Department that would cooperate with the attorney general and land officials in combating illegal activity. But with such pervasive chaos and bloodshed, no one had time for the trees. After reviewing internal correspondence from the department, historian Roza el-Eini summarizes: "With the outbreak of the Arab Revolt on 18 April 1936, the Forests Department lost access to many of its reserves which were in remote hill areas and as with Agriculture, it saw the destruction of much of its work. Licensing became difficult and over-cutting went uncontrolled."[104] Key achievements from fifteen years of work by the Forestry Department—like the Balfour Forest near Nazareth—were simply burned to the ground.[105]

The British eventually made a strategic decision in the *White Paper* of 1939 to capitulate on matters of Jewish immigration and land ownership to contain the violence. They came to see forestry as a program fueling Arab antagonism that needed to be curtailed. After the first year of insurgency, the British Colonial Office convened a fact-finding mission in Jerusalem to consider the situation and recommend appropriate measures. It was headed by Lord William

Peel, the grandson of former prime minister Sir Robert Peel and a senior politician in his own right. Despite the formal Arab boycott, the six members of the commission met dozens of witnesses from all sides. Besides being the first body to recommend "partition," the "Palestine Royal (Peel) Commission" also considered forestry policy in Palestine. Ultimately it gave afforestation a lukewarm endorsement, calling for a more conciliatory approach:

> The Commission fully realizes the desirability of afforestation on a large scale of a long-term forest policy, but, having regard to their conclusion as to the scarcity of land in the hills for the agricultural population, they cannot recommend a policy involving expropriation of cultivators on a large scale until other cultivable land or suitable employment on the land can be found for them. In the aggregate, however, a large amount of land is fit for afforestation but not for cultivation, and the Commission endorses a policy of afforestation of steep hillsides to prevent erosion, the prevention of grazing on land fit for afforestation, and, where practicable, the establishment of village forests for the benefit of neighbouring cultivators.[106]

Order was finally restored and the Forestry Department at long last was ready to exploit its independent status and new budget. But by then British energies were focused on World War II. The Forest Service continued to design ten-year plans, complain about the unavailability of qualified forestry staff,[107] conduct professional training programs, haggle with bureaucratic rivals over decentralizing its work, and mostly present a stiff upper lip of "business as usual." But these were not usual times for the world, and Palestine was no exception.

Due to a Mediterranean blockade, wood imports into Palestine during the war period were significantly curtailed. The only alternative was local forests, whose poor quality meant that logging was even more extensive than anticipated. The British military was the largest consumer, after which came local industries that swelled to meet demand for war-time products. Although it was illegal, army acquisition agents frequently purchased from merchants who would buy contraband wood from fellaheen; these procurements included timber from olive trees, chopped down for a quick sale.[108]

The Forestry Department created a special war-time Utilization (or "U") Section that was given the task of "reducing the wood resources of the country as possible."[109] It would scamper to identify trees located outside forest reserves to supply the seventy-five hundred tons of wood needed by local match factories, or help create a new wartime plywood industry that relied entirely on local trees. Forestry officers took to their new role of advocating timber efficiency, but the exigencies of wartime trumped any real long-term silviculture considerations.

In a six-year report for the 1939–45 period, Goor, by then acting conservator of the forests, summarized the turbulent decade: "The Department of Forests was formed on 1st April 1936 and published its first report in 1939. The whole of this period was one of civil unrest and no serious development of the work was possible. This report again covers a wholly abnormal period, that of the War years. Although the Disturbances ceased in 1939, reduction of staff and a great burden of new work has meant that few fresh advances have been made; in many fields it has been difficult to retain pre-war levels of achievement."[110]

The immediate postwar years were actually highly productive ones for forestry. For example, government seedling nurseries raised almost seven million plants, three times the levels of the 1930s. In 1947, 656 hectares of forest reserve lands were planted, primarily with pine trees, a bumper crop by Mandate standards.[111] But the era of British forestry in Palestine was drawing to a close.

An Unfinished Mandate

When the issue of "Palestine" was submitted for consideration to the newly established United Nations, Great Britain never actually expected the UN to support so willingly Jewish or Palestinian nationalistic proposals. This is clear from the general historical record. It also emerges from the extensive report that the Mandate government prepared for the associated Anglo-American Committee of Inquiry in 1946, entitled *British Mandate: A Survey of Palestine*. The survey constitutes a three-volume, three-thousand-page summary of the state of affairs under the Mandate, including twelve pages that focus on forestry and soil conservation and its future challenges.[112] It reads like anything but a eulogy.

The Forestry Department compiled a concise retrospective review that is a treasure trove of data regarding the British forestry experience. On paper at least, the cumulative numbers are impressive: Some 166 forest reserves on 64,400 hectares were "gazetted" between 1926 and 1928. A few of these would be disqualified in the demarcation process, but by 1946, some 56,534 hectares enjoyed reserve protection, almost a quarter of them established after 1939. Almost 10 percent of the reserves were designated as "closed" to all grazing and timber activity.[113] In the south and along the coastline, trees had helped stabilize the itinerant sand dunes that bedeviled local residents. But there was still dissatisfaction with overall performance.

Thirty years after the Great War, the British remained extremely judgmental of Turkish land stewardship, but the survey does not avoid criticism of the Mandate's own failures in management: "In recent years a large oak forest clothes the hills south and south-west of Hebron until destroyed by overcutting and overgrazing. In various parts of Samaria the ruin of similar forests has been completed by the same agencies since 1933. Most of the pine and oak forests of Carmel have been lost during the past thirty years while the destruction of natural scrub forest in the Galilee is still progressing."[114]

Most of all, the Forestry Department was looking ahead with a clear agenda, stating its future priorities:

(a) The training of staff so that an effective instrument of forest policy can be built up.
(b) The improvement of communications and accommodation so that control of the forests may be made more effective.
(c) The consolidation of the reserves by the straightening of boundaries, the elimination of enclaves and the settlement of disputes.
(d) The large-scale closure of reserves, their management and improvement, and their eventual use for agricultural purposes or as permanent production forest.[115]

It is telling that "Engaging the Arab communities of Palestine" was not on the list. In that sense, the British Mandate sported a colonial forestry service to the very end. One can only imagine how the landscape might have benefited had an effective effort been launched, from the first days of Herbert Samuel's administration, to engage the hearts and minds of the local Arab population in support of afforestation.

All the same, Mandate foresters today have fans in unexpected quarters. Yoav Sagi has been a leader at the Society for Protection of Nature in Israel (SPNI) for some fifty years. During this time he has been a fairly consistent critic of Israel's forestry practices. When asked about the British Mandate's planting, he is quick to change his tune:

> In designing an afforestation strategy, you have to have a sense of proportion. Plant a grove. But if you open it up, it will be much richer and provide a home for birds, butterflies and plants. It becomes part of something larger. If you go to the old British forests, say in Nes Harim or in the Meron area, you can see such the results of such an approach. They worked in diverse patches. Today, we actually try to emulate the British approach in our work in the Yad Hanadiv private reserve. Of course we need to let natural succession take place.[116]

One of the few surviving foresters from the period, Dr. Yerahmiel Kaplan, worked in the parallel JNF forestry efforts at the time and continued on as an Israeli forestry researcher for decades. He has no problem acknowledging that the forests planted by the British foresters were better than JNF stands of the time, citing their superior capacity for thinning and more optimal spacing.[117]

If today's usually divided foresters of the JNF and the nature advocates of the SPNI can both offer praise, then the Forestry Service and Forests Department of the British Mandate must have done something right. The actual statistics of the period show that there has been a bit of glorification about the British commitment to tree diversity. Between 1920 and 1947, the Aleppo pine was surely its predominant tree species of choice. All told, it made up 50 percent of the trees planted, with Italian cypresses comprising an additional 20 percent.[118] And the Mandate's insouciant experimentation with exotic plants would create an invasive species headache that still has not been resolved in some parts of Israel. But the arboreal mix was far richer than the great monocultures that would characterize Israel's forestry in the 1950s and 1960s. The trees and offspring of the original British forests serve as a model of sorts.

Most of all, the British forestry adventure was significant because it represented an abrupt departure from the previous two thousand years, It showed that trend need not be destiny. In dozens of degraded corners of the land, Mandate foresters proved that though the ancient hills had for centuries been without vegetation, they could be restored and, with proper care, blossom. Determination, when combined with professionalism, overcame the inhospitable soils and climate. These achievements fired the enthusiasm and contributed to the confidence of the next wave of afforestation crusaders, as the new State of Israel set about writing the next chapter.

4

Enthusiastic Saplings

And when you enter the land, you will plant.

—Leviticus 19:23

Collapse

Like most of the snows that fall upon the wooded Jerusalem hills every few years, the flurries provided an ephemeral change of color and excitement for children who equate even imperceptible levels of snow with missing school. The 1972 event coincided with a cold snap so that even Be'er Sheva, in Israel's sunny south, was briefly wrapped in white. Ultimately, it was hardly a remarkable Mediterranean storm—nothing remotely resembling a blizzard. No one ever imagined that it might bring ruin in its wake. But it did.

As one drives from Israel's coast to Jerusalem and begins to ascend the inclines of Judea, one passes through Sha'ar HaGai—a Hebrew translation for the Arabic Bab el-Wad, the "Gate to the Ravine." This strategic pass between the hills, roughly halfway from Tel Aviv to Jerusalem, was the site of fierce battles in Israel's War of Independence as Jewish forces attempted to break the siege on their newly declared capital. (Eventually, the area was secured from the Jordanian forces by a young lieutenant colonel named Yizhak Rabin, who would go on to conquer other military and political objectives.) The area was home to an adolescent forest, planted between 1927 and 1937 by the Mandatory Forestry Service, which was later expanded by its Israeli successors. Soon Sha'ar HaGai became one of the notable woodlands lining the highway linking Israel's two largest cities. Hectares of dense pine forest spread a lush green blanket across the hills. More than just a fetching recreational destination, it offered proof of regeneration.

There were those who had already noticed that the stands hiding behind the vigorous trees lining the roads were not doing well. In the summer of 1968, in order to meet the increased traffic coming to see the united capital after the Six-Day War, the highway from Tel Aviv to Jerusalem was widened to four lanes. To make room for the asphalt, the outer row of trees was removed, revealing a sickly, wilting stand of "scarecrows."[1] Four years later, even the more robust, edge branches had dried into a desiccated, reddish hue that was brittle and weak. Snowflakes clung to the boughs and began to collect. Like the straw that breaks the proverbial camel's back, thousands of pine trees simply collapsed under the weight.

JNF foresters from the period found themselves under attack—with the media vilifying what appeared to be systematic incompetence.[2] The director of JNF's Central Region, Haim Blass, convened a press conference to assuage the concerns of the press and its charges of ecological disaster. These Aleppo pine trees were fifty years old, he explained. They had reached full maturity, completed their role as pioneers, and were in their final stage of an ecological succession. The forest would be just fine. But the politicians didn't buy it. They continued to drive by the crumpled trees on their way from Tel Aviv to the Knesset and were uncomfortable with the sight and the symbolism of the moribund landscape. It was as close to an ecological trauma as the nation had ever undergone.

The decrepit state of Sha'ar HaGai's pine forests was well known to the foresters who held general suspicions about the culprit, but lacked empirical proof. A fourteen-person, blue-ribbon, expert committee was convened in 1974, to analyze the problem. It was chaired by Hebrew University botany professor Abraham Fahan. The committee listed twelve possible causes for the atrophying of Sha'ar HaGai's trees but was loathe to indict a definitive "smoking gun." In March 1975, Hugh Wilcox, the renowned American plant pathologist from the State University of New York, appeared before the committee. Only after Wilcox had confidently shared his unequivocal diagnosis, was the committee convinced.[3] Wilcox attributed the tree losses to infestation of the Israeli pine bast scale, a well-known local pest.

The Israeli press was quick to publish the findings.[4] There were a few dissenting voices like Haifa University professor Ze'ev Naveh, who posited that the problem was probably linked to air pollution.[5] Others blamed the rocky soils.[6] Dr. Joseph Halperin, a leading Israeli researcher in forest entomology, boycotted the subsequent "arboreal pathology" discussions in protest of the committee's excessive reliance on international, nonentomological expertise.[7] But the scientific consensus before and after the subsequent investigation was clear, placing the blame on scale insects.[8]

When you plant a monoculture crop, conditions for pests are stable: they enjoy homogeneous cover and a bountiful food source. This is true in forests as well. In general, the greater the diversity of tree species in a woods, the less frequent and less severe the pest outbreaks.[9] As a standard textbook on forestry explains: "A forest with a mosaic of age, classes and species is more resistant to insect attack, especially disastrous outbreaks, than are even-aged forests."[10] This is especially true in pine plantations,[11] whose simplified ecosystem offers fodder for pests, encouraging their immigration and procreation. At the same time, the habitat and environmental opportunities for natural predators is reduced.[12] In a word, the hapless pines became an all you can eat buffet.

The JNF had noticed as early as 1935 that almost three-quarters of the pine trees in its flagship forests were dying.[13] The few entomologists working in the nascent Israeli academic institutions of the time were pressed into service, immediately identifying the culprit: a tiny scale insect. Scale insects are miniscule parasitic bugs that adhere to plants and live off the plant's sap. The damage they inflict on trees is often mistaken for a disease.

It has been said that there is usually an inverse relationship between the size of an organism and its Latin taxonomical name. That is surely the case for the *Matsucoccus josephi Bodenheimer et Harpaz*—with *Josephi* conveying a dubious taxonomic honor to JNF chief and pine champion, Yosef Weitz. (*Bodenheimer* and *Harpaz* refer to the scientists who first studied the pest.) Quite a mouthful, international literature simply calls the bug "the Israeli pine bast scale."[14] Some thirty species of scale insects have been identified thus far globally. All share common traits of being "monophagous"—that is, they develop exclusively on a single or a few closely related pine species. The *Matsucoccus josephi* has eyes only for the Aleppo (Jerusalem) pine.[15]

In their initial, larval stage, these tiny bugs emerge as yellow, elliptical crawlers. Thriving between the months of January and May, they waste little time before snacking on the pine boughs, swelling so much that their legs and anten-

Matsucoccus josephi Bodenheimer et Harpaz, the Israeli pine bast scale that devoured the dense, monoculture pine-planting strategy that characterized Israel's first thirty years. (Photo courtesy Zvi Mendel)

nae cease to function as they mature into brown-colored adults. The females never grow longer than 2.7 millimeters and males are but half that length.[16] As the insect munches through the bark, it excretes poisonous saliva, which serves to disrupt water transport to the tree and essentially paralyzes any new growth.[17] The result appears as a "drying up" of the affected branches. For vulnerable young trees whose upper boughs are easily reached, the end comes quickly.[18] Scale infestation also makes healthy mature pines far more susceptible to the pathogen *Sphaeropsis sapinea*, a deadly fungus. This "*Sphaeropsis* blight" penetrates the wounds on the bark created by the scales and slowly disfigures the branches, weakening the entire tree.[19]

Even before the perils of monocultures became axiomatic to ecological thinking, Professor Hillel Oppenheimer, a leader among the pioneering cadre of agricultural researchers in the country, had no hesitation in stating the practical implications of the *Matsucoccus* epidemic. In 1939 he wrote the forestry managers in the JNF:

> With this it is clear to me that the disease is raising serious doubts about the soundness of the basic forestry system in practice until now. After the disease has spread to such a great extent, I doubt if we can continue to successfully grow forests comprised only of densely planted Jerusalem pines. It is essential to plant the trees at appropriate distances and to thin them every several years. And we must, with greatest seriousness, consider whether it is possible to:
>
> a) exchange the Jerusalem Pine with another such as the *Pinus brutia* or the *Pinus pinea* etc.
> b) consider whether the time hasn't come to return to experimental plantings of mixed forests as you did previously, if I'm not mistaken, in the Herzl Forest in Ben Shemen and in other places.

> You need to try to mix the pine with other deep-rooted varieties such as oaks and carob that should be planted, of course, several years before the pines.[20]

It is worth noting that in successional forests, the order is usually reversed, with the shallow-rooted conifers preceding the deeper-rooted broadleaf species. Nonetheless, the clear call for an end to monocultures as early as 1939 in an internal JNF report is remarkable.

Forestry researcher Israel Gindel, a contemporary of Bodenheimer, also touted the advantages of mixed stands, comparing lessons learned from pest epidemics in German monoculture plantations to the resilience of the mixed stands in France. He wrote Weitz in 1938, calling for fundamental changes in the assemblage of new plantings: two-thirds broadleaved species and one-third conifers.[21] But the warnings went unheeded. Weitz felt that after twenty years of experimentation, the best available knowledge incontrovertibly pointed to single-species, densely planted forests. The rations for the *Matsucoccus* scales soon grew exponentially.

Matsucoccus species can be found outside of Israel, in countries like France and Spain, but do not appear as devastating to Aleppo pines there. It is not clear why the *Matsucoccus josephi's* effect on young Aleppo pines in Israel is so lethal. *Brutia* pines, in contrast, exhibit greater biotic and abiotic tenacity and hardly appear to be bothered by the scales. Genetic weaknesses may be part of the reason. Aleppo seeds were imported into Israel and did not coevolve with the local varieties of scale. These pines may enjoy genetic resistance to insects prevalent in the western Mediterranean, but be highly susceptible to the eastern Mediterranean varieties they encounter in Israel.[22]

Whatever the explanation, the fact remains that despite hopeful proposals for biological control via pheromones and advances in chemical controls, entomologists remained helpless against the giant waves of irrepressible scales. All the frustrated foresters could do was watch the trees die and try to plant more smartly in the future. Zvi Mendel, who for the past thirty years has been Israel's leading etymological forestry researcher, explains: "The sole control method in forest practice is to prune the lower branches of young trees (from 3–9 years)."[23] This presumably should be done at ages four to five and subsequently at ages nine to ten. The opening exposes the pine bast scales' eggs to sunlight, which dries and destroys them.[24] Stem injection for trees in parks and camping sites, a costly intervention, was never proven to be effective.[25]

The *Matsucoccus* was not the only insect to discover Israel's new pine monocultures. Pine processionary moths (*Thaumetopoea wilkinsoni*), first identified in 1937 in Samaria, are not only problematic for trees, but also for people. When the moths are in their caterpillar larval stage, they gather in nests high in

Pine processionary caterpillars find another victim. (KKL-JNF Photo Archives)

the treetops and feast on pine needles. The pest's name derives from the caterpillars' tendency to "march" in a procession through the forest, protected by a hairy exterior that causes acute irritation to human skin. As pine trees began to fill the Jerusalem corridor after independence, the crawler population overwhelmed the fledgling forests and spilled over into Israel's central coastal zone.[26] Caterpillar dermatitis and conjunctivitis soon reached epidemic proportions among humans.

In 1958 even the Israeli military was forced to take notice when it canceled maneuvers in Hulda Forest after six hundred soldiers required medical attention following contact with the ubiquitous caterpillars.[27] This meant war! In 1959 a full-fledged eradication campaign was waged against the pests by a consortium of government ministries and the JNF. It involved stem injections with insecticides as well as aerial spraying with endosulfan, a highly toxic chemical, and diflubenzuron.[28] The population immediately was cut in half and by the end of the second year infestation had dropped by 90 percent.[29] But the chemical bombardment killed not only the caterpillars: many of the natural predators of the *Matsucoccus* were exterminated, setting up the next pest crisis in the pine forests.

In retrospect, the decimation of Sha'ar HaGai's pine trees was a highly predictable disaster waiting to happen. The snowstorm in the Jerusalem hills in 1972 can be seen as a watershed date, when an entire approach to planting forests collapsed along with the lifeless trees. It would take the better part of twenty years for the message to sink in with foresters in the field and translate into an operational shift to greater species diversity, lower planting densities, and a preference for the more resistant *Brutia* pines.

Even today, JNF foresters have not fully internalized the fact that thinning trees at a relatively young age is not a luxury or recreational preference but a management necessity. Indeed, if a pine tree grows up under crowded conditions, by the time it is thirteen years old it will probably be too late to thin the tree to strengthen it, and the tree will never really recover from the associated "high-density" stress.[30] The Jerusalem pine debacle jumpstarted a rethinking process.

Even the most ecologically devout contemporary foresters working in Israel still salute the first generation of Israeli foresters and are indulgent about the vast stands of pine monocultures that they planted. This unique cohort was fiercely dedicated to its difficult work and driven by the original ideological and scientific orientation of the JNF. Nevertheless, the triumphant propagation of Aleppo pine trees across Israel during this period, before their ultimate fall, was possible because of political support and pressure applied at the highest level.

The Prime Tree Planter

On August 18, 1949, Yosef Weitz, the most senior manager and forester at the JNF, was preparing materials for an upcoming trip to the Galilee. Without warning he was called to Prime Minister David Ben-Gurion's temporary office in the Jewish Agency building in Jerusalem. The last of the four armistice agreements officially ending Israel's War of Independence between Israel and its neighbors had been signed with Syria at Rhodes in July 1949. A miraculous military victory had been achieved; the borders of the new state were more or less defined and democratic elections had been held. Yet the country was in dreadful condition.

The battles had taken six thousand lives—1 percent of the Jewish population, but a horrendously higher proportion of males between the ages of eighteen and thirty. Military expenditures were estimated at five hundred million dollars, creating debt that could not be paid back because the country's economy was in ruins. Citrus groves, which had always generated foreign currency, were decimated.[31] The Dead Sea potash plants were badly damaged during the war. Exports of produce to neighboring Arab countries, an important component of the balance of trade, was now unthinkable. And to make matters worse, the needs of the country were growing rapidly. Every month an additional eighteen thousand new citizens arrived, many of them traumatized Holocaust survivors coming from the fifty-two displaced persons camps throughout Europe. An additional seven hundred thousand Jews from Arab countries were on the way.

Weitz and the prime minister were of the same generation and made of the same sociological cloth. Only four years separated them. Both were prototypes of the idealistic, pragmatic, self-educated, humorless, talented, and totally focused leaders that made the "Mapai," the Labor Zionist party, the dominant political force in the country. A labor union leader for many of his early years, Ben-Gurion's only real manual job, "the happiest years of his life," was farm work at Sejira. Weitz was brought on as farm manager there a few years later. It might be assumed that the prime minister of a newly declared country, with existential military concerns and seemingly impossible social and economic challenges, would be absorbed with matters of state and domestic crises. But Ben-Gurion had trees on his mind, so he summoned his long-time colleague.

Weitz arrived at ten o'clock in the morning to find the prime minister alone in his office with a notebook. After remarking how long it had been since the two had last met, Ben-Gurion got to the point: why had it taken so long to respond to his recent memorandum outlining a vision for planting trees in the new country? Weitz apologized for the delay and noted that Amihud Goor at the Forestry Department in the Ministry of Agriculture was also reviewing the memorandum. They would convene a committee meeting during the coming week to respond. Ben-Gurion was clearly disappointed by what he perceived as a runaround: "I understand from your comments that you don't see the potential for any great momentum as I described in my letter." Weitz responded: "Jumping from 'zero' to that sort of height is impossible and may actually be dangerous." Ben-Gurion began to pepper Weitz with professional questions: the appropriateness of different species, the availability of seeds and saplings for planting, the number of work hours required, and constraints associated with the seasons. He also asked about the appropriateness of soils in the hills, the planes, and the valleys—even in hyperarid Eilat and the Arava Valley. The prime minister wrote down each answer in his notebook and told Weitz how important forestry was for the building of the land.

Weitz confided to his diary: "He sees it as the end-all. 'A year ago, the war was before us and all our efforts were focused on winning it: to defeat and to conquer. Now, all of our efforts are directed to the building of the land. If we are to conquer the soil—only the forest, only the tree will lead us to our goal. Accordingly, we need to think big: half a million dunams will be planted this year as well as next year.' I explained to him that this was not within the realm of possibility. Even a smaller percentage would not be possible during the

“A billion trees for a new nation” (1958): Prime Minister David Ben-Gurion plants one. Yosef Weitz, the “Father of the Forests” (with glasses, unruly hair, signature mustache, and uncharacteristic smile) looks on. (Edgar Hirschbein, KKL-JNF Photo Archives)

coming year because we don't have any saplings. He queried me about their number and said: 'If that's the case, we will concede with regards to 1949–1950. But why are we not able to plant half a million dunams?'"

Ben-Gurion proceeded to discuss the Jerusalem corridor. "'How many dunams for afforestation are there?' he asked. I told him: '400,000 dunams.' If so, these we must plant in 1950–1951,' and he began to explain to me why it had to be in that place. 'Indeed, we have many enemies. Nations and religions rise up against us to drive us from these parts. But if we can bring 20,000 Jews there, and transform the moribund landscape and revive it all, and if Jerusalem will grow—they will never get us out of here.'"

Weitz responded that this was all well and fine, but that planting trees was a biological process. Ben-Gurion accepted that there were "*natural factors*" that they would not be able to overcome: "We won't be able to plant in the summer—rather only in winter. To this I surrender. But the human, technical factors? The budget? These we must overcome. Look here: We had no Hebrew army, and did we not draft in a short time one hundred thousand men to the military? Every Jew that moves to Israel will need to go immediately to the work of afforestation. And tin cans for plants can be prepared, and flower pots can be made by the millions and money will be found" (emphasis added).

Weitz agreed on the conceptual level but objected to the dimensions that Ben-Gurion proposed. In practice, it was impossible to implement Ben-Gurion's vision. There would be room for some expansion. And planting a million trees per year was certainly feasible immediately; perhaps in two years' time, ten million, even twenty million—but a hundred million? Weitz also insisted that forestry be coordinated with other manual-labor projects; otherwise the workers would find themselves unemployed for four to five months of the year.

The conversation went on for two hours, but Weitz wouldn't budge, and offered no solid figure for the coming year's planting goals. He summarized the discussion with the general observation that the prime minister was given to grand thoughts that might lead to great things if they could be brought down to earth and made functional.[32]

Three months later, none of Weitz's pragmatism appears to have gotten through to Ben-Gurion. At the opening of the second Knesset Ben-Gurion thundered:

> We must plant hundreds of thousands of trees on five million dunams—one quarter of the area of the state. We must wrap all the mountains of the land and all the slopes in trees—all those hills and rocky soils that no farmer can cultivate the sands of the valley and the coast, the plains of the Negev east and south of Be'er Sheva. We will not be faithful to one of our chief objectives for the state—making the wilderness bloom—if we limit ourselves to the needs of the present

> and to endeavors whose return is only immediate; if we don't, in our generation, build enterprises that will offer a blessing to all future generations.[33]

Part of Ben-Gurion's optimistic program was an extension of his bombastic, visionary persona. Part of it was based on a complicated but abiding affection for nature, especially the kind that was nurtured through human determination and ingenuity. Yet another part of it had to do with Ben-Gurion's confidence in the operational abilities of the JNF as a no-nonsense-get-the-job-done partner he could rely on. In what many still consider the organization's finest hour, once this public company received its marching orders it prepared to shift into high gear.

A Jewish National Forestry Agency

Since 1961 the Jewish National Fund (JNF) has enjoyed sole authority to conduct afforestation activities in Israel. This makes Israel's forestry experience idiosyncratic, not only substantively but institutionally. The JNF is run by a board of directors who adhere to Israeli corporate law and its forestry division is not a part of the executive branch or subject to government policy decisions. To be sure, it is not a profit-maximizing entity, its board of directors is democratically elected, and it operates according to what it believes to be the public interest. But neither is it an NGO or a nonprofit association. Although its Hebrew title, Keren Kayemeth L'Yisrael, can be translated literally as the "Sustainable Fund for Israel," it is unquestionably a *Jewish* corporation.

From a strictly legal perspective, since its establishment in 1901 the JNF is owned by the World Zionist Organization, a body overseen by a Zionist congress that convenes every four years as an international, ersatz parliament for the Jewish people. Technically at least, the JNF's stockholders are none other than the entire Jewish people. The JNF's landholdings, according to its bylaws, are owned in perpetuity by Jews around the world and its land development and forestry work is conducted on their behalf. In theory, a Jew in Melbourne, Australia, or Teaneck, New Jersey, has as much influence and representation regarding forestry policies and programs as a resident of Haifa or Tel Aviv. For better or worse, this is no ordinary forestry service.

The institutional history of the JNF is somewhat arcane and has been previously recounted in detail.[34] Originally established to serve as the land-acquisition arm of the World Zionist Organization, the JNF soon took on a variety of tasks associated with building a physical infrastructure for a Jewish state in Palestine. This included real estate purchases, agricultural assistance programs, and even orphanages. Forestry was something of an afterthought.

Appalled at the conditions he found during his 1898 visit to Palestine, the founder of the Zionist political movement, Theodor Herzl, conferred with Professor Otto Warburg, a botanist and active Zionist.[35] Herzl called for the planting of ten million trees there, but never found the time to actually promote this agenda. Only when Herzl died in 1904 did the JNF finally plant a "forest" in his honor. In fact it was more of an olive grove, which foundered in the sticky soils of the Ben Shemen region. The forest was soon supplemented by grapes along with almond and apricot trees in a one hundred and forty-hectare fruit orchard.[36] World War I was not kind to the trees of Israel, and Herzl's forest essentially disappeared following the 1915 invasion of locusts. The insects were followed by subsequent waves of Turkish, German, and ultimately English soldiers who turned the grove into campgrounds. Only a tenth of the original woodlands had any trees left at all after the war and these were the odd Aleppo pine, cypress, and Australian casuarina.

With a British mandate that encouraged afforestation, the JNF decided to try again. By 1922 it had planted sixty-five thousand trees in the renewed Herzl Forest. Other JNF afforestation projects would soon follow, all characterized by the provisos of agronomist and senior JNF official Aqiva Ettinger. Ettinger stipulated that the trees be non-fruit-bearing varieties, requiring little maintenance, and that they not be planted in the "fertile valleys" which were prioritized for agricultural cultivation. These forest stands were associated with new Zionist settlements as a way of asserting sovereignty over recently purchased lands that were too steep or rocky to be farmed for crops. To a large extent, the early JNF forests served as public works projects to provide minimal employment for the new communal kibbutzim that frequently lacked adequate agricultural work for an entire community.

Analyzing the difference between the forests of the JNF and the British during this period, Tel Aviv University researchers Nili Lipschitz and Gideon Biger conclude: "The difference between the goals set by these two bodies is clear: the Zionist Organization regarded afforestation as a tool for enforcing its ownership over fallow lands, and as an economic stronghold for new settlements. The British regime used trees in its battle against the movement of sand and soil erosion and planted forests in order to enrich ground water reservoirs. The JNF was mainly active in the mountain regions, while much of the British efforts were on the sand dunes."[37]

After his extended 1939 visit, U.S. soil conservation expert Walter Clay Lowdermilk was diplomatic in contrasting the two systems of afforestation: "There is a considerable difference between a 'forest area' as the term is used by the Palestine government, and a 'forest' in the usual sense of the word. A 'forest area' is a stretch of land closed, at least theoretically, to goats and fuel gatherers in order that opportunity may thus be given for the growth of native vegetation and trees. On certain of these areas the government has planted new trees, some of which have reached a considerable size. The forests of the JNF are planted more densely and more nearly approach our conception of woods. While there may be various opinions regarding these two systems there is no doubt that both government and Jewish organizations are doing excellent work in this important field."[38]

For the JNF, the thirty years of British rule until 1948 can be seen as a "pilot" run for Zionist forestry practices and its leading practitioner. Yosef Weitz joined the new JNF Land Development and Afforestation Department in July 1919 at the behest of its director, Akiva Ettinger.[39] Ettinger provided much of the conceptual framework for JNF's tree plant program that Weitz would eventually expand, particularly after taking over as head of the department from 1932 until the 1960s. For the half-century he was affiliated with the JNF, Weitz retained his signature upright, lean appearance with wire-framed glasses and full mustache. His work ethic was legendary; his subordinates always knew he would be the first to arrive and last to leave. Although he was hardly a chummy sort of boss, most of his workers were completely devoted. When he visited a forest on a field visit, regardless of its location in Israel, he was always there by 6:30 A.M.[40] Even his critics acknowledged that underneath his crusty sometimes priggish affect was a romantic who absolutely loved trees. Relying on his own keen intelligence and meticulous attention to detail, he developed the afforestation methods that would transform the countryside of the new nation and presided over their implementation.

Forestry in 1948 was still a matter overseen by the Ministry of Agriculture. As the Mandate's chief forest curator, Amihud Goor was a natural choice to continue in this position. David Ben-Gurion decided to bet on more than one horse. In May 1948, a year before his famous meeting with Weitz, Goor was also summoned, this time to the prime minister's home in Tel Aviv, where Ben-Gurion wrote him a check for one thousand pounds and told him to run the new state's Forestry Department in his old Mandatory office building and to hire whomever he chose.[41] Rather than resolve the institutional rivalry between the "Zionist" and the "government" forest agencies, Ben-Gurion perpetuated it, allowing it to simmer for more than a decade.

And so it was that the new state and its ambitious forestry agenda had two charismatic individuals vying for control. Each represented very different sociologies and perspectives on forestry. Goor was eight years younger than Weitz, a native of Palestine who grew up in Mikveh Yisrael, the leading agricultural training center in the country. To the extent that the Yishuv, the Jewish community of Palestine, had an aristocracy, Goor was a member. By contrast,

Weitz was born in czarist Russia in 1890 and moved to Israel when he was eighteen. Goor's father was a highly regarded teacher in an agricultural high school, while Weitz's father helped supervise the farms of Russian noblemen and manage their forests of Scots pine trees.[42] Goor fought in the Jewish Brigade during World War I. Weitz sat out the conflict and worked as an agricultural laborer and manager. During the 1920s, Goor attended the world's most elite universities and attained advanced degrees in forestry. Weitz never attended college but was fluent in several languages and spent a great deal of time reading. Goor was a civil servant for the British Mandate while Weitz became a central figure in a Zionist bureaucracy that came to view the British government with suspicion and eventually animosity. Goor attended professional conventions, traveled extensively throughout the world, and served as an advisor to international agencies. Weitz left Israel rarely, preferring to attend to the day-to-day minutiae of settling a country. Both were extremely smart, completely dedicated to the establishment and nurturing of a Jewish state and its forests. Before Israel's independence, they had cooperated on any number of projects, including discussions in 1941 about establishing a university forestry degree in Rehovoth.[43] But they had very different styles.

It was clear that the illogical, bifurcated system of forestry that emerged after 1948 would not last. A 1952 intervention to get the two to share power in a unified committee completely failed. In the end, only one would remain standing.

By the 1950s, Weitz's authority within the JNF on matters of forestry was completely unquestioned. Haim Blass, who worked at the JNF for forty years during this period, recalls his old boss with fondness and devotion: "Yosef Weitz was my Rabbi and teacher. His personality was absolutely dominant in the JNF. He was practical and considered by many to be a hard man. You had to be hard in those days. There was so much going against you that there was no choice but to be aggressive and stubborn. There was no bureaucracy or support staff. And everything was new."[44]

Initially, the two forestry services set out planting trees at a ferocious rate. Notwithstanding the presence of British infrastructure and expertise which remained with the government, the JNF foresters quickly took the lead and never lost it. The graph below shows the scorecard from a decade of parallel work: JNF forests filled 18,000 hectares, government foresters established a third less.[45] During the country's first decade, without the benefit of heavy machinery, they planted together an area that constitutes a third of all forested lands.[46] The numbers do not fully tell the story, nor can they faithfully portray how dramatically transformed the landscape was in the center and the north of Israel. Within ten years, the countryside looked entirely different.

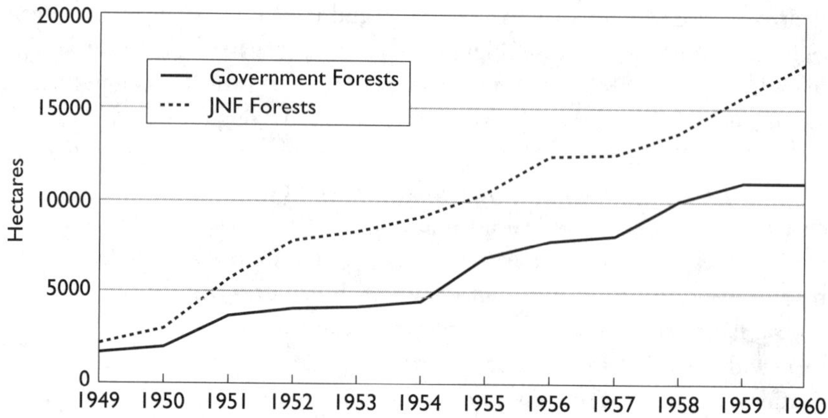

Comparison of tree planting: JNF versus Israel Government Forestry Department: 1949–60 (in hectares) (Source: *Israel Statistical Abstract* [Jerusalem: CBS, 1961], as quoted in Zvika Avni, "Government Afforestation in the State of Israel: An Epilogue," in *Forest,* 5–6 [2004]: 5)

In theory it was not a race, but there was more competition than cooperation between the two agencies. During the first decade of Israel's independence, more than anything else the scope of tree planting reflected unemployment patterns rather than institutional capacity. With all due respect to the prime minister's high minded rhetoric about "future generations," a high percentage of Israel's new citizens found themselves without work, living in tents, disappointed and frustrated. Ben-Gurion understood that a single shovel could provide jobs for two people on back-to-back daily shifts. Like Weitz, he strongly believed that the new immigrants' self-respect would be fostered if they worked for their pay rather than living on the dole. Until alternative livelihoods could be generated, more than five thousand workers were hired on a daily basis to plant trees.[47] The labor was overseen by both JNF and government managers.

The work was punishing, but the old-timers who remember the period unfailingly praise the dedication of the immigrant foresters, practically none of whom had any previous experience in foresting or even agriculture. Unemployed immigrant men would congregate in the early morning in city centers and pile into military-style transport trucks, returning in the evening, exhausted from the heat and the hard work, bearing new blisters and calluses. To some extent, these reminiscences are colored by nostalgia and selective memory. Correspondence from the period suggests that output was often low and that laborers did not believe productivity should be linked to the salary received. Much of the JNF's budget came from donations abroad. Many immi-

Immigrant workers from Arab countries planting the Martyrs' Forest in 1952, commemorating European Holocaust victims. Signs mark groves according to countries from right to left: Lithuania, Italy, Latvia, Denmark, Austria, and France. Nursery in background. (Abraham Malvaski, KKL-JNF Photo Archives)

grant laborers felt that the funds had been contributed on their behalf and that they were doing the JNF a favor by working at all. Weitz felt otherwise.

JNF decided to change these norms and base its salaries on actual planting performance or to pay on a piecemeal basis. (When the work required precision, workers could be paid by the hour.) Not surprisingly, this policy was highly unpopular among the new immigrant workers, who were primarily from North Africa, Yemen, India, and eastern Europe. Frequently there were strikes. This situation existed until the mid-1960s, with thousands of otherwise unemployed working as day laborers planting trees. Workers with physical constraints were paid lower, part-time salaries (five hours per day) and given lighter tasks (e.g., nursery work, pruning). Ultimately, most of these salaries were covered by the Ministry of Labor.[48]

Mordechai Ruach was a new immigrant from Egypt whose resourcefulness and diligence were impressive enough to land him a permanent job at JNF. Eventually, he would become its forestry chief. Ruach explains that because the work was so arduous that JNF foresters had a schedule with fifteen days "on" and then fifteen days "off." Israel's young forests were planted on "marginal lands," which was a euphemism for steep and rocky terrain. "We planted forests at impossible angles. Immigrants from Kurdistan and Iraq—we took huge risks. Today they are afraid to let people climb up to those heights. Not even Arab workers. Slowly but surely we started getting machinery to reduce the manual labor. Eventually we got a small tractor which made the work much easier."[49]

Although the government and the JNF selected their laborers from the same reservoir of unemployed men, the contrasting woodlands reflected the ideologies of their senior managers and the resources of their respective agencies. On the one hand the density in the JNF forests was far higher. Goor's government foresters planted trees with an average of three meters of space between the trees—amounting in practice to thirteen hundred trees per hectare.[50] This was far lower than the JNF standard, according to which trees were never separated by more than two meters and density typically reached four thousand trees per hectare.[51]

Yerahmiel Kaplan, a JNF forest manager during this period, believes that part of the reason that JNF forests had densities even higher than this standard was the pay framework that existed, especially before the creation of the state. Group contractors such as kibbutzim were paid according to the number of saplings planted rather than by the land covered, creating an incentive to pack the parcels with pine.[52] Over time, this would be detrimental to long-term forest health, as the impassable stands were rarely if ever thinned properly, leaving crowded and enfeebled trees.

In general, the JNF had more money than the government Forestry Department; its workers received a supplement beyond the very modest public work stipend that welfare provided. Often, the government was forced to hire JNF "rejects." During this period, neither agency enjoyed a permanent, highly trained staff: as in most financial transactions, they got what they paid for.

Conflict between the two forestry agencies soon grew very personal. Although their initial relationship was reportedly cordial, as institutional rivalries hardened they became strained.[53] By the mid-1950s Weitz had resolved that the JNF needed to take over the government apparatus and become the country's sole forester. His 1957 diary records several meetings held to that end. Weitz's primary "co-conspirator" was close friend, powerful politician (and future prime minister) Levi Eshkol, as well as Ben-Gurion himself. Beyond personal connections with the top level of the Labor government, Weitz could also bring to the table JNF's capacity to independently generate income. Its twenty-odd offices around the world were staffed by skilled fundraisers, adept at engaging hundreds of thousands of generous supporters of Israel.[54]

For a government perennially strapped for cash, the advantage of having a forestry agency that didn't fight for allocations in the annual, budget squabbling rituals was compelling.

These circumstances ensured the outcome of the power struggle even before it began. On March 31, 1958, Weitz recorded a meeting with Eshkol; the minister of agriculture, Kadish Luz; and Goor during which the minister suggested leaving research, enforcement, and timber exploitation in government hands and transferring all other forestry work to the JNF. Neither of the two adversaries supported the compromise: Weitz was fearful that any government involvement in forestry would spoil its appeal as a fundraising gimmick, while Goor called for "a single forest department—but governmental."[55]

There were certainly substantive reasons for Weitz's "power grab." Weitz felt that survival rates in the government forests were greatly inferior to those in the JNF, as indeed they were. The government trees were not up to the specifications set during the British Mandate. Unlike the Mandatory Forestry Department, Israel's Agricultural Ministry lacked sufficient funding to maintain a permanent team of field workers. They had neither time nor resources to replant when trees did not take. The ministry could maintain a permanent staff of only 98 workers, with the actual planting, thinning, and pruning done by 220 nontenured laborers.[56] Workers and laborers were paid very little and were often devoid of skills and motivation, so the quality of work was inconsistent. There was another problem too: the Forestry Department director was never around.

Goor's success as an expert emissary and advisor for the United Nations was considered a national asset by Israel's Foreign Ministry. Its diplomatic staff encouraged Goor to accept the growing number of invitations from around the world to consult in sundry startup silvicultural ventures. Undoubtedly, Goor enjoyed the prestige and challenges associated with touring the globe and crafting major forestry initiatives. All of his trips were professional in nature and approved by the director general of the Agriculture Ministry. But because of his traveling, year after year he was not always present to oversee the critical planting season between November and March. In a 1958 meeting with Eshkol, Weitz was quick to point out the inadequacies of his institutional rival, highlighting: "the absence of a 'proprietor': they work without any planning or efficiency in their performance, due to the lack of basic funds. The budget is used for supporting personnel and overhead while actual forestry in the field is dependent on the Ministry of Labor public works for the unemployed."[57]

Goor was clearly frustrated at the unreliable flow of funds to his office. Although the finance minister, Eliezer Kaplan, was a personal friend and neighbor, Kaplan's job during the early 1950s was practically impossible. There was simply not enough blanket to cover all the feet that crowded the bed in the new, indigent state. A worker in Goor's office during that period recently penned this memory:

> The head of the office or rather, the department to which the office belonged, was a svelte aristocrat by the name of Amihud Goor, who, so it was said, was a world renowned expert on forestry. The peculiar thing about him was that in spite of his habitual gentle and courtly conduct, every morning he raised his voice while talking on the phone with the Finance Minister Eliezer Kaplan (who after Ben-Gurion was the most influential person in Israel). "Why," Goor would roar, "haven't the new immigrants who were employed as relief workers been paid yet? How dare you keep them in such a dire condition and let their families starve!" This ritual repeated itself with ever rising tones every morning for quite a while, at least for the time of my being employed there. I don't believe that anybody else ever dared to talk to Kaplan in such tones.[58]

It would seem that given the choice between international travel and haggling with the finance ministry about budgets, and an internecine turf battle that he was bound to lose with the JNF, it was easier for Goor to devote his talents to professional assistance overseas. Thus, the situation in his department went from bad to worse as the immigrants eventually found more reliable and lucrative employment. The government's forestry professionals had no one to do the actual fieldwork.

Eventually, Weitz went to the press with his allegations. A 1959 media report reads:

> Minister of Agriculture, Kadish Luz sent urgent instructions today to Amihud Goor, head of the Ministry's Forestry Division, to return to Israel from a South American tour to answer public charges of mismanagement of the state-owned forests. The charges against his department were made by Joseph Weitz, head of the development department of the Jewish National Fund. Mr. Weitz alleged that only about 25 percent of the 12,500,000 trees planted by the Ministry survived as compared to some 80 percent in the JNF forests. The Ministry of Agriculture conceded that a smaller percentage of the saplings planted survived, but argued that this was due to the fact that the forests being planted by the Ministry were—for reasons of economy—not irrigated as were the JNF plantings. Mr. Weitz had charged that the failure of the state plantings was due to mismanagement, disorganization and labor practices. Behind the charges, which have been attracting considerable attention in this election season, lay mounting friction between the Ministry of Agriculture and the JNF.[59]

It took a few years before the final deal was cut with the minister of agriculture and translated into a series of statutes that Eshkol (as the finance minister who replaced Kaplan) shepherded through the Knesset. The transfer culminated in a "covenant" signed between the JNF and the government on November 28, 1961. This covenant formally gave JNF responsibility for planting and maintaining Israel's woodlands, with a modest official ministerial oversight role for the Ministry of Agriculture, which maintained its forestry research operation at the Ilanot station. Long before the new division of labor was finalized, in April 1960, Goor tendered his resignation. The matter did not pass without controversy and public disgrace for the Weitz camp. There was a "hue and cry" after it became known that Yosef Weitz intended to appoint his son Sharon to be head of the new, unified Forestry Department rather than Goor. The act of nepotism seemed particularly crass in light of Sharon Weitz's lack of any academic training and Goor's unparalleled credentials and stellar international reputation.

Professional organizations remonstrated. The dean of natural sciences at Hebrew University, Shmuel Samburski, wrote an open complaint, decrying the decision: "In our State, until now we have enjoyed the great advantage of a single manager with both scientific knowledge and experience as is accepted in all advanced countries. Separating the positions and entrusting the operational post to someone without appropriate scientific background means a step back that will lead to friction and poor coordination—and to a drop in the scientific level of forestry in Israel."[60]

In his diary Weitz records that nine leading scientists signed a letter of protest to the JNF board of directors that was published verbatim in the leading newspapers. But he was unyielding: "I will not lend my hand to bringing a person into our club that knows how to plant but actually plants his own interests. As for Sharon, I have already said, that his work is 'work.'"[61]

Committees in both the Ministry of Agriculture and the JNF were convened to consider the matter, but Weitz was powerful enough to push through the appointment. Ben-Gurion himself was aware of the power struggle and felt uncomfortable with the outcome. On July 29, 1959, he wrote a letter to Goor that was leaked to the press: "Regarding the styles of planting, I am of course not an expert, but I know of your great expertise in this profession and if for some reason you have been distanced, or will be distanced, from this blessed work—no person will be sorrier about this than me."[62] It was little more than lip service. There are no signs that the prime minister saw fit to actually intervene.

Although the final agreement left a few forestry researchers in the Ministry of Agriculture with formal authority for overseeing the Forestry Ordinance, Weitz's terms for institutional surrender were in fact unconditional. The JNF brought over all permanent government foresters (Goor's deputy became Sharon Weitz's deputy) as well as the twenty tractors and seventeen cars that were used by Ministry personnel.

The appointment of Sharon Weitz was ostensibly a simple case of overt favoritism. In fact the controversy had repercussions within Israel's forestry program for years to come. An anti-intellectual, anti-academic bias characterized many of the state's leaders during this period, who had themselves given up university training in order to bear the burden of state building. This perspective was manifested in Israel's forestry program. Yosef Weitz actually valued academic expertise and found funds to send promising workers for graduate studies, like a doctoral fellowship in 1960 for Yerahmiel Kaplan to attend Oxford.[63] But Kaplan was anomalous.

During the first thirty years of forestry in Israel, professionalism was not a valued qualification. In 1958, the only worker in the JNF Forestry Department with any advanced training in the field was Moshe Kolar. Kolar was a competent forest engineer, but "due to his diffidence and consideration for others for the most part he avoided expressing nonconformist opinions on professional matters."[64] Common sense rather than scientific method drove decisions, and Sharon Weitz's appointment reinforced this organizational culture. With the equanimity of hindsight, one forestry researcher from the period charitably reminisces about the first generation of foresters: "They bequeathed us today's mature forests. One can compare them to the officers of the Israel Defense in the Independence War, most of whom had never learned in military academies but thanks to their dedication and sacrifice, the state of Israel was established."[65] Nonetheless, Professor Samburski's concerns were not without justification.

Weitz's cronies from the time are quick to praise Sharon Weitz as a pragmatic, experienced, hard worker. But fifty years after the "scandal" the old-timers at JNF still joked about the decision: "Weitz had three sons — the youngest and most gifted, Yehiam — was killed as a fighter in a commando raid prior to the War of Independence; the oldest had an impressive career as a regional planner in Israel's rural sector; and Weitz made certain that the middle and least talented son remained 'close by' so he couldn't get into trouble, leaving him the family forestry business." Sharon Weitz did not have his father's analytical, autodidactic gifts nor his openness to new ideas from subordinates. Even loyal family friends acknowledge that the son was "educated" through constant interventions behind the scenes from the real boss.[66] In retrospect, Sharon Weitz was probably the perfect middle manager to disseminate his father's JNF gospel and "one-size-fits-all" proclivity for the Jerusalem pine tree.

Planting Pines

Leah Goldberg's popularity as a national poet continues to grow, even years after her death in 1970. Born in Lithuania, she moved to Tel Aviv from

Berlin in 1935 at age twenty-four, having already completed a doctorate in literature. Goldberg was extremely erudite, and spoke seven languages. She started her academic career teaching in a Tel Aviv high school but quickly moved on to head Hebrew University's comparative literature department. Luckily, she found time not only to translate and analyze other people's writing, but to do a bit of her own. Unlike many of the "nationalistic" poets of her time, Goldberg's lyricism was more intimate. Nature was a constant focus, and she found inspiration in Israel's streams, flowers, and trees. Over the years, many popular-song writers and singers have made Goldberg's lyrics familiar to modern Israel.

Among the many themes that Goldberg's poems explore, especially in the face of the foreboding scenery in Goldberg's new land, is an immigrant's longing for a romanticized landscape of childhood. Even the arch-Zionist Ben-Gurion missed the verdant Polish scenery of his youth, confiding to the Knesset in a 1962 speech: "I visited Scandinavia this year . . . and was jealous of two things: the abundance of lakes and the abundance of trees. In Finland alone, whose territory is twenty times that of Israel, more than sixty percent of the land is forested. . . . I know we can't make lakes flow into our barren country . . . but there is no reason why our country can't be covered in forests with trees in every place possible — and there is no place where it isn't possible."[67] Since the advent of Zionism, conventional wisdom incorrectly assumed that biblical Israel had been as blessed with thick vegetation as the temperate European countryside. Restoring this lost mythical Eden to its previous glory was an aspiration shared by the foresters of the JNF, many of whom were of European origin.[68] One of Leah Goldberg's more famous poems, "Pine," is an ode to the tree which served as something of a vegetative bridge between European immigrants' native and adopted arboreal realities:

> Here I will not hear the voice of the cuckoo. Here the tree will not wear a cape of snow.
> But here, in the shade of these pines — my whole childhood reawakens.
> The chime of the needles: Once upon a time — I called the snow-space homeland,
> and the green ice at the river's edge — was the poem's grammar in a foreign place.
> Perhaps only migrating birds know — suspended between earth and sky —
> the heartache of two homelands. With you I was transplanted twice,
> with you, pine trees, I grew roots in two disparate landscapes.[69]

There was nothing coincidental about Goldberg's selection of this particular tree species. During her thirty-five years in Israel, there was no tree as ubiquitous or as symbolic as the "Jerusalem" or Aleppo pine.

Little at the start of the twentieth century indicated that this particular species would come to dominate the forests of the new State of Israel. The tree is native to Israel and Jordan, but only isolated relics survived, as captured on vegetative maps of Judea prepared during the extensive surveys of the 1870s.[70] They were also found sparingly in the Carmel and the western Galilee, but did not attract much attention from anyone except perhaps agricultural researcher Aaron Aaronsohn. In a 1913 piece criticizing Turkish forestry policy, Aaronsohn admired Aleppo pines' success among Templar plantings.[71] But the article was even more complimentary about a more popular, recently arrived tree species: the eucalyptus.

It would take a century from its "discovery" in Australia by Captain Cook in 1777 for eucalyptus trees to reach the shores of Palestine. But once the first seeds arrived at the Mikveh Yisrael Agricultural School, it was an instant success.[72] Within seven years, these trees can reach a height of eleven meters; one of the original Mikveh Yisrael trees, planted in 1884, was measured at 47.5 meters in 1940! By 1920, almost 80 percent of the JNF trees were eucalyptus, playing a key role in land reclamation activities. Draining the swamps through tree planting soon became an integral part of Zionist "lore" from the period, which was made out to be glamorous but was undoubtedly a punishing business: "The swamp was full of water snakes, but these were harmless. Leeches though, were troublesome at the time. On the orders of Rosnak [a JNF official] each planter would receive a daily bottle of brandy. Half-tipsy and stinking of swamp they worked well and the saplings prospered."[73] The JNF initiated a trial-and-error process, observing seventy-one species of eucalyptus. It eventually limited its plantings to three: river red gum (*Eucalyptus camaldulensis*), Tuart (*Eucalyptus gomphocephala*), and Swamp Yate (*Eucalyptus occidentalis*).[74]

During the early years of statehood, the military maintained a special unit for tree planting, primarily around highways and bases, with the objective of camouflaging armed forces' movements. The fast-growing Australia trees were a favorite specimen for this task.[75] But the eucalyptus did not thrive everywhere and not all of local foresters were fans of the Australian arboreal colonists. Goor reportedly grew less enamored and began reducing their range. This trend continues to the present. (Today there are less Israeli lands planted in eucalyptus than there were in 1960, as many trees were destroyed when highways were eventually widened.)[76]

As a young forester, Weitz sought a more reliable species for JNF's new forests. It did not take long for a winner to emerge. In the early JNF stands, pine trees consistently outperformed the eucalyptus on the rocky hillsides. The oaks, carobs, cypress, and fruit trees also could not compete. Pines are in fact an eclectic group of trees. They can be as tiny as three meters or as lofty as eighty-two. They all share certain basic features—scaly bark, cones, extensive "tap roots," resin, and needles. They are all evergreen. Although there are some outrageously old pine trees—with one California specimen reaching the incomprehensible age of forty-six hundred years[77]—for the most part, pine trees live about a hundred years.

Experts estimate that there are between 105 and 125 different pine species, depending on one's taxonomic criteria. The foresters of the JNF experimented with about a dozen promising varieties: the Brutia pine, Stone pine, Canary pine, Corsican pine, Monterey pine, and the Bishop pine to name a few.[78] It was the Aleppo pines, however, that grew the fastest, consistently proving to be the most adaptable of the *Pinus* species.

Aleppo pine is widespread throughout the Mediterranean.[79] First classified in 1778 as *Pinus halepensis,* the species ironically does not appear to be indigenous to the Aleppo area of Syria at all. Pines are mentioned only once in the Bible, yet the *Pinus halepensis* is undoubtedly native to Israel. There were times when its indigenousness was challenged, yet one need only go to Jerusalem today and see the immense pine trees at the Armenian Patriarchate or view photos of a recently deceased 250-year-old Aleppo pine in the courtyard of the Rockefeller Museum to realize they are not recent arrivals.

Just how prevalent Aleppo pines were throughout the land of Israel remains the subject of debate. Initially, the position of Israeli botany expert Professor Michael Zohary was widely accepted: "One must admit that this type of forest has occupied in the past much larger areas than at present. The reason of its reduction is two-fold: this kind of rendzina soil is arable and has, therefore, been largely captured by agriculture. Second, the pine tree is the only soft-timbered tree and has been extremely used by the local population."[80] Recently, analysis of past travelers' logs, palynological research (that identifies pollen samples), as well as excavations and dendroarcheological analysis of ancient wood items appear to corroborate an opposing view. While probably the dominant pine species historically in Israel, *Pinus halepensis* was a limited component of the original wood assemblage.[81]

Yosef Weitz was not one to quibble excessively over taxonomical nuances. Nor was he one to argue with success. By the mid-1930s, the Aleppo pine had been renamed "the Jerusalem pine" for Hebrew vernacular and made up about 98 percent of JNF stands. In a 1936 publication, Weitz animatedly justified his bias:

> And this pine, nicknamed the Jerusalem pine or according to its scientific name (*Pinus halepensis*) was recognized by scholars who decided that the pine's homeland is the western coast of the Mediterranean Sea and this is the coast that lies next to us. It is therefore a child of this land from its creation,

and as such well adapted to local conditions. At first when the land was blossoming, before the forests were eradicated, it was widely distributed, particularly in the hills of Judea from whence they would bring sacrifices. Its beauty warmed our forefathers' hearts, so much so that it served as a motif for decorating and ornamenting and was not unlike the other prized fruits—the grape, the pomegranate, the olive etc. The tree never stopped growing in our country, from time immemorial, and in the most far-flung places of Israel humans can find remnants of ancient pine forests. The qualities of the Jerusalem pine that make it so specialized as a forest tree are:

a) It adapts to different climates: from the Jordan valley, the wasteland receives its shade. In Jericho and Degania you will meet it. In the coastal plains and in the Sharon it will flourish and on mountains 800 to 1000 meters above sea level;
b) It does not discriminate according to soil-type. It is happy to blossom in sandy and organic soils alike and even on rocks it sends its roots to explode them and grab hold. It finds soils rich in lime to be most pleasant, so it can be planted in the most desolate places in the land;
c) It is easy to expand cultivation;
d) It grows quickly.

The Jerusalem pine is hard, and from this vantage point thought to be the best of the pines that can be used for furniture and building materials. The foresters know the advantages of the Jerusalem pine and so it has their blessings over other trees as the primary tree to plant in the new forests. It is not only a praiseworthy forest tree, but also an excellent tree for decoration and gardening. Its green landscape is always a dark joyous green, and its smell is sap still dripping. Its needles and its cones earn the tree a respected place in the garden near the house, in the public park, in the city streets and in the village. . . . The number of pine trees planted during the past decade is eight million.[82]

The, newly planted, orderly pine plantations stood in stark contrast to the chaotic, natural maquis scrublands with their gnarly oaks and impenetrable assorted shrubs. For Weitz, these natural woodlands were not only devoid of value but almost an affront to civilization. As late as 1974 he wrote:

The sorry state of forest blocks covering thousands of dunams is the result of a natural growth which has been completely neglected. The question is whether this natural state is desirable in a modern world and should be hallowed in the form of a reserve. Or should it not perhaps be regarded much as any other blind and destructive natural force, against which man must defend himself and if possible bring under control. . . . This natural forest reserve and others like it are indeed untouched by man—and they are also undesirable and of no

> use to anyone. But over the rest of the area, man could step in and convert the scrub into forest capable of functioning as such, possessing a natural beauty, providing nesting sites for songbirds and ranges for forest animals, as well as a place for man himself to wander and refresh his spirits.[83]

An Israeli timber industry was a dream harbored by the early foresters and encouraged by Yosef Weitz.[84] After all, notwithstanding the Middle Eastern dryland conditions Israeli yields in agriculture mirrored those of their old homes in Europe. There was no logical reason why wood production should be any different. Dr. Israel Gindel studied forestry in Warsaw, moved to Israel, and began working as a researcher for the Mandate and later the Ministry of Agriculture while completing a doctorate in Italy. His 1952 tome, *The Forests and Afforestation in Israel,* offered a comprehensive "state of the art" review (in Hebrew) about the existing methods for planting and managing forests. Like the foresters of his generation whom he advised, his economic expectations were quite clear: "With the increased sophistication of agricultural production, we need to add the afforestation division, whose establishment is one of the essential conditions for a healthy economy. This can be realized, especially in light of the novel inventions that offer the possibility of upgrading the qualities of weak timber."[85] Pines offered the most promising silviculture strategy.

The JNF did not just acquire a singular preference for Jerusalem pines. It also developed an almost industrial protocol for planting them that quickly came to replace manual labor, with its shovels, hoes, and pick axes. Initially applied on lands with dark, humus-rich, rendzina soil overlying soft bedrock, the mechanized assault on the new forestry plots was soon expanded to terra rossa soils with harder limestone and dolomite bases.[86] Most importantly, trees were raised to be tall and straight, enhancing their value as timber. This meant planting at extremely high densities, frequently five thousand trees per hectare.[87]

The main goal of site preparation was efficiency, and any competing flora in the "bare" shrub lands did not warrant consideration: At the start of the planting process, the ground cover would typically be burned to ensure eradication of existing vegetation. After the fires, large D-4 bulldozers (known as "rotaries") were brought in with a clearing rake and a three-tine ripper (gigantic steel claws) for "brush elimination."[88] When soils were deep, a subsoiler cultivation tool was used to break them up at depths of half a meter. Cultivation on shallow soils relied on disc ploughs attached to tractors that prepared the surface for planting. Trenches were dug and the soils were ground by routing. As years went on, herbicides joined the arsenal wielded to suppress native trees and shrubs.[89] Nurtured by such aggressive preparations, *Pinus halepensis* — and little else — thrived.

"Make way for pines": bulldozer preparing hillsides before spraying and dense planting of conifers in Israel's early monoculture plantations. (KKL-JNF Photo Archives)

In contrast to JNF afforestation practices, the approach of the government Forestry Department during the 1950s was markedly more diverse and less destructive, reflecting its British methodological roots and Goor's personal orientation. In spite of the war raging and prevailing confusion, during its first year of activity the new State of Israel's Forestry Department not only managed to plant 283 hectares of diverse trees, but also to write up a full report of its work. Small homogeneous stands containing two dozen tree species were established as distinct units in different parts of the country.[90] The patches were based on the underlying soil conditions, with the purpose of "studying the ecological and silvicultural requirements of this region." In addition, larger stands of evergreen oak and stone pine were planted.[91] It would take many years for these species-rich woodlands to be recognized as a more sustainable forestry model. In the annals of Israel's botanical history, the Jerusalem pine had won this round.

Under the best of circumstances, planting any conifer so single-mindedly would spell trouble. But unbeknownst to it, the JNF had bet its forest's future on genetically flawed material. The Jerusalem pine debacle of the 1970s and 1980s was not just a story of monocultures' inherent susceptibility to pests; there was also a hereditary component to the pathology.

Pine trees grow from seeds. The DNA contained in the seeds released by the pinecones of a forest determines the characteristics of the next generation.

When afforesting "barren" lands with depleted seed banks, seeds need to be introduced from outside the forest. Once the seeds are planted, it takes years to find out what sort of trees they produce.

Thus, only after much of the damage was already done did it become clear that Jerusalem pine forests were collapsing throughout Israel because they unwittingly had been selected for feebleness. In hindsight, rather than the stock of trees being built up on basic notions of "survival of the fittest," it seemed that they were planted with perpetuation of the weakest in mind. JNF senior forester Gershon Avni explains that in order to meet their ambitious schedules and limited budgets, British Mandate foresters purchased most of their seeds abroad. After all, there weren't many trees that survived the Ottomans in Palestine, and few conifers were available for samples. Efficiency and price drove procurement decisions.[92] The JNF collected some of the seeds it needed from the maturing Mandatory forests, perpetuating its eclectic sources that ranged from North Africa to eastern Europe. Seeds were collected from trees that were not necessarily impressive in terms of their shape, growth, or resistance, but that were easily accessible to climbing by aging foresters.[93]

This ignored a key principle of forest management: seeds need to be selected so that they that have similar "provenances" or locations of the population from which they are collected. It simply makes sense to rely on trees whose genetics evolved under local conditions. Bringing seed or root stock from outside a provenance increases the risk of planting weaker trees that grow slowly and are more vulnerable to attack from insects and pathogens.[94]

Another paradox is that trees under stress or that are dying, perhaps anticipating the end is near, generate copious amounts of pinecones and seeds during their final years.[95] The evolutionary impulse to pass on genetic material is a powerful one. A flourishing pine tree, by contrast, focuses its finite energies on foliage and growth, with less production of fruits and seeds. It provides relatively modest numbers of cones from which foresters can mine seeds for future woodlands.[96]

Finding millions of seeds from such robust specimens is a tedious task. For ordinary workers tasked with collecting seeds, *the path of least resistance* literally led to trees *with less resistance.* Moreover, when the JNF decided to fully embrace *Pinus halepensis,* the shortage in mature local Jerusalem pines was counteracted by the attractive prices offered by western European seed companies. The fact that the evolutionary forces that shaped their seeds were entirely different than those of the Middle East was not considered. During the 1950s and 1960s, quality control for genetics in pine species was not exactly a salient factor in the tender. Weitz and his foresters imported *Pinus halepensis* seed from Europe, and inadvertently began to perpetuate weaker genomes.

Professor Gabriel Schiller, of the government's Volcani Institute of Agriculture Research, is one of Israel's only full-time silviculture researchers. With no academic training in forestry available in Israel during the 1960s, Schiller received a scholarship from the Swiss government to complete a graduate degree there. Upon his return, he was rejected by the JNF Forestry Department as overqualified, forcing him into a career of research. For more than forty years afterward, Schiller enjoyed the role of academic gadfly, openly expressing his exasperation at the systemic incompetence he found in the JNF Forestry Department. He was among the first to chronicle the genetic debacle:

> The Mandate foresters brought seeds from around the Mediterranean. These are places that often have a totally different ecology than that of Palestine. We live in a fairly unique place. It's very hot here. The soils for the most part are horrible. I don't have to tell you how great the challenge is. In the other countries, no one even attempts to try to forest these areas. When they come visit they can't understand why we invest so much in these forests. So the pine trees that survived in Israel over the years, naturally evolved and adapted. They should have been the genetic base for the new forests. Instead of taking advantage of these robust local trees, the British bought cheap seed from oversees. (If you go to the national archives you can still see the receipts from the Paris seed company where they bought the *Pinus halepensis* seeds.) The JNF continued it. I don't know what they were thinking.[97]

Schiller's subsequent genetic research would help correct the problem and offer important insights to afforestation programs internationally.[98] Given the climatic similarities, not surprisingly, Aleppo pine seeds from Greece show greater pest resistance and so were increasingly integrated in plantings.[99] All the same, scientific insights could *not* uproot the many millions of trees that were already planted with defective French seeds during Israel's initial pine fervor. They were there to stay. Nor, could it curb the natural appetite that the *Matsucoccus josephi* had for them.

The new forests soon made enemies other than pests and disease. During the 1950s the Society for Protection of Nature in Israel (SPNI) emerged as the largest NGO of any kind in the country, establishing a national network of educational field schools.[100] Gidi Ne'eman is a distinguished professor of ecology at Haifa University. His scholarship is a natural extension of the formative experience he enjoyed in the military during the early 1960s, when he worked in the Galilee out of the SPNI's Field School on Mount Meron. Ne'eman smiles during an interview as he recalls the nocturnal retaliatory raids that were made by the young nature lovers who uprooted newly planted pine trees they saw as trespassers, sullying their beloved natural landscape:

> We saw the foresters as the enemy, a total catastrophe. They would say: "You protect nature. Our job is to make forests. Don't tell a farmer how to grow pears; don't tell us how to raise trees." I remember at Har Hazon, not far from my present home, one fine day they showed up and burned all the natural vegetation. Then they returned with bulldozers, dug up the entire northern side, and planted pine trees. There had been beautiful oak and *katlav* [Greek strawberry] trees there. Wildflowers. All destroyed. That's how they worked in those days. So we fought back.[101]

As luck would have it, the chief JNF forester for the north of Israel, Tuvia Ashbal, had a daughter who worked as a volunteer at the field school, so attaining intelligence was not a problem. She would leak the site of tree plantings which could then be easily sabotaged. This confrontation was settled decades later when the JNF came to rethink its planting methods. Other conflicts proved more difficult to resolve.

Forestry Ideology and Nationalism

The pine monocultures and the passions associated with their promotion during the first twenty-five years of Israeli independence reflected an ideological perspective. Three different kinds of motivation drove JNF forestry policy. The first was a fundamentally romantic desire to restore a homeland perceived as neglected and in an advanced state of decline. Planting trees was a noble venture, part and parcel of fulfilling a solemn commitment to a beloved homeland. "It seems as if God has covered the soil of Palestine with rocks and marshes and sand, so that its beauty can only be brought out by those who love it and will devote their lives to healing its wounds," wrote Israel's first president, Chaim Weizmann.[102] This vision of ecological restoration was also present among many British foresters and senior officials. Yet, their impulse had neither the historic sense of self-righteousness nor the territorial ambition of Zionist afforestation ideology.

The second type of motivation was related but more rational and utilitarian in nature. It focused on what today is called "ecosystem services." Afforestation reduced erosion, improved the climate, harvested rainwater for groundwater storage, and provided "cultural services" such as recreation. Here, there was no meaningful difference in perspective between British foresters and their Israeli successors. For instance, parallel to his government post in forestry, Amihud Goor headed the agricultural ministry's Soil Conservation Department. During the 1950s, the two programs were inseparable.

The third type of motivation was linked to Jewish nationalism. Here JNF's purpose and operational objectives were fundamentally different than the British forests department. With much of the world skeptical about Jewish sovereignty over the land of Israel—and local Arab populations vehemently opposing it—trees figuratively were drafted into a Jewish civilian "militia." Forests communicated permanence, prosperity, and control. To be sure, farming could send a similar, no less powerful message, but many areas were unsuited for commercial cultivation. This "nationalistic" function remains a defining element of Israeli afforestation to this day.

Technion University researchers Shaul Amir and Orly Rechtman explain: "Protection of national lands via afforestation became an important task, especially in security-sensitive areas, where it was also used to camouflage military units, installations and movements. . . . Due to the special role of afforestation in developing the country, policy-making in this regard was mainly limited to activities related to implementation of national goals."[103]

It was this assertive side of local forestry that made it the target of animosity and violence. Zionism was not the only national movement in the land of Israel. The JNF's trees were an anathema for Arab nationalists from the advent of Jewish immigration to Ottoman Palestine. When given the chance they struck the trees down.[104] Predictably, in response Zionists redoubled their afforestation efforts. The phenomenon is captivating enough to have spawned a few books. Geographer Shaul Cohen describes in his *Politics of Planting* the politics that were quickly integrated into the unofficial rules of the contest.[105] Across the land of Israel, trees became analogous to soldiers, or what Irus Braverman has called "planted flags."[106]

The politicization of trees before 1948 was aggravated by signals sent out by the legal system. To preserve order, the British Mandate extended the provisions of the Ottoman Land Code from 1858. In retrospect, these rules served to encourage lawlessness in and violence to the forests.[107] Article 103 of the Land Code recognizes the legitimacy of ploughing and cultivation for anyone in need of open spaces—known as *mawat* lands. These are vacant areas "such as mountains, rocky places, stony fields . . . and grazing ground which is not in possession of anyone by title deed, and lies at such a distance from towns and villages from which a human voice cannot be heard at the nearest inhabited place."[108]

Even more important, article 78 of the code guarantees that "everyone who has possessed and cultivated land for ten years without dispute acquires a right by prescription . . . he shall be given a new title deed gratuitously."[109] This provision applies to *miri*, or state lands, which comprised the bulk of Palestine's territory. Land that remained fallow or in a natural state was "fair game" for squatters. Arab cultivation could challenge ownership and repossess Jewish holdings. Zionists came to see this as a "use or lose" test that they needed to win to hold on to their hard-earned lands. Trees were often the best way to assert ownership. The game was afoot.

Before Israel attained independence in 1948, many of the first JNF forests were on plots next to agricultural settlements like Mishmar Ha'emek, Kefar Hahoresh, Gineigar, Ma'aleh Hachamisha, Neveh Ilan, or Kiryat Anavim. Planting trees was often no more than a precautionary act to ensure preservation of Jewish property rights on rocky, undulating parcels.[110] Haim Blass, a forester during the 1930s and 1940s, is clear on this point: "During the interim period, 'holding the land' was a key goal so that it shouldn't go back to foreign hands. Those lands that were abandoned actually ended up reverting to Arab ownership. No activities could achieve this as inexpensively as forests. Just a year or two's work, and the trees didn't need any more help."[111]

The "politics of planting" took on new forms after the establishment of Israel. Despite its quasi-official status in afforestation, the JNF remained the operational arm of the nongovernmental Zionist movement that faced a new list of national trials. The JNF contributed actively to addressing many of Israel's collective national challenges like unemployment, poor infrastructure, and soil degradation. It was also expected to play a central role in the battle for Jewish sovereignty over the lands of the state. Just as trees had been successfully pressed into service to assert ownership prior to 1948, they were once again considered a critical tool in resolving the question of abandoned Arab lands.

Arab and Jewish narratives about the origins of the refugees who left Palestine during the 1948 war are of course entirely different. Arabs claim mass deportations,[112] and Israelis describe voluntary departure from a war zone at the behest of the Arab political leadership who promised a triumphant return with conquering Arab armies.[113] Regardless of the cause, the results are not disputed: before the war 850,000 Arabs lived in Palestine; after the war only 160,000 remained.[114]

According to one study, 418 Palestinian Arab villages were abandoned during the 1948 war.[115] Many houses were soon demolished and nature was encouraged to reestablish its hold. Some 43 percent of these abandoned villages are now surrounded or overrun by nature reserves, parks, or forests.[116] This is no coincidence. The JNF planted forests on roughly eighty-six of these depopulated villages, most of which were razed entirely prior to the planting.[117] While Israelis may not have precipitated Arab departure, they surely were not sorry about the outcome. The country was not about to encourage the return of hundreds of thousands of people who not only called for the elimination of the Jewish state but continued to actively pursue this goal.

Planting trees not only transformed the landscape aesthetically, it also effectively erased the evidence of Arab presence and asserted the biological stamp of the land's new owners.

Two additional factors, one institutional and one personal, made the JNF's work an even greater subject of controversy. Palestinian refugees had left behind considerable land holdings, much of which was quickly declared by the government to be "absentee property." Israel's political leaders were keenly aware of the political sensitivity associated with state nationalization of disputed real estate and the country's vulnerability to international pressure. They happily assented to two major land deals, in which the government sold 2.2 million dunams of absentee lands to the JNF. By 1954 the JNF had tripled its real estate holdings. The JNF now owned 13 percent of the country's territory. It is worth noting that JNF donations really did pay a market rate for these lands. Moreover, most of the JNF forests are *not* planted on these disputed lands nor on those that were purchased before 1948. (Afforestation typically took place on *national* lands that were subsequently zoned to that end.) Nonetheless, to Palestinian Arabs and their sympathizers, these developments solidified the JNF's reputation as an active conspirator in the postwar land heist by an enemy state.

As the most senior and prominent JNF official, Weitz played a central role in the land acquisition of the 1950s. But, it was his activities prior to this, during the War of Independence, that make him such a reviled figure for Arabs in this context. Weitz was one of three members of an officially recognized "transfer committee" that in 1948 assessed the feasibility of a population exchange between Arabs in Israel and Jews in Arab lands. The forced 1947 migration of 6 million Hindus and Sikhs to India and 5 million Moslems to Pakistan at the end of British colonial rule provided a model that was fresh in the world's minds. Officially the committee was tasked with assessing the refugee situation and making recommendations. Weitz strongly opposed the return of Arabs who had fled during the War of Independence and sought to further encourage Arabs to leave JNF-owned lands. As a lobbyist, his voice was heard. Ben-Gurion's diary records Weitz's constant pressure to do more to harass and stem the tide of "infiltrators" who sought to reclaim their homes.[118]

It is important to also remember that Weitz was one of the few Zionist leaders who vociferously supported full compensation for Arabs who abandoned lands in Israel. His sense of culpability to the refugees even led him to travel to the province of Mendoza in Argentina, to seek alternative agricultural lands on which interested Palestinians might resettle.[119] Weitz never advocated forced expulsion. Despite an express cabinet decision forbidding such payments, immediately after the war he continued to purchase Arab lands and assist those who wished to move to Lebanon and Jordan.[120] While Weitz favored population exchange as a logical part of the final settlement, the formal recommendations of his committee called for full and equal rights for Israel's Arab citizens.[121] Needless to say, none of this did very much to improve his or the JNF's standing among the Arab communities of the area. This acrimony often extended to the forests themselves, and remains an unfortunate legacy of this formative period in Israel's forestry history.

In short, part of the zeal behind the JNF planting program was to contribute to the resolution of the country's military conflict. This was in no way the primary motivation for afforestation and only a minute fraction of JNF woodlands are actually located on abandoned Palestinian villages. Nonetheless, there was a recognition that trees offered an important means for staking the new country's territorial claims, limiting the expansion of Arab towns internally and protecting its borders in future negotiations. The JNF saw this as nothing less than a righteous mission in which the Aleppo pine's robust and fast-growing qualities made it the perfect arboreal partner.

Trial and Error

Gershon Avni joined the staff of the JNF at the start of the 1970s, just as the doctrine of monocultures was being questioned. He was wont to say that "revolutions don't happen overnight—especially in forestry. In agriculture, you can plant tomatoes every three months and try new things. In forestry it takes 20 years to see if something works."[122] It is well to look charitably at a great deal of Israel's first forty years in afforestation as an experiment. Not only did it take decades for the results to arrive—but sometimes foresters didn't even know what questions they should be asking.

Indeed it was a time of considerable experimentation, with hypothesis testing in many areas. Most of the research was not controlled and surely did not take place in a laboratory. Haim Blass recounts the first time a JNF nursery faced snow in Haifa.

> I was director of the Haifa region of the forestry division in 1950 and responsible for the largest nursery in the country, located in Kiryat Haim, producing one million pine trees a year. Today that's a lot of trees, but in those days, it was simply enormous. That year was one of the few times in recent history that it snowed in Haifa. Now the snow fell just after the seedlings had sprouted. I was terrified that the snow would wipe the entire crop out. It was very early in the morning and I was the only JNF worker with a four-wheel drive vehicle that could make it through the snow. No other workers could get there. I furiously tried to uncover the seedlings as fast as I could, but there were hundreds of

> thousands of them and I realized it was an impossible task. How much could I do? So I went home completely dejected. Well, it turns out that the few seedlings that I uncovered suffered and many died from the much colder ice that came afterward, while those safely buried under the blanket of snow did just fine. But how was I to know that then? Everything was so new to us. We just had to use our best guesses and hope.[123]

Then there was the brief foray into carob trees. Based on a recommendation from local agricultural researchers who sought a replacement for imported animal feed, considerable stands of carob trees were planted by both JNF and government foresters. A whole program emerged, replete with nurseries, irrigation techniques, and extension services to local farmers. Twenty thousand dunams were planted, but soon this turned into a debacle. In September 1959, twenty years after the original research, Ra'anan Volcani, head of the government's agricultural research center, sheepishly explained that when the experiments were repeated a very different picture emerged: "It soon became evident that there was a great deal of difference between carobs and barley as cattle fodder and even more so, as poultry feed. This difference lies not only in their respective nutritional values, but also in the presence in carobs of a factor or factors which adversely affect or retard the development of young chicks, and which also reduce the milk yield in dairy cows."[124] Goor wrote an angry letter of censure, but given the apparent toxicity to livestock, the project quietly was discontinued. This is the nature of trial and error on a national scale.

During this period, basic forest recreational infrastructure specifications had to be invented: "I built the first forest picnic area around 1960 in the President's Forest near Eshtaol," recalls Blass. "Of course we weren't sure how to do it. Should we spread the tables out or put them near each other? Or should we encourage people to be alone in nature? (It seems that Jews prefer the togetherness.) The result of course was something in the middle."[125]

Even the methodology for pine monocultures was being honed. Weitz describes standard procedures in a controlled experiment to assess the success of *Pinus halepensis* as nursery saplings versus seeds:

> In the experimental plot (1963) the ground was prepared in the usual way for planting in this area. This included:
>
> 1. complete eradication of all woody vegetation, perennials and annuals by means of a D4 tractor;
> 2. plowing with a rooter to a depth of 40 cm.; and
> 3. stone clearing and making ridges about 30 cm high and spaced 3.20 meters apart, using a Fordson tractor. As the plot was rectangular in shape, the ridges were made to follow the length of the plot.

> Broadcasting seeds took place and contrasted with planting. The Aleppo pine was the fastest growing conifer (versus Stone, Calabrian-Brutia, Canary and Monterey pines). But after the first year it was necessary to plow between the rows and remove the weeds and thin the shoots out. The same for year two with the rows hoed. In the third year, treatment was plowing and pruning the lower lateral growth. This took a total of 102 man days and 17 tractor hours and proved to be less cost-effective than planting saplings. The (cost) disparity would still be 50% in favor of planting.[126]

In a scientific experiment, determining the proverbial "dependent variables" typically drives results. In the real world of forestry, picking the units that are to be assessed involves values. Not surprisingly, measures of success in Israel during these years were economic and agronomic—not ecological.

Shimon Ben Shemesh was Weitz's long-time personal assistant during the 1930s and 1940s and then served for many years as JNF director general. He recounts that as early as the 1940s there were debates among JNF staff about the relative virtues of diversity. Two streams emerged in the debate as to whether pine monocultures were appropriate. Weitz insisted that the workers adopt the view that forests be seen primarily as an economic resource with efficiency the paramount priority.

Ben Shemesh was sympathetic to this perspective: "Weitz was not opposed to nature like some people try to make him out to be. Sometimes, personal animosity distorts what people say. Weitz wanted to develop the country and ultimately the opposing position wanted to leave it as it was. If you ask me, they would have been happy leaving us back in the Middle Ages. So Weitz attacked that view whenever he encountered it." Evidence actually suggests that Weitz harbored Romantic inclinations as well. In a 1945 lecture he waxed poetic: "A mixture of colors, shadows and sounds make up the forest. . . . man finds peace and an atmosphere that bonds him with the Creator. . . . In the forest—man finds nutrition of a sort that can be more important than actual food!"[127]

Looking back, Ben Shemesh believed that the two streams eventually reached the same point of departure for future policy. After the initial pine planting, it was time to diversify the forest. It is only natural that the original generation of foresters takes a somewhat self-serving view of history.

As a forestry researcher for the Ministry of Agriculture for some fifty years, René Karschon monitored local forestry practices up close. Born and raised in Germany, with strong international training, he was a thorough and solicitous man, maintaining good relations with his colleagues at the JNF throughout his career. Ever diplomatic, he was fond of confiding to friends that, in German, the word *enthusiasm* has a negative connotation. Mistakes are made in the throes of overzealous excitement.[128]

In the aftermath of Israeli independence, the JNF afforestation program had the best of intentions. As a second generation of forests fills in many of the original pine stands planted during Israel's early years, it is important to acknowledge how important it was to seize and save those huge swaths of open space: not just from rival ethnic squatters, but in the race against the rapacious forces of development. The trees certainly sprinted off the blocks—but this strategy had a price. After all, afforestation is more of a marathon. The loss of the young Aleppo pines to *Matsucoccus josephi,* like the collapse in Sha'ar Hagai or the reliance on genetically feeble tree specimens demonstrated the perils of enthusiasm. Because trees were planted on the worst possible soils, a short life span would be expected even if optimal diversity and spacing had been pursued. Present estimates suggest that some forty thousand hectares of conifer monocultures were lost to pests before Israel finally decided to change the way it planted forests.[129] Considering the densities of the time, that is a lot of trees. Yet, precisely because they so believed in their mission, Israel's foresters were serious about learning the associated lessons. The journey from aggressive monocultures to diversity and sustainability had begun.

5

Sustainable Forestry

Let the land produce vegetation: seed-bearing plants and trees on the land that bear fruit with seed in it, according to their various kinds. And it was so.

— Genesis 1:11

Trees without Timber

In 1981 Haim Zaban became head of JNF's Land Development Authority, overseeing forestry, and made a surprising discovery that changed forestry in Israel forever: "I did a 'back of an envelope' calculation that was very simple. It showed that the cost of cutting and clearing a tree was more than the economic value of the tree as timber. I was all in favor of continuing forestry—just not producing wood. It didn't make sense."[1]

Historically, forestry programs around the world were established to harvest timber. Trees provided lumber, paper, furniture, medicines, fruits, dyes, chemicals, weapons, statues—and many other products. Maintaining a steady stock of wood constitutes a basic survival strategy for many cultures that are dependent on trees for raw materials, fiber, and food. Efforts to ensure timber supply were part of public policies adopted in Europe during the Middle Ages. Robin Hood legends might be reinterpreted by casting the sheriff of Nottingham in the more favorable light of forest ranger and conservator.

Modern forestry is generally thought to have begun during the 1660s in France. Jean Baptiste Colbert, the prime minister, discovered that French forests could no longer provide the navy with high-quality timber required to expand the country's naval fleet. Initial strategies to address the problem were primitive, involving staggered allocation of designated forest lands for harvesting over a multiyear schedule. The need to beat the British and Spanish armadas at sea opened the government's eyes to its forests' finite dimensions and the need for systematic management.

Over the subsequent century, German scientists accrued a critical mass of knowledge about afforestation, establishing a formal, academic forestry school in Freiburg by 1795. The department survives to this day as part of Dresden University.[2] During the nineteenth and twentieth centuries a new applied discipline of "silviculture" slowly emerged, dedicated to renewing and managing tree cover. Even as the field became increasingly ecologically sensitive, "sustainable" techniques were still designed for profitable and lasting wood harvesting.

Ever since the British Mandate, forester managers in Israel also assumed that timber was their ultimate raison d'être. Omri Bonneh, among Israel's most experienced forest managers and experts, recalls that when he began as a JNF forester, old-timers told him that one of the reasons for pruning pine trees was to generate wood of sufficient quality to make violins.[3]

Suddenly, this axiom was no longer valid. To be sure, dryland forestry provided a broad range of benefits. But in a land of limited rainfall and poor soils, a profitable timber industry was not one of them. Zaban's perspective took a while to sink in, as top-down reforms often do. In the early 1990s, the Forestry Department contracted to supply a new Israeli particle board factory with wood. Unfortunately, it did not meet distribution schedules; Israel's forests couldn't deliver,[4] and world prices slumped.

By 1990 the JNF Forestry Department codified the new perspective in an *Updated Program for Management and Treatment of Planted Forests in Israel*. Drafted by the country's top managers, in consultation with leading academics, the goals of forestry were redefined. *Environmental benefits* — including landscape improvement, remedying environmental and aesthetic hazards, soil preservation and sand dune stabilization — topped the list of forestry objectives. *Social benefits* — like hiking, recreation and tourism followed. *Economics* appeared last, almost an afterthought (alongside grazing and "protecting national lands").[5] By 1992, in public presentations, JNF's head of forestry put wood and timber at the very bottom of the list, as a by-product of forest management rather than an objective function to be maximized.

The 1990 program offered a frank assessment of limitations:

> Timber is a product with an economic value. This value is dependent on the existing supply and demand in the timber market. The primary market for timber from JNF forests until now has been particle board factories, a most limited market, both in extent and in the price it is willing to offer for wood. There are other uses for JNF timber: sawmills (for shipping pallets, A.T.), heating, etc. These possibilities are not exploited as they might be, something that prevents appropriate prices for lumber and limits the general quantity of local wood that the market can absorb.[6]

Israel's experience in fact was paralleled by enough countries for the UN Food and Agriculture Organization to redefine forest plantations in 2005 as bifurcated systems: *Productive Plantations,* established predominantly for provision of wood, fiber and non-wood products; and *Protective Plantations:* planted predominantly for provision of services like soil and water protection, rehabilitation of degraded lands, combating desertification, and so on.[7]

Previous forestry programs in Israel relied on extremely dense plantings, assuming they would be thinned and pruned to produce maximal yields and profits. But this assumption was no longer considered compelling. Pine stands were to be thinned according to a seven-year cycle, not to mention two to three prunings of low tree branches during the tree's formative years. In practice, neither happened very often. Subsidized labor was no longer plentiful and the JNF Forestry Department was outsourcing much of its fieldwork to contractors. Managers knew exactly how much the work cost—and chose more pressing priorities. As the JNF was run by politicians with a Zionist rather than an economic or ecological agenda, political pressure focused solely on new planting. The forestry budget primarily went into expanding forested lands and the associated maintenance requirements were usually neglected. When it did happen, thinning mostly took place near major roads to make forests accessible.[8] The remaining woodlands were unkempt and dense. The time for change was at hand.

Even under the new orientation, Israel's forests produced 10 percent of the country's wood requirements. In 1990, Israel set a target of harvesting a hundred thousand tons of timber per year—a goal that was met for eight years, until production levels slumped, with present production stable at thirty thousand tons per year.[9] This could not be attributed to new accounting criteria or revised projections for timber quantity. The entire orientation had shifted.

Freed of any illusions, the shackles of yield intensification and profit maximization that had once constrained managers were gone. New policies and practices coalesced in every realm of forestry: matching tree species with soil type; diversifying plantings; reducing stand density; encouraging natural regeneration; preventing fires; increasing handicapped accessibility and integrated pest management. Naturally, changes were easier to envision than implement and the new strategies brought many technical problems. But they also opened the door to creative solutions. The policies offered the promise of forests that more closely resemble Israel's original woodlands; more effectively prevent erosion; are more resilient to pests, droughts and fires; are easier to hike in; and are more picturesque.[10]

New Faces in the Forest

One could argue that a key factor in JNF's transition to sustainable forestry was Haim Zaban's aversion to having a personal driver. Zaban remains one of Israel's most innovative and bold agricultural economists. In 1960, fresh out of graduate school, he started up the ladder in the Ministry of Agriculture. Farmers at the time were traumatized by JNF's indefatigable planting, alarmed that the new woodlands would usurp agricultural grounds, rangelands, and water resources. The minister of agriculture still retained formal statutory authority over forestry. To mollify his concerned constituents, Zaban was appointed head of the agricultural planning department, authorized to review and approve all JNF planting plans.

Zaban's eyes twinkle behind his spectacles as he recalls the haggling that went on: "If the JNF had a budget for ten thousand dunams, they would invariably ask for permission to plant twenty-five thousand dunams." Using age-old bargaining strategies from Middle Eastern bazaars, foresters more or less got to plant whatever they wanted.[11]

This wasn't the only context in which he became acquainted with JNF personnel. Watershed management and soil conservation naturally required coordination with the Forestry Department. As an economist and public servant, Zaban was concerned about glaring inefficiencies he identified in resource utilization for water projects. The JNF leadership, however, appeared largely indifferent. In one case, Zaban approached the agency's strongman, Yosef Weitz himself, showing him how a proposed JNF dam for the ephemeral Shikma stream made no sense economically. Zaban suggested that a cost-benefit evaluation might be in order. Weitz smiled and explained: "I have money; I don't need to do an economic analysis." The dam was built.

In 1981 Zaban received a call. It appeared there was political support for him to become director of the Land Development Authority, JNF's opera-

Haim Zaban in 1982, soon after starting a quiet revolution that professionalized the country's foresters and introduced sustainable forestry practices to Israel. (Joe Malcolm, KKL-JNF Photo Archives)

tional arm, responsible for forestry. "Come quickly," outgoing director, Meir Shamir told him. "Otherwise, Ben Shemesh [then the JNF CEO] will grab the position for himself!"

Zaban didn't hesitate to seize the professional opportunity, but soon realized he would have to make meaningful changes in personnel if he was to implement the many reforms he envisioned. The JNF's labor union to this day wields remarkable power—and imposes its views on most personnel appointments. Bringing in new blood is no easy matter, and the politically elected board of directors is typically disinclined to drag the organization into the throes of a general strike over nuances in résumé quality and professional qualifications.

And so Zaban began to sneak idealistic and able men with relevant university training into his refurbished forestry organization. He was particularly interested in finding workers inclined to study problems analytically rather than automatically accepting the way the system worked. The labor union, naturally, was totally opposed: "Over our dead body are you going to bring new academics for management jobs into the JNF," the union informed him. "If you want university graduates, we'll send our people to get degrees. Then you can hire them."[12]

Multiyear, internal training programs for staffers were established. But Zaban did not want to wait for the field workers and bulldozer operators to complete degrees in agronomy, geology, and horticulture. He began to fund external academics such as Yossi Riov and René Karschon on the condition that they devote research time and resources to the applied questions that JNF faced. This approach fell far short of the changes he had in mind. Zaban found that there were few jobs for which he enjoyed unrestricted authority to hire without consultation. Nor was it easy to create new positions. But there were a few exceptions. His personal driver, for instance, was an appointment he made independently, as it required extended intimate contact and personal compatibility. To Zaban, the very idea of a personal driver seemed absurd and wasteful. Zaban decided to continue driving himself while hiring young college graduates to take a fictitious chauffeur's position. He would give them sundry forestry assignments and let them build up practical skill sets. When he would post a new professional position, the driver could compete for the job as a bona fide JNF employee. Based on the academic training criteria he tailored, the driver's appropriate CV had no trouble winning the tender. Then Zaban would hire another "driver."

> There were those who talked to me about 'transparency' in management; they were right in theory. But if I wanted to upgrade the staff, I simply didn't have time. I hired academics as secretaries too. Mordechai Ruach, who joined the JNF planters as an immigrant from Egypt in the public works program thirty-years earlier, turned out to be a very talented fellow. On his own initiative he had acquired a bachelor's degree. As soon as I could, I made him head of the Forestry Department. That assuaged the union a little bit.[13]

When Ruach stepped down he was succeeded by Menahem "Nachscheh" Sachs—a Ph.D. ecologist from the government's agricultural research center. Sachs was a particularly unusual "acquisition," who actually managed to enter the Forestry Department through the "front door." Zaban brought him on in 1981 to provide academic support for Mordechai Ruach, with the immediate mission of overseeing the country's twelve tree nurseries. The dispersed and inefficient nursery system was a remnant of JNF forestry work during the Mandate, when travel was difficult and nurseries were established contiguous to major afforestation projects. Unencumbered by traditions and preconceived notions, the ebullient Sachs reached the conclusion that one national nursery would be quite enough. Given the shockwaves and human perturbations closing eleven nurseries would cause a somewhat ossified organization, Sachs was told to be realistic. He ultimately succeeded in cutting the number down to three. When Ruach retired, Zaban had a replacement ready to go. Israel's Forestry Department suddenly was being run by a research ecologist—which was nothing short of a revolution!

And so through Zaban's seemingly chaotic methods, a cadre of "young Turks" came on board. Most of them graduates of the Hebrew University Faculty of Agriculture—as there has never been an independent forestry degree granted in Israeli universities. A surprisingly high percentage of today's senior managers in JNF began as Zaban's "recruits": Itzik Moshe, who has overseen the JNF innovations in arid land forestry; Israel Tauber, who introduced geographic information systems into the planning process. David Brand, present head of forestry; Omri Bonneh, who ran JNF's northern region for two decades before serving as chief scientist. The list goes on and on.

Israel Tauber remembers how he came to be Zaban's driver: "I was finishing my studies in soil and water at Hebrew University and heard that JNF was looking for a soil surveyor. So I started buying materials for sampling and driving around in my old Fiat testing soils for them. I didn't have a contract and they probably paid me no more than the costs of the gasoline, if that. But it opened a door that allowed me to start working for Haim Zaban directly as his 'driver.' It was like being in school again. I began to see the big picture."[14]

Itzik Moshe, another recruit from the period, recalls the cultural clash between the academics and the veteran work force. "I was only twenty-seven when I started work. Pretty much all the foresters were in their fifties and sixties. In those days there were hundreds of workers planting trees every day. I was probably the only forester at the time with a university degree. But I had to work with these people, and show them that I wasn't some pretentious academic, but just a simple guy from Ashkelon who grew up in the same poor neighborhoods that they lived in."[15] In retrospect, it did not take him long to gain his colleagues' confidence and redefine the way trees were planted in Israel's southlands.

The dramatic transformation that took place in the JNF since Zaban's tenure can be attributed not only to the resourcefulness but to the stamina of these young workers. Environmental NGOs around the world are unable to pay workers well, so they often suffer from a corresponding "revolving door" phenomenon. Young people have a good time at their NGO posts and gain experience from them, but after a few years seek "real jobs" with pensions, health care coverage, and paychecks that cover the costs of living. Israel is no exception to this phenomenon. Because of the revenues from its land holdings, JNF can offer its workers competitive salaries. This has sustained the idealism of Zaban's talented foresters, allowing them to build lifelong careers in which Israel's forests became a personal mission.

A New Green Gospel

For Menahem Sachs, forestry constituted a second career after reaching department chair in the government's Volcani Institute of Agriculture Research. He would joke that his "previous life" should at least make him immune to accusations of being addicted to trees. Sachs's academic colleagues thought he was mad when he answered Zaban's call, leaving academia behind to start anew at the JNF nurseries, without pay, "to see what forestry was all about." With a professed goal of "moving ahead while making the minimum mistakes possible," Sachs liked the idea of joining an organization that was fundamentally a development agency.[16] He minimized errors by opening up the organization to multidisciplinary consultation and peer review.

Sachs decided to turn the absence of a JNF research department to his advantage. Unlike the U.S. Forest Service with its thirty thousand employees, he ran a small national forest service, with few workers formally trained as foresters, an anti-intellectual heritage, and minimal internal research capacity. The void enabled him to seek advice from outside experts who were not locked into age-old biases or given to pressures from intellectually inflexible or tightfisted bosses. As head of the Forestry Department he embraced three basic principles: first, he had a duty to be ecologically responsible; second, he knew surprisingly little about the natural processes associated with Israel's unique forestry conditions—but wanted to learn more; finally, he was committed to seeking out people who could help the Forestry Department in this journey of discovery.

It was this ethos that led him to Professor Imanuel Noy-Meir, challenging him to take a year off and play the role of ecological conscience for the country's woodlands. Born in Argentina, Noy-Meir moved to Israel as a child and was considered a brilliant but introverted ecologist. His work in the Hebrew University botany department led him to explore the effects of grazing on grasslands, whereby he developed a model of vegetation dynamics that remains highly influential to this day. Noy-Meir took a sabbatical and agreed to review Israel's forestry program. He presented his findings in an internal 1985 lecture, later summarized as an article in a now defunct German forestry journal, *Algemeine Forst Zeitschrift.* His perspective provided an academic paradigm for the operational changes that took place during the Zaban era.

Noy-Meir began by summarizing the five essential complaints of environmentalists who disparaged conventional Israeli foresters and pine forests: (1) Monocultures created a visually monotonous landscape. (2) Native vegetation suffered from the planting of a single species—creating "pine deserts" beneath the shade of the conifers that were barren and boring. (3) Tall, dense pine forests were an emulation of Europe and inappropriate for Israel. (4) Jerusalem pine trees were only suitable for growing on rendzina-type soils. And (5) dense pine forests did not allow for pastures to grow, sabotaging potential grazing.[17]

Rather than refute these points, Professor Noy-Meir implicitly accepted their legitimacy, preferring to understand the events and experiences that led to the decision that favored dense Aleppo pine monocultures. Noy-Meir concluded that it came about through a thoughtful process based on cumulative local know-how that evolved from systematic observations and controlled experiments. The system was not an application of inappropriate theories imported from overseas. Noy-Meir was actually sympathetic to the logic behind the seemingly "anti-ecological" practices that had successfully "afforested large areas in a short time at a low cost." The trouble was, he explained, in the long-term, the strategy was doomed to failure:

> A high initial density of the forest populations affords an advantage in competing trees from other species, but it also increases the competition between trees from the same variety. The more dense the forest, the less light, water and minerals are available for each tree, and therefore, the growth rate of each individual tree is slower. In the dense stands, trees tend to be smaller, weaker and more susceptible to diseases, pests and adverse climactic conditions. . . . The negative influence of density on the wellbeing of trees is small during the first years after planting, but constantly increases as the trees mature, the tree canopies take full advantage of all the available light and the roots occupy the total root space.
>
> In planted forests it is difficult to rely on the process of self-thinning as an efficient regulator of population density and to expect it to bring about natural selection in favor of the genetically most successful trees. . . . Ecological considerations lead us to expect that uniform and dense planting will accelerate the development of the forest during its initial years, but will adversely affect its quality and healthy during its mature years, leading to a shorter life span.[18]

While acknowledging the contribution of thirty-five years of afforestation to the country's landscape, it was clear that existing woodlands also created problems that foresters in 1985 needed to tackle. The most pressing immediate task was to reduce tree density and make a much more serious commitment to thinning existing stands. The problem is that thinning young forests is never cost-effective in dense stands with low precipitation, as the trees removed are skinny, with negligible economic value. But it needed to be done. For many of the country's pine forests, the diagnosis was grim: it was already too late to upgrade tree robustness by thinning. Many stands were beyond remediation. Noy-Meir posited that the longevity of many Aleppo pine trees would remain low, less than half *Pinus halepensis*'s usual 100- to 150-year life span. Pest outbreaks would remain constant.[19]

Despite this pessimistic projection, his analysis ultimately contained a silver lining:

> "Pioneer plants" is an ecological term that refers to plants which are the first to populate and occupy areas upon the removal of the previously existing flora (as a result of land slides, fire, felling). These plants flourish quickly in the open and produce seeds rather early on. However, their life span is

> relatively brief. Over a period of time, these are gradually replaced in the process of succession by slower-growing plants which take full advantage of the soil and light. These establish the permanent plant population. In many cases, these plants cannot survive in the open and only the pioneer plants create the shade, moisture and soil conditions indispensable for the germination and establishment of the permanent plants.[20]

The fact that there were forests around the world where pines played the role of "pioneer trees" gave a sense of order to what might otherwise be considered egregious ecological folly and waste of resources. Noy-Meir was encouraged by anecdotal locations within Israel, where "the slow process of replacing a mature planted pine forest by oaks that have been growing in its shade is currently underway." He also cited research suggesting that oaks are more successful in the shade of pine trees and under pine needles than in open spaces. Noy-Meir concluded that in a small, crowded, increasingly urban country, the recreational role of forests as a home for hiking and nature would only grow. This had operational ramifications:

> In the past the success of afforestation was measured by the size of the forest (the number of hectares and trees). In the future, its success will be evaluated by the quality of the forests created, in environmental terms and/or economic contribution per hectare. If in the past, the demand was to produce forests fast—in the future it will be to create long-lived forests which do not have to be replanted every few decades. . . . A forest may be long-living if its constituent trees are so or if it fosters processes of natural turnover in the population through regeneration. A forest that exists over a long period of time as a result of such processes is a permanent forest. If in the past the Forestry Department's goal was to create the pioneer forests, the goal for the future is to bring about permanent ones.[21]

Most foresters in the JNF had never heard of Imanuel Noy-Meir before he was recruited for the sabbatical, and he remains largely anonymous today. But his involvement offers a rare example of a dispassionate and ingenuous academic's ability to articulate a critical sea-change in policy.

The Transition

Even before Noy-Meir's analysis, the transformation from monocultures to mixed stands had begun. Committees convened after the Sha'ar HaGai episode and recommendations were made. The immediate response in the 1970s after the crumpling of the Jerusalem pines was a shift to other, more resilient *pine* species. By 1987, conifers still made up 65 percent of the trees

Thirty years later, in 2012, Omri Bonneh visits the Carmel foothills and the first multispecies forest he ever planted. (Photograph by Alon Tal)

planted, but the Jerusalem pine accounted for only 11 percent, with a variety of other nonnative species (*Pinus brutia, Pinus pinea, Pinus canariensis*) and cypresses filling out the ranks. Oak trees and eucalyptus were roughly 8 percent of planted trees and there was modest experimentation, integrating fruit trees.[22] In 1990 the national benchmark for "broadleafed species" was upped to 25 percent.

These changes had little to do with ideology or aspirations to restore authentic ecological assemblage and a lot to do with reduction of pest risks. In a 1989 article Zaban wrote matter-of-factly: "Various pests and arboreal diseases have damaged many of the common forest trees and necessitated the introduction of alternative varieties."[23] Dr. Omri Bonneh was among the influential members of the committee tasked with redesigning forestry practices and was even blunter in his pragmatism: "Planting a mixture of coniferous and native broadleaf species is aimed at preventing the loss of complete plantations in the case of a species-specific damaging agent. In case of a heavy die-back of a particular species, the remaining species will ensure the sustained functionality of the forest stand."[24]

But this constituted a transitional stage. A new, ecologically driven ideology that idealized diversity and eschewed homogeneity began to push the organization to an even more dramatic transformation. Just as Weitz waxed eloquent about pine trees' virtues, the new forestry gospel saw variety as the heart and soul of sustainable silviculture:

> The mixed forest captures the eye from the perspective of landscape and in so doing embellishes the environmental and the social benefits it provides. In addition, its appropriateness for grazing rangelands increases. The endemic, broadleaf trees are not temporary, and they have the ability to develop nicely in the shade of the conifers. Thus, it is possible to reach a mature forest where

> the upper level is held by conifers and the undergrowth flourish in natural woodlands. These trees' longevity and their ability to regenerate after a fire turn them into one of the lasting treasures of the land.[25]

While the pathology of monocultures led to their "fall from grace," it is well to recall that Aleppo pine trees always grew in Israel. Yoav Sagi has been a leader in the Society for Protection of Nature in Israel (SPNI) for fifty years and a consistent critic of JNF forestry practices. One of the things that most angers him about local forestry history is that it gave *Pinus halepensis* a bad name: "With the right distribution, it's actually a very nice tree."[26]

Historically Aleppo pines were relatively rare, primarily relegated to the tops of ridges. Aviva Rabinovitch was the chief ecologist at Israel's Nature Reserve Authority during the 1970s and 1980s, and arguably the single most vociferous opponent of accepted forestry practices. Her doctoral dissertation focused on the natural distribution of soils for Israel's indigenous trees. She surmised that the Aleppo pine trees only grew naturally on the soft rendzina soils found on the hillsides.[27] Today, the pine trees are confounding her painstaking research and sweeping hypothesis by successfully establishing themselves across Israel's woodlands on a variety of soils.[28] Rising above the Mediterranean oak scrub, they naturally emerge as 15 to 20 percent of the assemblage in renewed forests after fires. This has surprised the experts, who long postulated that left to their own devices, successional processes in woodlands would lead to a diverse broadleaf assemblage dominated by oak trees. "When the shade tolerating shrubbery is left alone, the rejuvenation of the pine on the spot will be extremely difficult, so that the pine forest will disappear and on its stead the evergreen maquis will appear."[29]

A well-funded, highly motivated, afforestation bureaucracy was not about to wait centuries for nature to take its course. Open spaces in a crowded country need to be managed thoughtfully and professionally. The operational goal became to accelerate natural processes and to expedite the rich assortment of trees in new forests. This "recruitment" of native species into monocultures' existing understory in order to improve the genetic, structural, and functional diversity is considered a key component of sustainable management in forest plantations.[30] As it turns out, adding additional species is not without its problems.

To imitate nature, foresters interspersed a healthy mixture of tree sizes and species throughout newly planted parcels. During a twenty-year period beginning in the 1970s, the number of tree species raised in nurseries for planting steadily increased from fifteen to sixty.[31] Yoram Goldring, an ecologist working at JNF, completed a doctoral thesis in 1977 around the hypothesis that pioneering coniferous species could improve the establishment of broadleaf native trees and enhance their growth.[32] While he found practically no oak trees growing in young pine stands, they appeared as understory in many older stands. Theoretically, it seemed that with a little assertive intervention, oaks could easily be recruited Unfortunately, however, the trees did not cooperate.

When broadleaf natives—such as oak trees or terebinths—were randomly distributed among pine seedlings, they could not compete with the conifers. If they did manage to find sufficient light to survive, native broadleaf trees grew painfully slowly and remained as suppressed understory, fifteen years after planting. Techniques were eventually developed for transplantation of mature oak trees that yielded impressive survival rates of 96 percent.[33] But this was an extremely expensive measure and infeasible on a macrolevel—a feasible solution would have to rely on acorns.

The nuts coming from Israel's common oak tree (*Quercus calliprinos*), like most acorns, contain a single seed, enclosed in a tough leathery shell, maturing a year and a half after pollination. Israel's acorns tend to be on the small side, with a length and diameter rarely exceeding two centimeters. In retrospect, their difficulties in getting established in Israel's pine forests should not have been surprising.

The forestry literature reveals a well-documented paradox associated with recruiting new types of plants into existing monocultures. It translates into a practical management dilemma: seeds or seedlings can find available sites to germinate either in open spaces or in forest shade. Neither option is perfect; forest canopies simultaneously have a positive and negative effect on growth. On the one hand, many tree species have an easier time getting established in shady sites. On the other, they pay a price for the additional moisture and evapotranspiration advantages. While surviving the first year, they languish as a result of the diminished lighting.[34] Moreover, a physiological tradeoff has been hypothesized, suggesting that the better plants are able to cope with shade the less they can handle drought conditions, which are not uncommon locally.[35]

These generic dynamics were examined systematically in Israel by Ben-Gurion University graduate student Shani Ben-Yair. Her research sought to characterize optimal conditions for recruiting oaks into established pine stands. Working in diverse plots in the Massuah Forest in the Judean "lowlands," she found that acorns planted under a protective forest canopy significantly outperformed those planted in gap areas. While the initial survival rates in the woods were high, subsequent tree development was poor and the new oak saplings remained puny and vulnerable. In the open sites, development was vigorous, but only a modest proportion of acorns successfully germinated.[36]

Mature pine stand in the Maneshe Forest. Note rich understory that thrives after trees are thinned. (Pnina Livny, KKL-JNF Photo Archives)

In the real world, outside of controlled experiments, tree regeneration is limited by the availability of seeds, not only planting sites. Israeli *Quercus calliprinos* acorns, it turns out, do not drift far from their mother trees, unless rodents transport them or they roll down a slope. Moreover, under field conditions, they are only viable for seven to twenty-eight days, although they can last longer when protected by forest canopy.[37] During this time there are plenty of hungry foragers roaming about, especially in the shaded areas, seeking the nuts. Rich in protein, carbohydrates, and fats, along with many minerals and vitamins, acorns provide critical food for birds and rodents, as well as for larger mammals such as deer and wild boar. In forests, foraging animals are better protected from predators; moreover, with mother trees nearby, they enjoy a much larger supply of nuts. By way of contrast, open gaps in the forest are relatively free of foraging animals and serve as relatively "safe sites" for acorns, even late-buried ones. When combined, these factors led Ben-Yair and her advisors to the conclusion that oak recruitment in forest gaps was a more promising long-term strategy than recruitment in heavily shaded stands.[38]

Even twenty years before this empirical demonstration was published, JNF practices had already made this shift intuitively. New forests contained parcels with different tree types, together forming a patchwork quilt with an assortment of species. The layout made for more efficient management, with uniform weed control, irrigation, or polypropylene tree shelters designed especially for a given patch. This sort of "separate-but-equal" affirmative action improved the survival rates of the broadleafed natives. Prior to that, many of the valorized broadleaf species simply could not make a go of it in areas where rainfall drops below four hundred millimeters.[39] Recent experience shows that establishing Tabor oak trees is more successful when direct sewing of acorns is accompanied by supplementary practices, such as using tree shelters, plastic sheeting, and a nutrient containing polymer (Barbary Plante). Mulching can obviate the need for irrigation during the summer months.[40]

The management reforms were also manifested in a softer and gentler approach to the planting process. For instance, in areas with modest rainfall, soil conservation and ecological integrity received greater attention. Native trees were meticulously preserved while preparing the ground for seedlings, and the wholesale burning of undergrowth phased out. Responding to criticism about the damage caused by breaking up terra rossa soils through deep subsoiling, protocols changed to minimize visual and ecological impacts. Hulking bulldozers were replaced by more meticulous "point preparation" machinery. At first augurs were pulled by small-wheeled tractors, but even this caused unnecessary injury to lands. Eventually, JNF planters employed excavators with pneumatic hammers that could break up even basaltic soils and boulders, shattering bedrock up to depths of sixty centimeters. Even on steep terrain, the hammers showed reasonable maneuverability and left much of the natural soil and vegetative cover intact. In a happy coincidence, the gentler planting techniques were not only easier on existing ecosystems but they also produced improved seedling survival rates.[41] Each climate and terrain required its own planting methods. For instance, preparation of lands in the semi-arid Negev necessitated entirely different techniques than steep, rocky hills in the Galilee.

Pesticide use is another area where decades of habits began to change. By the time the reforms began, Israel's forests had a fairly toxic baseline. In 1960, the JNF introduced simazine into its chemical arsenal. This herbicide is particularly effective against broadleafed weeds and annual grasses. Sprayed from airplanes over large swaths of lands, it did a superb job of keeping weeds at bay, saving thousands of worker hours. But it was a highly toxic poison. Not only were there concerns that the chemicals would run off and bio-accumulate in aquatic organisms frequenting local streams and lakes; it was known that deleterious health effects in humans were also likely. For example, a high incidence of mammary gland tumors in female mice appeared after exposure to simazine, suggesting endocrine toxicity.[42]

Once again, forestry policy moved in stages toward sustainability. At first ground crews used backpack sprayers to ensure that native trees and vegetation patches remained free of the chemical.[43] Cognizant of European policies regarding the chemical, the JNF board's Sustainable Development Committee decided in 2005 to phase out use of the use of simazine in Israeli forestry programs as well. Since then, when JNF needs to resort to herbicides, it uses the more benign and costly biodegradable alternatives.

Other upgrades in basic planting practices were implemented. For instance, planting conifer seedlings used to take place with their bare roots exposed. As forests were planted on increasingly marginal (and less hospitable) lands, saplings had to be tougher than ever. The critical factor for success was an ability to develop extensive root systems. Tree nurseries began to maximize a parameter known as *acorns' root growth capacity* (RGC). This required planting in containers that hold soil and nutrients, giving saplings a better micro-environment for getting started.[44] And with the results of many years of research and application, the quality of the seeds steadily improved.

No area of change was as dramatic as that of tree density. In 1974, Weitz summarized Israeli planting practices of the time: "For the purposes of obtaining cordwood etc. one may conclude that Aleppo pine should be planted as close as possible." Weitz rejected conventional notions that density led to crooked trees, and attributed any kinks and bends to genetic predisposition. Accordingly, he prescribes no less than 2,500 pines per hectare as an optimal planting density.[45] In practice during the 1950s, density was even greater—reaching 2,890 trees per hectare. When one considers the percentage of stone cover on many rocky planting sites, the effective density was higher still.[46]

The 1990 JNF best-practices guide called for a completely different strategy. It assumed that due to high expenses, especially on rocky and steep sites with no accessibility to mechanical forestry, there would only be a single thinning for pines that reach a height of 2.5 to 3.0 meters, between the ages of eight and twelve years. Presumably by then, trees would be less vulnerable to damage from grazing animals. With this in mind, a new density standard in rocky areas was set at a mere 450 to 650 trees per hectare—an order of magnitude lower than Weitz's reference point—with "no limitations on subsequent thinning."[47] By way of reference, in semi-arid regions in Spain, Aleppo pines are typically planted at a density of a thousand trees per hectare, albeit, typically only half of the saplings survive.[48]

The dramatic reduction was made easier because of steady improvements in seedling quality, weed control, wheeled tractors equipped with augers and excavators, maneuverability in point-site preparation, fencing, and tree shelters. The proximate reason behind the decision to lower densities was im-

Considerable efforts have been invested in ensuring a more robust seed stock. Upgraded jujube (*Ziziphus spina-christi*) seeds collected at the Beit Nehemiah National Seed Center. (Shimon Lev, KKL-JNF Photo Archives)

proved tree health. Yet, there was also an implicit assumption that greater spacing offered additional benefits, such as natural generation of broadleaf species on the exposed forest floor.[49] Of course actual field conditions drove final decisions. Trees in drier climates with less access to water or soil nutrients required even lower densities. The trend toward more dispersed planting, begun in the 1980s, continues to this day.[50]

Looking back on this historic transition, there are experts who believe that initial obsessions with dense Aleppo pine stands were fortuitous. One such apologist is Hebrew University forestry professor emeritus Yossi Riov, who offers an impassioned defense of early management decisions: "The cover of the pioneer species improves the habitat (adding organic materials, softening the soils, allowing development of helpful micro-flora and shading). This facilitates development of the second generation that can be quite different than that of the first generation. Relevant criteria should consider the qualities of the species and their suitability for afforestation. Pines are a robust pioneer species while most of the broadleaf varieties are sensitive to environmental conditions and prefer shade for initial development. The very low rate of survival of planted broadleafed species proves this. I have no doubt that thanks to the pine we succeeded in establishing the first generation of forests in Israel."[51]

An American Makes Israeli Salad

Muddling forward, the next generation of Israel's forests looked very different. Paul Ginsberg's experience as a forester in the JNF personifies the institutional and substantive transition that Israel's forestry underwent during the 1980s and 1990s. An American from New York, in 1980 Ginsberg came as a volunteer to work in the cooperative agricultural community of Kibbutz Sasa on Israel's mountainous northern border. He soon fell in love with the romantic country, the scenic settlement, and one of its residents. Already envisioning a career as a forester, he was urged by local experts to acquire his professional credentials abroad, so he returned to the United States with his girlfriend and completed a master's degree at the SUNY College of Environmental Science and Forestry in Syracuse, New York.

Not long after he and his bride returned to her kibbutz, he answered an advertisement for a forester's position in the northern region. There were thirty candidates for the job, and some concern that as a nonnative, his Hebrew was not yet up to the task. But in 1985, senior JNF management was looking for professionals and almost no one in the country had a bona fide forestry degree, much less from a leading American university. Ginsberg was accepted on a trial basis that stretched on for twenty-one years, during which time he was given responsibility for overseeing all Galilee forests.

His first job was to manage the woodlands around Baram, near his home on the Lebanese border. This is about as remote a region as exists in Israel. In the absence of a forest manager, personnel had been transferred to other locations. During his first two months on the job, Ginsberg's only employee was a Bedouin watchman, Ali Hamdoon, Without a staff to oversee, Ginsberg was free to spend his initial days wandering the woods now under his supervision. After familiarizing himself with the many plaques demarcating the magnanimity of donors from around the world, he began to map out the state of the forests and work that needed to be done in each stand. By the time JNF workers from the Safed region showed up, Ginsberg had a detailed program for pruning and thinning.

Ginsberg is a bit sardonic when he describes the veteran JNF workers who did not quite know what to make of their young American boss. This was before the shift to outsourcing and subcontracting in most forestry fieldwork. Many of the workers were the same Jewish immigrants from Arab countries who during the 1950s had planted the massive pine monocultures and had stayed on as JNF workers. "A white Ashkenazi boy, without a yarmulke was not especially well received," Ginsberg recalls with a touch of bitterness. "All these guys knew how to do was plant. Their view basically was that you weren't a real forester until you had planted a hundred thousand trees. Six times that year I wanted to quit. But I stuck at it and insisted on changing the orientation."[52]

Having done undergraduate work in environmental studies and planning, Ginsberg was well aware of the ecological critique of JNF's "green deserts" and of the political capital that critics reaped from passing judgment on its forests. So he was pleasantly surprised to discover that reality was actually quite different than the lifeless stereotype. Under the Brutia and Jerusalem pines planted during the 1950s and early 1960s was a rich understory filled with oaks, teberinths, and redbud trees:

> You could see the principles of forestry at play—the competition for light, usually described in native forests. When the forest is dense, oaks don't really develop and appear more as a high shrub. But in those places where cover was not full, the light gaps simply pulled the oaks up and they were as tall as seven or eight meters high. I studied in New York where there's a maple broad leaf forest. But I could see the exact same principles at play in the Mediterranean. It empowered me to apply these dynamics.[53]

Ginsberg was convinced that with the right interventions he could strike a balance between planted and natural forests. The two could coexist and even thrive alongside each other. In 1989 the Israeli government initiated a public works initiative in response to a wave of unemployment caused by massive Russian immigration. This provided Ginsberg with a platoon of field workers and a chance to put his ideas into practice in the Safsufa Forest. Combining the manual laborers with a subcontractor he had brought in to help with thinning, he began to remove copious loads of branches and wood boles. At once he could both thin the forest to reduce fire hazards and improve native vegetation. "At that point, basically there were no visitors to this area of the forest. Really, there were no recreational facilities to speak of. For my own selfish reasons, knowing what happened to forests when they are developed 'recreationally'—the damage to the trees, the litter—I wanted to keep it as my own secret garden."

Ginsberg's vindication came when he hosted the foresters from the central region. Gershon Avni, the seasoned JNF central regional director joined his forestry staff on a field trip to see what was new with his colleagues up north. They were not anticipating the experience they would have when they entered what can best be described as a natural cathedral. Ginsberg recalls them walking reverentially through the towering Brutia pine trees, embellished by natural Mediterranean woods. "Gershon Avni paid me the highest compliment I think I have ever received. He said: 'You know they asked Michelangelo—"How do you know how to make such beautiful statutes from a simple piece of rock?" And Michelangelo replied, "It's in there all the time—you just have to know what to take away."'"[54]

With support from the higher echelons of the Forestry Department, Ginsberg continued to develop what he describes as an "Israeli salad": "My approach to planting was to randomly mix trees together, with the idea in the end of having a two-layered forest: a high canopy of conifers and a low canopy of broadleafs. Developing the next stage of forests is a balance between science and art. It's like an artist with a palette. With a scientific approach, you could shape it into anything you want. I saw these newly formed forests as a personal Zionist contribution."[55]

Divine Favor for Forest Planting

Even among Israel's idiosyncratic environment and planning elite, Moti Kaplan is an unusual personage. He grew up in an ultra-orthodox family and still retains a strong connection to the world of Torah scholarship in which he excelled as a youth. Opting out of a natural rabbinical career, Kaplan chose to study geology. After graduating Hebrew University, he landed a position at TAHAL, a water-management consulting firm.[56]

By the middle of the 1980s, the JNF was beginning to understand that it was running out of space for planting forests and needed to be more sophisticated about matching its trees with underlying geological formations. Kaplan was hired by Haim Zaban to collect aerial photographs and identify appropriate, available open spaces across Israel for future forestry projects. Kaplan returned with a series of locations throughout the country that were not being farmed yet were home to natural woodlands showing signs of degradation. He suggested that the JNF forestry branch focus its efforts on renewing these native maquis and scrublands.[57] At first there was not much enthusiasm for his approach, but Kaplan has a stubborn streak and can be highly persuasive. Backed up by dozens of detailed maps and a gift for mastering technical details, he held his ground and sought allies for a strategy that would maximize open space preservation for the future. Eventually, Haim Zaban took note.

For some time Zaban had been frustrated by the lack of a statutory master plan for afforestation in Israel on which he might base long-term strategies. In theory, a national blueprint for woodlands should long since have been completed. When Israel's Knesset passed the Planning and Building Law in July 1965 it created a platform for systematically coordinating rapid national development.[58] It had taken twelve years to enact a law to replace the British Mandate's 1936 Town Planning Ordinance. The new system was innovative and democratic for its time—encouraging public participation by mandatory notification of proposed building initiatives and facilitating the filing of legal objections to development plans.[59]

The system is built on a three-tiered system of planning commissions with a National Planning Council at the top of the pyramid. Membership is composed of representatives from government ministries as well as a range of constituencies—from green organizations and engineering guilds to young people, mayors, and women. Its mandate is to coordinate land use in the country by reviewing and approving national master plans and their by-laws. In theory, the council functions as a marketplace where competing interest groups make their case and bargain over whether a piece of land should be zoned as an industrial park, a nature reserve, or a highway. After zoning decisions are made at the national level, six regional planning commissions translate the broadly sketched borderlines and principles into regional plans. Finally, 116 local planning commissions convert the regional plans to an even higher level of resolution as local outline plans.[60]

Once a master plan is approved it has the force of law. While broad policy issues are debated ferociously by the council, the thirty-two members do not function as a working group. Plans are drafted by the staff of the Interior Ministry's National Planning Administration or by professional planners it hires. These specialists wield tremendous influence. Much of the environmental protection existing in Israel is the result of specific provisions and maps prepared by planners who compose national master plans.[61] The national plan for forestry constitutes an extreme case.

In 1976, Israel's National Planning Council called for a national master plan for Forest and Afforestation (Master Plan 22), entrusting the task to the JNF, the Ministry of Interior's Planning Administration, and the Ministry of Agriculture. The JNF commissioned an initial draft, circulated in 1980, from eminent urban planner Shlomo Aronson. The plan engendered such fierce political resistance that it was quickly dropped.[62] Soon thereafter Zaban began his tenure overseeing JNF land development. He understood that the lack of a comprehensive planning scheme perpetuated disarray, narrow thinking, improvisation, intuitive decision making, misplaced priorities, and ultimately . . . mistakes.

Many foresters understood the benefits of demarcating the borders and configurations of forests, especially when facing pressures from rival land users. But JNF's CEO, Shimon Ben Shemesh, was disinclined to follow that course. The organization had gotten used to operating with little external scrutiny and its leadership enjoyed the flexibility and spontaneity that its special, poorly defined status provided. It might seem appropriate to plant trees today, and later make sense to develop the same parcel. What was the point of straightjacketing decision makers with procedures and bureaucracy? The message from Ben Shemesh was *not* to cooperate with the planners from the Interior Ministry when they asked for technical assistance. The Minister of Agriculture, Arik Nahamkin, was also disinclined, concerned that the plan might whittle away his formal ministerial authorities regarding the forests.[63] Without an institutional sponsor, the draft forestry master plan became an orphan and was soon forgotten.

Dramatic changes in Israel's demographic reality, however, were becoming readily apparent, and the implications for forestry planning could not be ignored. In 1989, a massive migration of one million Jews from the former Soviet Union began to reach Israel's shores. The country faced an acute housing shortage accompanied by an existential crisis: after working tirelessly for the removal of restrictions on Jewish immigration, Israel would not be able to provide the immigrants, their oppressed brethren, a place to live. How then could Israel claim to offer, a haven for the oppressed? The government responded by removing many of the traditional constraints on development.[64] In the resulting tumultuous speculation and land grab, open spaces and particularly forest lands were sacrificed.[65] Under these circumstances, a national plan to consolidate and protect the country's forests made sense.

National programs in other areas had already been approved and the country was filling up. Lands that might have been forests were already zoned for power plants, military training, quarrying, farms, and cities. A new CEO took over at the JNF for a brief tenure — former general Ori Orr. He was more open than his predecessor to a master plan. Dina Rachevsky who headed the Planning Administration at the Interior Ministry approached him and after ten minutes agreed that the time had come. But the ministry had neither the resources nor the expertise to take on the huge task of identifying available lands and designating which forests were appropriate. This time, JNF volunteered. Zaban was at the end of his eight-year tenure, but managed to commission Moti Kaplan, the young maverick geologist, for the job. The choice for such a formidable task was not self-evident. But Kaplan had already prepared preliminary maps, and was a lot cheaper to hire than well-established private planning offices. And everyone agreed that he was very bright.

For Kaplan the project was the ticket he had been waiting for to leave his safe Tel Aviv job and set up his own shop in Jerusalem. The marching orders he received were to map out the sixty thousand hectares of land that were home to the existing woodlands. From the first days of JNF afforestation in 1921, the official vision of total future forest coverage was limited to 3 percent of the country or about sixty thousand hectares. This objective was never officially revised, and by the 1980s there was an unspoken consensus about it.[66] In its day, the British Mandate's Forestry Department reached more ambitious conclusions, suggesting 15 percent of Palestine, or four hundred thousand hectares, suitable for afforestation.[67] In Israel's early days, Yosef Weitz opted for the more modest target.[68] The sixty thousand hectares of woodlands were mostly planted and Kaplan believed they would surely be preserved, regardless of his efforts. He saw Master Plan 22 as a chance to save additional countryside that would otherwise be lost to development. In 1989 few Israelis understood how fast the country's open spaces were dwindling. That year Kaplan wrote an article and brought together Zaban and SPNI nature advocate Yoav Sagi as coauthors. In retrospect, the publication offered an indication of how ambitious National Master Plan 22 would eventually be.

Kaplan started by redefining conventional notions of forests. His proposed plan recognized eight different categories of forest lands. Some of these are ordinary enough: *existing planted forests* referred to the stock of large coniferous areas, home to most of the country's wooded sites for hiking and picnicking. *Natural woodlands for conservation* included outstanding scenic areas, where endemic trees were either in a successional state or contained threatened species that needed special protection. *Proposed planted forests* identified remaining areas available for planting conventional forests, and *natural woodlands for cultivation* were not yet rehabilitated natural lands that could be used for extensive recreational use or even timber production. Other forest categories, however, were much more irregular.

Existing forest parks, for instance, indicated rangelands with as little as one hundred trees per hectare. Typically these would include a few scattered Tabor oaks, pistacia, carobs, and olive trees. *Proposed forest parks* were selected precisely because they were devoid of any vegetation at all, even shrubs. Here, scattered drought- and salt-resistant species were to be planted to enhance grazing pastures in the southlands. *Coastal forest park*s went one step further, denoting many of the beach regions south of Tel Aviv. There were few trees in these sandy areas, but Kaplan's plan called for planting scattered sycamores, figs, and dates. *Riverside stream plantings* targeted areas bordering the intermittent streams that flowed when seasonal rains filled them each winter. The restored vegetation was required to conform to the natural, riparian character.[69] This was no blueprint for future afforestation; it was an undisguised attempt to preserve the country's open spaces.

By the time Kaplan had completed his draft Master Plan 22 in 1990, it included 180,000 hectares of disparate forest lands, rather than the 60,000 hectares Zaban had originally envisioned. Before taking the draft to the National Planning Council, Kaplan had to present it to the JNF international board for approval. The architect of the original plan from the 1970s, Shlomo Aronson, was invited to attend. He was asked about the prodigious size of the plan's proposed forest lands. Aronson shrugged: "Moti must be confused—that can't be the number." Kaplan is convinced that the JNF board members themselves did not believe the number either. After all, it meant that the land resources they oversaw would triple overnight. But Kaplan begged: "Please approve the number so when the negotiations begin, I'll have a comfortable level to come down from."[70]

On August 3, 1991, the plan was ready to be shown to the National Planning Council. Not surprisingly, many eyebrows there were raised when they saw just how ambitious the foresters' appetites were. Kaplan still gets excited when he recalls the heated discussions at the time and what it took to push his plan through its initial review:

> Coastal Forest Parks were a case where I had a particularly hard sell. It was easy for me to see then that much of the country's natural sand dunes had disappeared and that only a few reserves remained. These lands were pretty close to "beachfront property" and under high demand. I knew that if their new status became conspicuous, it would spell trouble for the plan. But I had to try and save as much as I could. I came up with this concept of "light yellow" shading. You see, this was before the days of color printers, so I had to color in all the maps by hand. In the map's legend I used a very "light yellow" to indicate the lands set aside as coastal forest parks. To tell you the truth, I imagine most of the people who glanced over the maps really didn't notice the marking.[71]

(When Kaplan proudly digs out an original map with its "light yellow" coastal parks, the shading really is close to invisible.)

The plan was approved "in principle," but then the National Council sent the draft out for comments to regional and local planning committees, government ministries, and other stakeholders. This is precisely the stage at which the plan's previous incarnation had met an untimely death. This time, however, Kaplan's energy, obstinacy, and intelligence changed the dynamics.

The job of lobbying for the plan was too much for an single individual, so a committee was convened for the task. In the ensuing twenty years, Moti Kaplan has been involved in most national planning initiatives, but in those days he was still an unknown geologist. Someone of stature was required to give the plan greater respectability. Aronson agreed to have his name added to the maps, lending them an air of propriety. Ilan Be'eri, a career JNF forester, agreed to serve as the in-house "partner in crime, Be'eri," handled the very complicated logistics of dissemination, presentation, and communication with the many interested parties. Be'eri's agreeable temperament and in-depth understanding of forests and geography proved to be invaluable. "My only argument with Ilan Be'eri during the whole process," recalls Kaplan, "was that his modesty was so exaggerated that he refused initially to have his name included among the authors of the plan."

Looking back, much of the credit for the blueprint's ultimate survival goes to the personality of the late David Nachmias. Nachmias had been a senior official in the Jewish Agency when he was recruited by Zaban, whom he ultimately replaced in 1989 as head of the JNF's Land Development Authority. Nachmias's contribution was important enough to move Kaplan to dedicate his recent book about the principles of Master Plan 22 to his memory:

> Nachmias was an Egyptian Jew—an extremely clever and shrewd man with a great heart: a very unique person. He believed in the program and was willing to make the considerable investment in aerial photographs, photocopying and fieldwork. Sometimes, forests had burned down and didn't appear in the aerial photographs. You had to go into the field, make the measurements, and reinsert them in the maps. Nachmias could see the big picture, and he understood that the plan had potentially historic implications for Israel's future landscape.[72]

The road to approval involved a seemingly endless litany of petty demands. Kaplan's tactics were simple: "I would just give in to whoever argued with me. Today it sounds crazy and even dangerous. A charlatan could have been given the task and it would have been disastrous. But I think I had good intuition at the time, and that I managed to preserve as much as possible." For moral justification, Kaplan relied on the great rabbi Moses Maimonides and his famous ruling that it was permitted to sacrifice one Sabbath in order to save a thousand Sabbaths in the future. "The mayor of Shlomi in the northwest Galilee came to us and said: 'My city is suffering from the Hanita Forest which is preventing our development.' It was a lovely, mature stand of woodlands. But I immediately said: 'Well then, we'll take it out.' People in the JNF were very unhappy about that, but I told them: 'What do you want? To fight for three hundred dunams here and there and sabotage the plan, or get a 1.6 million–dunam package?'"

No little amount of persuasion was required: "The Ministry of Agriculture was very unhappy with the designation of so much land as Forest Parks, which it felt would come at the expense of the country's grazing livestock. I explained to them that not only would they get what they wanted for their clients, but that the plan would actually upgrade the land's carrying capacity. In general, that was my approach trying to overcome the opposition during the prolonged period of notice and comment. I gave everyone pretty much everything they wanted."[73]

Other agencies filed objections. For instance, the Housing Ministry was concerned the plan would slow their emergency campaign to build apartments for the wave of immigrants arriving from the former Soviet Union. The Ministry of Defense, concerned that training grounds for military maneuvers might be affected, delayed progress for a considerable period. And innumerable rural and urban governments were concerned that the plan would stymie future expansion. There were instances when Kaplan decided to wage a fight. To overcome an unreasonable demand, sometimes he would call in heavier fire power, such as the charismatic minister of environment at the time, Yossi Sarid. Sometimes he relied on his own resourcefulness. The Jerusalem corridor was one such case.

The Mateh Yehudah Regional Council contains much of the Judean hillsides surrounding the city of Jerusalem. The mayor of the council, Meirka Weisel, was a hero from the Yom Kippur War who constituted a formative adversary. Weisel entertained a vision of creating six Beverly Hills–style communities on empty lands that were within his council's jurisdiction. The lands were completely desolate, without a single tree. But Kaplan insisted that the open spaces not be fragmented and that the lands be included as *planned forest parks*.

He brought the National Planning Council members on a special field trip to the rural expanses outside of the capital to decide for themselves. There he recited the famous Saul Tchernichovksi poem from the 1930s: "Oh, my land, my homeland; rocky bald mountain." Then he railed at the planning commissioners: "There you have it: Rocky bald mountains. Do you want the country to still have lands that look like that so that future generations will be able to appreciate Tchernichovksi?" They were convinced, and even without a trace of foliage the planned forest parks remained protected.

Preparation of the plan produced one of the more fascinating political episodes of the period, bringing together the foresters at JNF with their traditional adversaries at Israel's Nature Reserves Authority. The two rival agencies had a long history of competition and distrust, which was as much an ideological conflict as it was about turf. During the steering committee meetings, debates over the precise zoning of future forests grew so vociferous that the director general of the Interior Ministry, Amram Kalaji, who sat down the hall, burst into the conference room to make sure that no one had been attacked.[74] From its inception in 1963, the Nature Reserve Authority staff waged a determined battle to preserve habitat for all creatures of the land. Their approach was draconian: declaration of a reserve effectively terminated any development in and around new scenic reserves and wildlife sanctuaries. It did not take long for municipal politicians to become parsimonious in responding to the Nature Reserve Authority's requests. From the perspective of local jurisdiction, nature reserve lands were perceived as lost forever.

JNF forests were another matter entirely. As a Zionist settlement corporation, its perspective was unapologetically anthropocentric, allowing its foresters to be flexible to a fault in accommodating development. Every rural area in the country over the years had enjoyed considerable economic assistance and technical support from the JNF settlement budget. Kaplan saw this good will and historic cooperation with local governments as an asset and an opportunity. He approached Dan Perry, the head of the Nature Reserve Authority, about several places that had been previously envisioned as nature reserves, but that remained undeclared due to intransigence among the local governments or the military.

Perry had a long history of animosity with the JNF, including a number of disagreements over land and wildlife management. Kaplan asked him whether those lands would ever become nature reserves, and Perry acknowledged that for the foreseeable future it was unlikely. So Kaplan proposed a deal. For now, the *would-be reserve lands* would be included as the forests in Master Plan 22 and would be under JNF control. But a provision would be added to the master plan allowing for their reclassification as nature reserves through a simple mechanism that circumvented the usual approvals required from local governments, the Defense Ministry, and so on: "He gave them to JNF as a 'deposit,'" Kaplan said. "We sat down and wrote the provision together. It's my favorite one in the plan." As the years went by, the JNF was as good as its word, and most of the "deposited" lands have quietly been returned and reclassified following the cursory approval of a regional planning commission.[75] As a result, critical habitats and scenic treasures were saved through their temporary designations as forests.

On August 2, 1995, National Master Plan 22, with its long list of revised maps, countless revisions and compromises, was finally presented for final authorization by the National Planning Council. It included a grand total 162,000 hectares of forest lands, over a hundred thousand more hectares than had originally been intended. It still required the approval of the cabinet. A long list of pressing matters required government approval and it was not at all clear when forest planning would make it onto the agenda. The minister of the environment, Yossi Sarid, lobbied effectively, and a cabinet subcommittee approved the final draft of the plan on November 1, 1995, even as the minister of housing opposed and filed a former "reservation." This meant that the master plan would need formal cabinet approval fifteen days later. In retrospect, it was one of the very last decisions that Yizhak Rabin's government would make. Ehud Barak, the newly appointed minister of interior, voiced objections to the plan's flexibility and the provisions that allow a percentage of some forest lands to be easily rezoned. But his senior staffers convinced him not to rock the boat. On November 16, 1995, just days after the assassination

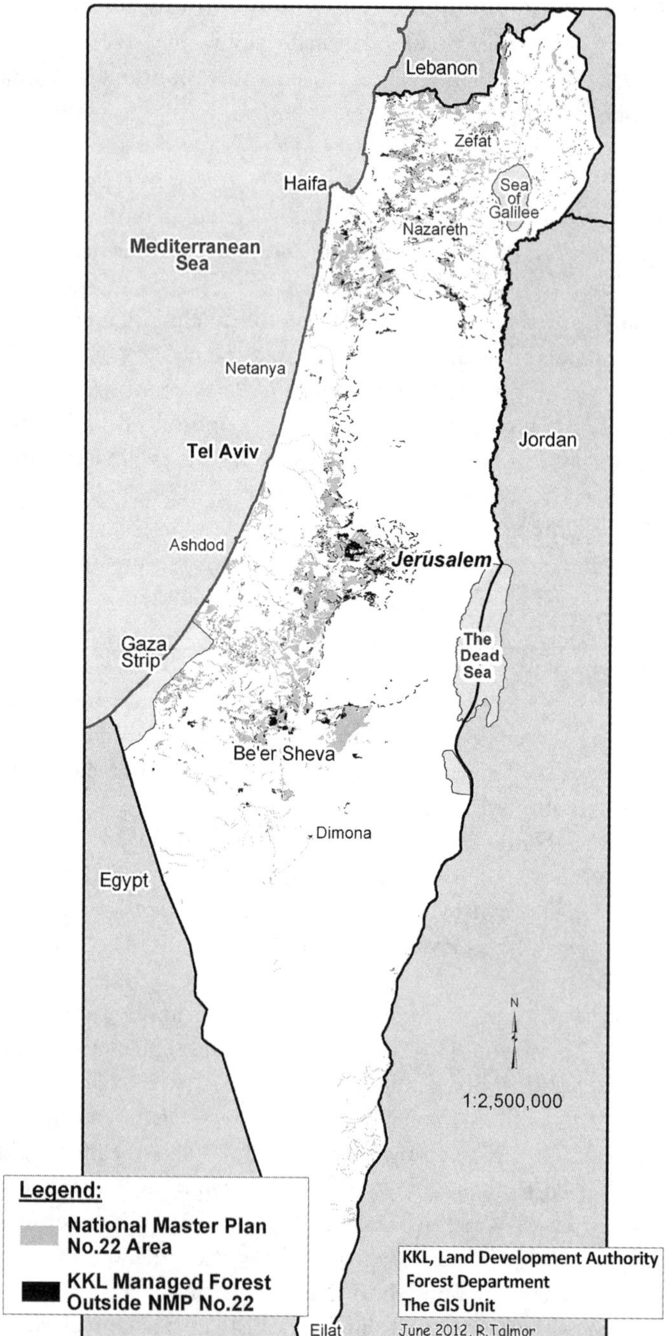

Israel's 180,000 hectares of zoned and unzoned woodlands. (KKL-JNF GIS Unit)

of the prime minister, a dazed Israeli government gave Master Plan 22 its rubber stamp.[76]

When the plan passed, no one was more surprised than Kaplan himself. Reflecting on the experience twenty years after he began the project, he expressed his wonder with a mystical religious expression that acknowledges fortuitous circumstances: "It really was a 'time of divine favor.' Today, a plan of these dimensions would never have a chance of being approved."[77]

Sustainability in the Wind

By the end of the millennium a new vision and commitment to sustainability had come to inform Israel's forestry programs. The transformation that took place in the JNF and its foresters was not just a change in ecological literacy or economic ideology. It was a cultural shift that led to greater openness and an ability to listen. When Azariah Alon went with Zaban on a tour of the forest in the Gilboa hills across from his home on Kibbutz Beit HaShita, the great nature advocate complained that trees blocked all views of the valleys below from the trails and roads. Visitors could drive across the mountain and never see the surrounding landscape. He demanded that the JNF cut open "windows" along the route for the scenery to peak through. In the past, Alon had almost gotten into fist fights when trying to engage obtuse foresters about such matters. Zaban remembers his response. "Well—he was right. So we did it."[78]

It wasn't just that management practices were now more open to friendly—and sometimes unfriendly—suggestions. There was a move afoot to open up the forests so that people could spend more time there. By the end of the 1990s, Israel had roughly 278 square meters of woodlands per person.[79] This fell short of the three hundred cubic meters per capita standard that Zaban had set himself a decade earlier. But it was more than a critical mass. With all their drawbacks, the conifer monocultures were sited on some of Israel's loveliest rolling hills and scenic slopes. They had grown rapidly and provided rich canopies and innumerable picnic sites.

Forests were suddenly recognized as an appealing recreational resource that needed to be accessible to a general public that increasingly sought respite from the intense reality of its sweltering cities. In a small country, most citizens lived near some woodlands and visited in increasing numbers on weekends and holidays. Entry into Israel's forests remains free, but has enormous value for all.[80]

When Haim Zaban passed over the reins to David Nahmias in 1989, Israel's forestry program was totally different than the one he inherited in 1981. Its operational assumptions were altered. So were its practices and the professional backgrounds of the workers it hired. Change was most conspicuous, however, in the new forests themselves, which looked completely different. With the passage of the new national master plan in 1995, there was now a formal mandate and road map for the nation's diverse woodlands. In a very short time, Israel's forestry program had almost become transparent, with past plans and future initiatives theoretically available for review on the Internet. As the JNF would soon learn, being part of the national planning system also made its forestry practices accountable to the public not just in the abstract sense of the word but also in Israel's Supreme Court.

In a sense Israel's forestry had gone from a start-up entrepreneurial venture to an established national undertaking, ascending a steady and steep learning curve. The drama of restoring woodlands to the land of Israel entered a new phase. The challenge now was to avoid complacency and ossification by remaining nimble, creative, and innovative. For the problems of managing forests in a crowded, dry country would only grow more complicated.

6

Dryland Forests and Their Natural Enemies

The desert and the parched land place shall exalt—and the prairie shall rejoice, and blossom.

—Isaiah 35:1

Southbound

It is symbolic that a geologist spearheaded the remarkable afforestation ventures in the semi-arid Negev regions of Israel. Not changes in the genetics of trees but changes in the formation of the landscape and flow of water in the desert produced vigorous woodlands in these dry regions. Yitzschak "Itzik" Moshe has done more to redefine the way that trees are planted in Israel than any other individual since the 1980s. A modest, smart, and remarkably good-natured man, no forester in the country is better respected internationally and more liked by his colleagues. It is ironic that the new forests he created in Israel's drylands are the subject of such intense controversy.

In 1982, Itzik Moshe wasn't sure whether he would ever earn a living with his geology degree. After completing his studies in Jerusalem, he returned to

Itzik Moshe, who created a new system of arid forestry for Israel's southlands based on ancient methods of water harvesting. (Photograph courtesy Itzik Moshe)

his coastal hometown of Ashkelon, and cobbled together some ad hoc jobs doing surveys for soil conservation projects. The southern region of the JNF was looking for a soil conservation expert, and Moshe was happy to be offered a permanent position at the regional headquarters in Gilat. It has been his base of operations ever since.[1]

Initially, the JNF did not have a clear idea of how to combat the persistent soil erosion in the south. Management was acutely aware of the steady loss of the rich, dusty loess soils and the woeful fate of earlier afforestation projects in the drylands of the Negev Desert. David Nachmias was head of JNF's southern region at the time and told his new employee: "Drive around, look closely and start to think about how you can help me." One of the reasons that Itzik Moshe is so appreciated by his peers is his unusual ability to listen. When he asked his colleagues he heard any number of analyses about the problem. Moshe then put his mind to the task of stabilizing the slopes in dryland watersheds. It was immediately clear to him that forestry would only succeed if it was part of an integrated soil strategy:

> I wasn't a biologist. But already I saw that the tree, water, and soil are all part of the same picture. One of my jobs when I started was serving as representative of the Forestry Department with Israel's Pasture Authority. The loss of biomass was evident everywhere. If you want trees in the deserts to survive, you have to make sure that the surrounding lands are treated properly. You have to pay attention to the land. For instance, I noticed that wherever the slopes faced south, there basically was no vegetation. The sun hit the land

> harder there and soil was lost. Northern slopes, with less evaporation, had more land cover. If you want to succeed at planting trees, you have to think about the soil and water.[2]

Steadily, he began to learn why some forests flourished others did not, and how to overcome the stress imposed on trees by a hostile, desert environment.

A few years into the job, he received a phone call from his boss's boss, Haim Zaban, then head of JNF's Land Development Authority. Zaban told him to prepare a strategy for the next thirty years for water and soil conservation in southern woodlands. Moshe was not yet thirty and was relatively new as a forester. Only later did he realize that the plan he was to design was breaking new ground and did not rely on conventional forestry practices or protocols. Indeed, dryland countries at the time such as South Africa rarely attempted plantations of drought-resistant pines or eucalypts in areas where annual precipitation fell below eight hundred millimeters.[3] Even China, which "pushed the envelope" considerably, encountered difficulties when it extended its northern forest shelter belt to areas with four hundred millimeters of annual rainfall.[4] Trees need water, which is why deserts don't have very many of them. But planting had already begun in Yatir where the average rainfall only averaged 255 millimeters a year and frequently was much lower. So the usual precipitation barriers imposed by the rainfall gradient were already broken. The first generation of Israel's arid forests was somewhat unprecedented, offering many lessons.

By the 1960s, Yosef Weitz was looking for an ambitious afforestation project in the south. Part of his motivation involved land reclamation and part was a desire to assert a clear sign of Jewish sovereignty and control in a largely unsettled region. Weitz had been warned by any number of experts that low rainfall made the thirty square kilometers of land lying on Israel's border region south of the West Bank inappropriate for planting his signature Jerusalem pine forest. Never one to let the caution of experts get in the way of grand national visions, Weitz embraced the challenge.

Yatir lies in the so-called transition zone between the semi-arid Mediterranean climate and the arid climates of the desert. Typically there are no rains during the seven-month summer season, and drought conditions in winter are not uncommon. The land was hilly and highly eroded from centuries of overgrazing. Weitz gave the order: Some forty million trees were planted between 1964 and 1969—most of which were Aleppo pines.[5] During the initial years, the saplings were not watered, with runoff on the hillsides sufficient to allow the roots to maintain minimal soil moisture.[6]

Abed Abulkian, a Bedouin Israeli and a second-generation Yatir resident, is

Planting in Yatir, an arid, austere region that, as of 1965, received a yearly average of only 255 millimeters of precipitation. (Yehuda Hanegbi, KKL-JNF Photo Archives)

the district forester responsible for those trees today: "Yosef Weitz approached my father and told him: 'I want to give you a forest. It will be for grazing. It will provide shade for your sheep. But it won't belong to you.' The lands were in horrible shape. Jordanian traders and herders would cross over the border and their flocks would graze the land mercilessly. It was just completely overrun. The new forest stopped that."[7]

Two decades later, when Itzik Moshe began to map out Israel's forestry strategy for the southlands, he had the benefit of twenty years' experience and the wisdom of hindsight from Weitz's massive experiment. To begin with, he saw that the forest was losing large amounts of water, generating considerable runoff during rain events. At the same time, during drought years, the parched pine trees struggled to survive. Utilizing the runoff appeared to hold the key to healthier woodlands. Moshe had always been a great enthusiast of archeology. Without any established international best practices available for such arid conditions, he looked to the region's ancient agriculture for inspiration.[8]

Before converting to Christianity in the fourth century, the Nabateans were an ancient people with a remarkably successful run of nearly a thousand years in the drylands of what is today Israel and Jordan.[9] Although originally nomadic, over time they developed ingenious farming techniques. To compensate for the arid conditions, the Nabateans would contour the dry undulated, countryside into a terraced system, building spillways that converted their watersheds into virtual funnels. Rather than course through the valleys, runoff during flash floods was controlled, cascading gently from terrace to terrace. The rainfall would percolate into the rocky lands, creating tolerable conditions for cultivation.[10] In other cases, runoff was diverted into increasingly narrow channels until it flooded the valleys, allowing for irrigation of fruit trees on the loess soil.

The same system seemed to make sense in the twentieth century for planting trees on the eroded hillsides of the Negev. Rather than choreographing an avant-garde ballet for the southern foresters, Moshe took traditional steps and created a new form of dance.[11] The new tree stands that Moshe designed are supported by natural runoff, through an extensive network of contour trenches or swales. (Swales are depressions in the ground, typically grassy, designed to collect storm-water runoff.) Using bulldozers, topsoil is moved from the sides of the contributing slopes and piled into soil mounds that stretch along the contours of the hills. (The earthwork, of course, causes considerable initial disruption of existing vegetation, but it recovers within two years.)[12] The idea is to slow the velocity of runoff so that it no longer tears the little remaining soil from the earth and cannot carry it away downstream. Low-density conifer stands (two hundred and fifty trees per hectare) are then planted and spaced along a terraced watershed—much like strip farming.[13]

In addition to the extensive landscape manipulation, the newer stands that Moshe planted were more diverse, with native, broadleaf species and fruit trees, planted in the valleys. Yatir was the prototype, but since the 1990s comparable projects were launched, transforming large areas of the northern Negev landscape. Relying only on the irregular winter rainfall, sprawling woodlands in semi-arid areas like Meitar, Dudaim, and Lahavim prospered, changing local assumptions about where the desert began.

Beyond the merits of diversity, the new generation of Negev woodlands appeared to be functionally more successful than the original model. When forester Abed Abulkian is asked about the evolution of the forest, he is quick to praise the new rain-fed forests:

> In the first days, the JNF came and planted pines and cypress and simply raced forward. They saw them as pioneer species and didn't really think about the genetic quality of the saplings. And at first they were planting 1800 trees per hectare. That's way too much. Nobody thought about sustainability. Since then, we changed everything and the trees are doing much better. Today local

> Bedouins have discovered the forest as a place to relax and enjoy nature with their families. We try to offer them a fair dissemination of wood from the branches that are cut that we don't need.
>
> The rest of the "people of Israel" work hard all week. We can offer them this wonderful place to come on the Sabbath to hike, to bicycle. The Israel Trail, the national hiking trail, cuts right across the forest and we have hikers who come through. We provide them with a shower and a place to sleep. I even allow occasional cultural events with concerts in the forest that can go until midnight. We provide the chemical toilets and the parking lots. So of course I don't regret the early planting. If Yatir had never been planted, we'd have nothing: any soil that would be left would be completely compacted.[14]

Some skeptics claimed that there was so much noise coming from the JNF propaganda machine that it was hard to believe the elated claims of success. So comprehensive research was conducted: it confirmed the good news.

The Yatir Forest offers almost a controlled experiment, contrasting the contour methods in the new green parcels with the older stands, as well as with the adjacent, unforested, open spaces, which serve as a "control group." A recently published, multiyear study run by an Israeli team of researchers from Ben-Gurion University and their Palestinian, Jordanian, and Turkish colleagues highlights the many benefits of Yatir's forestry systems. During the years of 2001 to 2005, for example, no runoff at all was measured leaving the subwatersheds of the forest. In other words, the trees and the terraced landscaping sufficiently slowed the flow after rain events to allow vegetation to fully utilize the rainfall.[15] By contrast, the nonforested, contiguous rangelands showed substantial and high-velocity discharges. These lands are acutely overgrazed, with little vegetative cover. Erosion in these treeless areas was visible even when flow was still low and considerable quantities of sediment were carried away each year. In contrast, the Yatir Forest virtually *stopped* all soil loss.

This was remarkable, because for as long as anyone has paid attention, the region exhibited a steady hemorrhaging of soil with its organic matter, nutrients, seeds, and promise of life. Experts confirm the particular vulnerability to erosion of soils in the Negev. Subsequent afforestation efforts in a variety of southern locations involving bank stabilization and watershed harvesting produced the same remarkable results. There is nothing inevitable about soil loss in drylands that trees cannot prevent.[16] Ecological *restoration,* however, must go one step further.

Rehabilitating soils after prolonged or acute erosion is important because the loss in land productivity frequently is a key factor preventing recolonization of degraded lands by the original shrub and vegetative species.[17] Conventional wisdom claims that afforestation is good for soils in drylands—but the

Yatir (2003), Israel's largest and most arid forest. (Benny Mor, KKL-JNF Photo Archives)

case of Yatir turned out to be extreme. Soil carbon content in the Yatir Forest more than *doubled* during its first thirty-five years.[18] Nutrients are also far higher in the forest than in the surrounding soils; moisture in the new swales is 20 percent higher than in land just outside the forest.[19] While there was concern that the forest canopy might block rainfall, it turned out that the forest floor and soils in unforested lands received the same amount of precipitation and showed comparable storage capabilities.[20] Recent research using remote sensing and field observations suggests that Israel's dryland forested areas have greater organic material, active carbon, and soil moisture even as the soil permeability is poorer with understory vegetation.[21]

The rain harvesting approach in semi-arid woodlands emerged as a new model for dryland forestry: "savannization." A new mosaic is created in the desert with water channeled to create vegetatively rich patches. Only 120 trees were planted on a hectare to create a "savanna-like" landscape. Sites typically contain thirty species of salt- and drought-resistant trees, scattered in patches that enjoy enriched water supply from contoured surrounding lands.[22] Dozens of shrubs and grasses have begun to take hold.[23] In 2013, the JNF forestry department prepared a detailed report in response to ecological critiques where it cites numerous empirical studies that confirm the relative botanical richness of the savanna system and its impressive contribution to soil reclamation.[24]

Planting in arid regions where annual rainfall is below two hundred millimeters definitely requires different practices. For instance, trees need to receive modest irrigation for three summers before taking hold on their own. Planting doesn't usually begin until February or March, when soil moisture is acceptable, as opposed to planting in rainier areas, where moisture is optimal in November.

Trees are also grown on extremely arid terrain in earthen formations called *limans*. There are two schools of thought with regards to the origins of the word *liman*. Some say it comes from Russia, where geographers borrowed the term from medieval Greek. It refers to a lake or estuary that is formed by blocking flow with sediment. Another view traces it to the Greek word for "port" as it appeared in the Jerusalem Talmud almost two thousand years ago in a reference to "the Liman of Jaffa."[25] In either case, when Israeli foresters started their first tree-planting foray in the desert in the early 1960s, they expropriated the term for modern Hebrew.

In arid and hyperarid regions, there generally is so little soil on the ground that erosion prevention is not a key factor in land-use decisions. The need to utilize every drop of available water, however, is an obsession. During the cooler months in the desert, there are occasional pulses of rain that lead to massive currents of rainwater in rivulets and channels. In the past, most of this water would settle into depressions or would flow until reaching flat terrain where it quickly evaporated. A smaller percentage formed coursing, ephemeral streams that surged into the Mediterranean or the Dead Sea. Limans showed that the "waste" of such precious water was not inevitable.

Limans are created by building earthen dams (reinforced mounds of soil) at the bottom of the desert channels to trap flash floods. This creates a pond or microcatchment basin whose size ranges between a thousand and six thousand square meters. An access channel (spillway) is constructed along the side of the dam, allowing excess runoff to continue downstream.[26] Then all that is necessary is to wait for winter rains to come. Once floodwaters are trapped, the water penetrates the soil, supplying trees whose roots can find moisture and nutrients a meter or two beneath the surface year round.[27] Sometimes a single rainfall event creates an amount of water equal to one thousand millimeters or more of rainfall. Limans are home to a variety of trees. Initially these were mostly drought- and salt-resistant eucalyptus species, with an occasional tamarisk or pine.[28] Recently, however, mulberry trees have been surprisingly successful in limans, as have carobs and sycamores.[29]

The first limans were built in the 1940s.[30] Based on their success, in 1962 JNF began to duplicate them. At that time most southern thoroughfares were yet unpaved, so the trip down to Eilat could take as long as two days. Part of the rationale for planting was to provide sunbaked patrolling soldiers, drivers,

Oasis: a liman harvests rare winter rains to support year-round shade. (Yossi Zamir, KKL-JNF Photo Archives)

and passengers with a shaded refuge when they crossed the desert. Before most of the trees could fully mature, however, the roads were upgraded, cars became air-conditioned, and the travel time had dropped to two to three hours. The number of travelers utilizing the limans also plummeted.[31]

Today, over four hundred limans dot Israel's deserts, still providing shelter from the relentless heat for camels, occasional hikers, and sundry other creatures. Local Bedouins enjoy the small but steady supply of wood and they forage for livestock.[32] Migratory birds make pit stops as they cross the vast desert. The overall hydrological effect of the limans is modest. Even during a rainy year in the most liman-intensive regions, only 1 percent of rainfall, or 860,000 cubic meters of water, is actually harvested.[33] Limans remain scattered and have not irreparably altered the local desert ecology.

By the time that Moti Kaplan prepared the maps of National Master Plan 22, he could rely on a variety of proven models for Negev afforestation developed by his old university classmate and friend Itzik Moshe. Considerable stretches (over 660 square kilometers) in Israel's southlands were designated as forests. This is more forestland than the Galilee and the Haifa regions combined.[34] Today, most of the land zoned as forests in Israel that remains unplanted is located in the south. As the Israeli government seeks to stem the tide of unrecognized Bedouin settlement into open spaces, it has increasingly turned to afforestation. Every year hundreds of hectares of new stands are planted under the scorching, dry conditions. But the trees don't seem to mind too much.

It is worth pointing out that, in the more temperate areas of Israel, woodlands flourished in the past. Technically, this would make many of the tree-planting projects attempted in Israel during the past century *reforestation* efforts. In contrast, no woodlands of any density ever grew in Yatir or in the arid Negev region. Most of Israel's dryland forests today are examples of *afforestation:* trees were never able to survive before the ancient techniques of water harvesting were brought to bear.

Desertification affects between 10 and 20 percent of the world's drylands.[35] With six to twelve million square kilometers of the planet already degraded, chronic land deterioration exacerbates poverty and famine making desertification one of humanities most severe ecological problems.[36] Israel's afforestation achievements suggest that arid soils can be regenerated relatively quickly.

Ultimately, word got out and Moshe became a highly sought-after expert in the international forestry community. He has served as a guide to diverse delegations of foresters who regularly visit Israel on professional tours, happily sharing trade secrets with them. He has lectured at UN sustainability and climate-change convention meetings, and often traveled to distant lands, offering advice about the latest techniques for dryland afforestation.[37] Even Muslim-dominated Indonesia, with no established diplomatic relations with Israel, has flown him out to bring the Israeli gospel of tree planting to their drier provinces.[38] But not everyone is enamored of the results, and there are those who ask difficult questions.

Biodiversity and Afforestation

Alon Rothschild is not particularly keen about Israel's afforestation experience in Israel's southlands: "The problem with forestry in Israel is that there isn't a clear distinction between means and objectives. Planting trees is a means to an end—not a goal, in and of itself."[39] Rothschild brings a scholar's intelligence and a redhead's forthrightness to his work, coordinating biological-diversity preservation efforts at the Society for Protection of Nature in Israel (SPNI), the largest environmental NGO in Israel. Relations between the JNF and the SPNI have long been acrimonious. Indeed, the SPNI was established in 1954 as a frustrated response to the draining of the Huleh wetlands. Despite intense opposition by the nation's incipient conservation community, the drainage of the lake and surrounding swamps proceeded apace. The country's largest and most distinctive wetlands were transformed into farmlands in the northern corner of the Galilee.[40] It was the JNF that initiated and implemented the project.[41] This set the tone for many years of conflict that was ideological, but of course, in many instances also personal.

With Israel's remodeled ecological approach to forestry, the two institutional foes might have entered a new era of reconciliation, or at least of cease-fire. After all, the SPNI should feel vindicated that much of its original critiques vilifying the aggressive, industrial planting practices implicitly were internalized by the forestry agency.[42] Densely planted pine monocultures may still survive, but they are relics of a different era and gradually are giving way to more diverse woodlands. Forest projects today have to go through a rigorous review process in the planning stage. Gone are the days when foresters dismissively rebuffed green concerns about habitat with assertions that they were *foresters* and not *zookeepers*.[43] It is well to accept victory graciously. In fact, there are areas where cooperation exists, and natural alliances form between the two old institutional foes. And yet, fundamental differences remain. Conservation advocates are paid to worry, and Rothschild has plenty to worry about.

In 2000, an international team of biologists led by Oxford University's Norman Myers identified the twenty-five so-called biodiversity "hot spots" of the planet. About 12 percent of the earth's surface contains 35 percent of the planet's nonfish vertebrates and 44 percent of all known plants.[44] One of these hotspots is the Mediterranean Basin. All of Israel is highlighted in the map. The local conservation community was delighted by this validating "shot in the arm." It now had a definitive, international analysis supporting its calls for improved preservation. There are those who argue that Israeli biodiversity is in fact quite normal and surely not extraordinary.[45] But it is remarkable that after centuries of deforestation, erosion, and hunting, the land of Israel remains blessed with 534 bird species, 115 mammal species, 103 reptile species, and 2,780 types of flowering plants.[46]

Recent trends did not appear encouraging and it was important to constantly take stock and set priorities. To that end, in 2002 local zoologists and botanists compiled an Israeli *Red Book*.[47] Like the international catalog the Red List, published by the International Union for the Conservation of Nature,[48] the Israeli *Red Book* contains lists of the country's threatened and endangered freshwater fish, amphibians, reptiles, and mammals. It suggests that much of Israel's biological heritage is being squandered.

The diversity of mammals offers one example. Reflecting the range of Israel's rain gradient and transcontinental location, the country is home to 115 wild mammal species, a number that is only 57 percent less than the number of wild mammal species endemic to California, even though the Golden State is fourteen times larger. Europe is three hundred times the size of Israel but only has two hundred mammal species. Sadly, twenty-five of Israel's mammal species are endangered and twelve more are defined as *critically* endangered. Only thirty mammal species enjoy stability and are not defined as threatened at all.[49] Five of the animals that survived the nineteenth century—including the cheetah and the brown Syrian bear—never got the chance to enjoy the protection of Israel's strict wildlife protection laws. By the 1950s they had been hunted out of the local landscapes. On the positive side of the audit, five species that were lost to hunters centuries ago—including the Arabian oryx and the lovely fallow deer—have been repatriated. But the overall picture shows two-thirds of Israel's mammals on an unsustainable track.

Birds face similar, albeit less extreme dynamics. Of the 534 avian species recorded in Israel, thirty-eight are classified as endangered; seventeen appear on international lists of globally endangered species; eleven more are extinct as breeders; and four appear to be gone for good.[50] Amphibians worldwide are in terrible shape,[51] and Israel is no exception. After the draining of 97 percent of the country's wetlands, amphibians' natural habitats have simply dried up. Of the six species of indigenous amphibians, no population is considered stable. The green toad is threatened; the tree frog considered vulnerable; two salamander species are endangered; and the Syrian spadefoot toad and the banded newt are defined as "critically endangered."

For fifty years, the painted frog, once a happy resident of the Huleh wetlands, was thought to have joined the ranks of the globally extinct until it was found by a ranger in the region in November 2011. Excited scientists called the miraculous sighting "the nature conservation equivalent of discovering the Dead Sea Scrolls."[52] All the same, the species' future is far from clear. Like Israel's other amphibians, the frogs literally have nowhere to go.[53] Reptiles have fared somewhat better: only a third of the 103 local species are endangered, with three on the list of local extinctions.[54] Fish have done worse—almost a quarter of endemic fish are endangered and five are already extinct.[55]

Rothschild refers to the *Redbook* as his personal Bible. It highlights his problem with Israel's forestry program: "The *Redbook* tells us what we need to protect. I sat with two senior forestry experts at the JNF. They had never even heard of the book. Perhaps we are responsible for this gap in knowledge. But to my mind, the afforestation in the south of Israel is being done without any awareness of the ecological consequences. Take the Be'er Sheva fringe-fingered lizard (*shnunit Be'er Sheva*). This reptile is extremely rare, and it only lives in Israel. When they did the last census in 2002, it was still surviving in the loess soils of the Negev but was defined as endangered. When new forests are planted, its habitat disappears."[56]

Rothschild represents a new generation of conservation advocates who don't carry the burden of decades of arguments and recrimination. He openly acknowledges that in several cases, his success in affecting public policy has been bolstered by JNF partnerships. Perhaps that is why his critiques emerge as substantive, unlike much of the squabbling and intermittent clashes in the past. Rather than being angry, he is disappointed with his colleagues.

Like many other conservation advocates, biodiversity serves as Rothschild's ultimate criterion for assessing the legitimacy of a forest. He sees no room for planting in areas that already offer a special niche to wildlife or contain important or rare indigenous plants. Along the dryland loess soils in the Negev countryside, for example, there are many such scrubland areas, naturally devoid of trees, but with a dense growth of bushes and shrubs. They can spread over long distances or be limited to a small patch of semi-arid open space that offers a mini-"sanctuary" for nature. Although the untrained eye might see the terrain as barren, in fact the shrub-intensive scrubland supports a reasonably rich variety of birds, insects, and rodents. Rothschild argues that the second that this land is covered by a canopy, it changes the balance. Here, a long-legged buzzard cannot successfully stalk and capture its prey. New populations of birds and other opportunistic critters begin to colonize. Water sources for the natural vegetation are usurped by the new trees that soak up every last drop.

The JNF has responsibility for 1.6 million dunams of land under the national master plan for forests. The question is: what tools does it bring to optimally manage these lands? Part of the problem is personnel and professional orientation. In the past, the JNF hired an ecologist who worked at a national level. Before he retired in 2008, there was an expectation by Dr. Zvika Avni, then head of the Forestry Department, to supplement the work of the national ecologist with a regional ecologist for each of the three JNF regions. Not only did this fail to happen, but Dr. Yoram Goldring, JNF's single staff ecologist also opted for early retirement. Finding a replacement was delayed for years because of a labor dispute. Assuming foresters were attuned to the questions of biodiversity, they would largely have to rely on their own intuition. The SPNI calls for the forestry professional to be supplemented by four to five ecologists, who will not only serve in an advisory role, but actually make and oversee the policies that determine the scope and type of tree planting that is conducted.[57]

"I know that it's a cliché, but this really is a case of not seeing the forest for the trees," continues Rothschild. "There is understory, there are shrubs, bushes, geophytes, and of course animals. Birds are nesting there. There is a bit of blindness when foresters look at the progress of the trees as the sole indicator of forest health. Compare the JNF's work in the forests to the way the Nature and Parks Authority operates. They also manage large areas of land. But they see a broader picture. Being aware of the ecological dynamics, they will coordinate with the Drainage Authority and reach agreement that bird habitat shouldn't be disturbed until after the nesting season in July. It's fine to cut the grasses along the banks of streams so that flow will not be impeded. But wait until after the nesting season. The JNF needs to start developing the same kind of sensitivity."[58]

No area of controversy remains as unresolved as the issue of afforestation in semi-arid zones where the battle lines are clearly drawn. Many nature advocates would like to call off the celebration.[59] In response to an early draft of the newly prepared *Bible of Forestry* management guidelines, the SPNI submitted a four-page list of comments and reservations:

> Planting forests in the arid areas causes the spread of Mediterranean species into the Negev (birds like Paridaes, Garrulus, and common black birds). These generalist species supplant the original wildlife that is specialized for this desert or the arid-fringe habitat. Outstanding examples of this involve the disappearance of the Be'er Sheva fringe-fingered lizard and the Houbara from the loess soils on the edge of the desert, because of (among other reasons) unregulated afforestation. Forests based on rainwater harvesting are a serious intervention into the natural habitat on the desert's edge, because they change the hydrology of the entire area. Typically they are filled with exotic species. Moreover the limans in the desert offer a "stepping stone" for ravens and other invasive species into the heart of the desert. Forests based on rainwater harvest damage the biodiversity of specialist species.[60]

In November 2012, the SPNI issued a more formal forty-two-page report charging the JNF with ecological negligence in its planting strategy in Israel's semi-arid, northern Negev region.[61] The authors amass findings from sundry studies showing a retreat among local animal species in response to the forest offensive. For instance, a variety of desert birds, especially those who nest on the ground among the shrubs of the Negev's scrublands (e.g., the black-bellied sand grouse) are crowded out by Mediterranean avian species, better evolved for forested regions once the trees are established.

The report also cites research published in the prestigious journal *Conservation Biology* that indicts Negev forests for creating an "ecological trap" for some lizard species, in particular the aforementioned Be'er Sheva fringe-fingered lizard. Ecological traps are formed when animals' evolutionary capacity is not sufficiently nimble to respond to rapid modifications of their habitats. In the present case, the planting of trees in the open desert terrain provide avian predators with innumerable perches from which to hunt. The trees increase the vulnerability of the lizards and their eggs to aerial predation. The lizards, unaware that their habitat is now of "lower quality," are caught in an ecological trap and their numbers decline.[62]

Some nature advocates even point to the biodiversity outcomes in the Yatir Forest study as a basis for concern. While the forest creates a wealth of ecosystem services, the overall number of flora species that inhabit the "poster boy" dryland forest is *lower* (seventy-nine plant species) than in the *non-afforested* lands surrounding Yatir (ninety-five species). Not surprisingly, the total biomass of the undergrowth is also much poorer, with the quantity of shrubs and herbs on the forest floor a fraction of those growing on the nearby open spaces. In fairness, however, when the seed banks of the two areas are compared, the gap drops to a trivial difference of 109 species versus 100.[63] This suggests that it is not availability of seeds but environmental conditions that determine which species actually germinate. Many of the indigenous plants have not yet been able to successfully establish themselves inside the forest.

There is, however, some good biodiversity news for forest fans. To begin with, forests can be designed with diverse patches that reflect site-specific topographic and drainage properties. In the Lahav Forest near Be'er Sheva, for example, pines are planted on the hilltops; jojobas, Atlantic terebinth, and other species on the slopes; and olive trees in the valleys. After many years of aspersions by naturalists, with Israeli conifer forests being defamed as "pine deserts" — it turns out that pine needles may not be so bad after all. Researchers expected pine needle litter to reduce germination due to its allopathic (chemical suppression) effects. In fact, it serves to protect understory, apparently by regulating temperatures on the ground. Whatever the reason, parcels with pine needles had more than 20 percent more species than uncovered areas (twenty-two versus eighteen species per sample germinated).[64]

Plants' inability to compete with trees for light was also not considered a factor in forest biodiversity.[65] Rather it is competition for water and nutrients between the herbs and the trees (which have a shallow root system) that probably causes the reduced number of shrubs and generally lower biomass of herbs on the forest floor relative to surrounding areas.[66] A recent study shows clear associations between species-richness in the Yatir Forest and topographic conditions and soil type. The steeper the incline of the land and the higher the carbon-nitrogen ratio, the greater the number of plant species in the understory — with stands ranging from six to sixty-five grassy species. The results suggest that careful planning of forests can ensure high biological diversity and biomass quantity.[67]

In more temperate regions of the country, newly planted forests have shown even better capacity for supporting biodiversity. For instance, a joint German-Israel team of researchers assessed the respective assemblages of saproxylic beetle in Aleppo pine stands and oak forests in the north of Israel. The expectation was that there would be little mixing and that the insects would be separated, with clear preference for the nonconifer habitat. In fact there was an excellent mixing and little difference in species richness.[68] Like many pines forests in the rainier, subhumid areas, these stands show clear signs of a successional process. The best explanation for the results was that young oaks were flourishing as understory under the pine canopy.

The situation is different in semi-arid regions. Present signs indicate that trees really only begin regenerating naturally when annual rainfall levels reach a minimum of three hundred millimeters. It is unlikely that any meaningful successional process has begun in most dryland forests. Veteran ecologist Uriel Safriel believes that because pine plantations are typically short-lived, the Yatir Forest will start to disappear in a decade or two. He reckons that the dying trees may successfully germinate seedlings in open patches where sunlight can reach them, assuming that livestock is excluded there. An alternative strategy might be to start a systematic program of harvesting trees and creating diverse, protected patches.[69]

This of course raises the million-dollar question regarding the sustainability of Israel's dryland forestry experiment: will the stands ever have the ability to regenerate naturally? A management program could be designed where patches are harvested and replanted periodically to create a diverse, multi-aged forest. But such a scenario is not optimal. The JNF's sustainability principles prioritize creation of ecosystems that regenerate without human intervention or investment. Conventional wisdom holds that Israel's forests south of the Kiryat Gat, Ashkelon, four-hundred-millimeters-per-year precipitation line do not regenerate without human mediation. Few seedlings sprout in such dry conditions, even among Aleppo pines, and those that do rarely survive the heat of late spring. Professor Yossi Riov and his students' research confirm the merits of several management interventions. For instance, the chances of natural regeneration typically are improved by superficial cultivation of the upper soil layer, prevention of grazing, control of ants and other seed-collecting insects, and thinning of mature trees for reduced competition and shade. The latter practice is particularly important for pine regeneration.[70] Such measures fly in the face of the basic impulse to let "nature take its course."

When ecology expert Professor Uriel Safriel oversaw the Yatir monitoring project, he spotted little regeneration of trees in the forest there, except for the odd pine sapling on the side of the road, "where there's always sun and water." Nonetheless, he did not rule out the possibility of regeneration. The dryness and the heat are key obstacles, and the relentless grazing regime makes reestablishing young seedlings practically impossible. The soil inside the forest, however, is growing richer. If conditions were optimal, Safriel believes that over time a second generation might emerge.[71]

In responding to criticism about dryland forests, Itzik Moshe relies on an historic perspective: "The question is: what's your reference point. I look at the traditional agriculture that existed in the desert. People have planted trees in the desert literally for thousands of years. That's the natural dynamic."[72] Millennia of anthropogenic activities undoubtedly extirpated the original trees and vegetation that proceeded human settlement. Even so, today's forests in Israel's southlands are not natural. This doesn't mean that diverse vegetation cannot thrive in them. In Yatir, for example, the colonization of twenty-seven plant species that are not indigenous to the area has been documented. The flora of Yatir and the southern forests is gradually becoming more Mediterranean than the Irano-Turanian (semi-arid) or Saharo-Arabian (arid) plants in the surrounding rangelands.[73] With Mediterranean habitats steadily shrinking, dryland forests may help preserve national species richness.

Ecologists generally assumed that, zoologically, forests would be "pine deserts." In fact a whole suite of animals have moved in—from gazelles, mongoose, hares, and wild boars, to foxes, wolves, caracals, jackals, and even hyenas.[74] Besides the shade and cooler temperatures inside the forests, the trees (and the JNF rangers) provide a safe haven for animals fleeing the many hunters who ignore Israel's tough hunting laws.[75] Some of the wildlife, like jackals, are new arrivals to the Negev. In and around these new desert forests, a novel ecosystem appears to be emerging. The question remains: is this a good thing?

The Invaders

Some conservation advocates not only seek to slow the spread of trees in Israel's southern landscape; in some cases, they want foresters to remove undesirable trees altogether. While gaps have narrowed, there appear to be real differences of opinion regarding the kinds of trees that have a place in Israel. Eucalyptus trees, for instance, have long been a subject of controversy. This genus of flowering trees includes some seven hundred species, almost all of exclusively Australian origin. Already in the 1940s, researchers working with the JNF tested roughly a hundred varieties of eucalyptus. They demonstrated a broad range of characteristics and abilities to adapt to local conditions.[76] Many were integrated into urban development projects of the time. Although never as excoriated as Jerusalem pines, eucalypts were the subject of considerable abuse by nature lovers. Not only their "impure" geographic origins were impugned, but a host of hydrological crimes and misdemeanors were attributed to them. Significant damage, however, has never really been documented.

There are numerous cases around the world where, over time, introduced species blended in harmoniously, even assuming great cultural significance. The Romans had a proverb: *tandem aliquando invasores fiunt vernaculi:* "In time, invaders become the natives."[77] Eucalyptus trees may not enjoy a vaunted status among ecologically orthodox Israelis, but they definitely have succeeded in acquiring their share of Israeli fans.

Naomi Shemer, perhaps Israel's most popular songwriter of the past fifty years, penned hundreds of tunes which immediately took their place in the secular liturgy of national folk songs. In one of her more popular ballads, the lyrics speak of the serenity of eucalyptus trees, emblematic of the timelessness of the Sea of Galilee landscape where she spent her childhood in the 1930s.

> When mother came here, young and beautiful, father built her a house.
> Springs passed, half a century transpired, and curls turned to gray.
> But on the beach of the Jordan, nothing has change: the same silence; the same scenery;
> The eucalyptus grove, the bridge, the row boat, and the smell of the salt plants on the water.[78]

Before she was murdered by terrorists in a 1977 attack, American-born nature photographer Gail Rubin created an astonishingly beautiful legacy of natural images from Israel. Her many photos of trees are almost entirely eucalypts.[79]

Today, a softening in positions demanding removal of all eucalyptus can be noted, and there is even a quiet acceptance among indigenous tree-hugging quarters that in some places the eucalyptus can be left standing. For instance in Hadera, where the trees made a modest contribution to swamp draining, eucalyptus groves have become synonymous with the local landscape. (In fact, most of the drainage work was not by the eucalypts "biological pumps," but old-fashioned drainage canals.) In some other places, however, the xenophobia remains categorical.

Rothschild would like to remove the eucalyptus trees lining the banks of the Yarkon River, which meanders through the center of Israel until reaching Tel Aviv's central park and the Mediterranean Sea. The River Authority there has begun a gradual program to phase out the eucalyptus trees and replace them with willows. This is primarily out of concern for the toxicity of the terpene and polyphenols present in the leaves that fall into the stream, and their effect on aquatic biological processes.[80] Cognizant that the local public has grown fond of the tall Australians, thinning and pruning are limited to no more than is necessary to let in sunshine and avoid hitting power lines. When the eucalyptus trees die, however, they are not replaced.[81]

In other parts of Israel, there is surely a place for eucalypts. It is a question of scale. When planting in the Negev was in its initial experimental stage, the harsh and dry conditions killed many local tree species or left them in dwarf

The Australian influence on Israel's forests: a eucalyptus grove and picnic site. (Pnina Livny, KKL-JNF Photo Archives)

conditions. Eventually they were replaced by more drought- and salt-resistant species such as the East African river acacia (*Acacia elatior*) and of course Australian eucalypts. Ecology professor Gidi Ne'eman has no trouble with eucalyptus plantings in the desert, in small numbers near communities: they provide excellent fuel for woodstoves and the trees don't expand their range in the desert. But he cautions about the their instability: "You never know when the eucalyptus is going to fall on the road. Even in Israel there have been tragedies of people hit by falling eucalyptus trees."[82]

As climate change reduces rainfall in Israel's southlands even further,[83] there will be a need to include a shift in tree species. For instance, the present Mediterranean cypress trees (*Cupressus sempervirens*) will need to be replaced by more resilient cypress alternatives like the Tetraclinis (*Tetraclinis articulata*).[84] The hardy eucalyptus's place in the local forest assemblage will probably expand.

Another benign "visitor" is the *brutia* pine tree. Its natural habitat runs from Georgia and Azerbaijan through Syria and Lebanon to the Mediterranean. Geographic patent rights, however, remain contentious. The common name for the tree is alternatively the "Turkish" pine or the "Calabrian" pine—from the Italian region where it was planted. Back in 1926, when JNF foresters first introduced the species to Palestine, they called it the Cyprian pine,[85] as it was thought to originate in Cyprus and still dominates forest assemblage there. Subsequent dendroarcheological analysis showed that *brutia* timber made its way to ancient Israel, probably imported from Lebanon for, among other products, coffins and boats.[86] Calabrian pines were less given to pest infestation and generally did well in Palestine: well enough for the British to begin importing *Pinus brutia* seeds in 1937.[87] By 1989, 18 percent of the country's conifer plantings were *brutia* pine.[88] Not only are they less susceptible to pests and better able to withstand intensive grazing, they are healthier in the high-density stands which characterize the older forests.[89] A recent random census of *brutia* trees planted in the Herzl Forest during the Mandate showed that at age eighty, they are still going strong and show no signs of faltering.[90]

Pine tree species can look rather similar. Today many people have difficulty distinguishing between the Calabrian, which purists claim as *exotic,* and Aleppo which are recognized as *native.* Indeed, for years, taxonomists considered them a single species. With proper coaching, it is easy to identify the darker, coarser needles of the *brutia* and the ninety-degree angle on which their stemless cones attach to branches. Using the Bible as a baseline, the distance separating the two sister species was never much more than a hundred kilometers, and *brutia* wood definitely was used in Jericho during the Roman period. With so many of today's "indigenous" Israeli Aleppo pines being descendants of seeds imported from France, attempts to disqualify *brutia* on grounds of its geographic inauthenticity seem petty at best.

There is a growing school of ecological experts who believe that there might be an excessive level of "fanaticism" associated with ensuring the authenticity of local species assemblages. Exotic species "per se" are not always a problem. To birds in Arizona, for example, the structure of trees, rather than their native, genetic purity appears to be the most important consideration for nesting.[91] Restoration projects that remove exotic tamarix trees that colonized there without replacing them with high-quality native trees may harm local bird populations. The southwestern willow flycatcher, an endangered American bird, has even come to prefer the tamarix as its preferred nesting habitat.

On a planet where humans have completely overturned natural balances, native pests may become just as pernicious as introduced exotics. The native mountain pine beetle is a case in point, killing more trees in North America than any other insect—local or exotic.[92] As the noted Harvard biologist Steven J. Gould observed: "Over time, the only conceivable rationale for the moral or practical superiority of 'natives' (read first comers) must lie in a romanticized notion that old inhabitants learn to live in ecological harmony with surroundings, while later interlopers tend to be exploiters. But this notion . . . must be dismissed as romantic drivel."[93]

Because of some deeply disturbing "horror stories," there is an element of hysteria in the policy discourse about invasive species. When local ecosystems are unprepared, the results can surely be disastrous. In Guam the notorious brown tree snake obliterated the local bird population.[94] Kudzu, an aggressive climbing vine was introduced in the Japanese pavilion in the 1876 Centennial Exposition in Philadelphia as providing excellent animal forage and ground cover against erosion. But it quickly grew out of control and now reportedly spreads across the United States at a rate of ten thousand hectares a year.[95]

The vast majority of exotic tree species, however, never pose any problems at all to indigenous organisms. Israel is full of trees which were transplanted and became an indistinguishable part of the local assemblage. The wormwood tree is a medicinal plant that was brought to the Holy Land by the Crusaders, which fit right in.[96] Indeed, the "sabra" cactus is a relatively recent arrival to Ottoman Palestine even as it soon became iconic for Arabs and Jews as the very symbol of their indigenousness. Recent findings indicate that many "invasive species" attributed to Mandate afforestation efforts actually preceded British rule in Palestine.[97]

A common estimate of the scope of the problem is Williamson's "tens rule": only 10 percent of introduced species will actually succeed in the wild; only 10 percent of these will become established; and of these only 10 percent will become pests. All told, on average, only 0.1 percent of all alien trees should pose a problem.[98] But when that 0.1 percent takes hold, it can be very problematic.

Israel's southern coasts offer a conspicuous example of biological diversity being threatened by nonnative trees. In recent years, there has been a greater sense of urgency. Ever since Moti Kaplan invented "coastal forest parks" to preserve Israel's beaches development, large areas of the shoreline fall under the dominion of National Master Plan 22. Technically, the JNF is responsible for vegetation. In many places along the seaside, with no help from foresters, trees are overabundant.

While British foresters may have been disappointed at the tenacity of some species they introduced, there was one tree that never let them down. The British brought the blue-leaf waddle and coojoong trees (alternatively classified as *Acacia saligna* or *Acacia cyanophylla*) from Australia in their war against the coastal sands of Palestine. The dunes' proclivity for shifting made life precarious for coastal residents and their farmland. It did not take long before the Forestry Department realized it had a winner.

Regardless of one's sense of aesthetics, the sheer tenacity of *Acacia saligna* is impressive. Growing almost an inch in a single a week, it only takes four years for the tree to fully mature. By that time it releases thousands of seeds that stay fertile and dormant underground for years. The trees are not particularly tall and never exceed eight meters. (They actually seem shorter due to their stooped-over willowlike posture.) Every year at the end of winter, they burst out with radiant yellow flowers that are comely enough to be grown in homes worldwide as ornamentals. Its black fruits are legumes that animals are happy to eat. Other benefits that the tree bestows include firewood, fodder, mulch, and of course sand-dune stabilization.

Amihud Goor, the forestry-science pioneer and Israel's first chief forester, included blue-leaf waddle planting among the best practices recommended in the 1955 dryland forestry manual that he prepared for the UN's Food and

Agricultural Organization. Goor's dry, technical language did little to hide his intense admiration for the hardy species, and for its impressive adaptive and reproductive capacities: "Widely planted in the Near East for sand fixation, erosion control and fuel, thriving on almost every soil, even slightly saline soils. Strongly resistant to drought, doing well even when there is less than 300 millimeters annual rainfall. Useful as windbreak in arid areas and for the road sides. The best species to use for binding moving sand. It coppices freely."[99]

For many years, this remained the prevailing view among Israelis who had grown fond of these colorful Australian arboreal immigrants. But this position soon changed 180 degrees as the trees became the bane of nature lovers' existence. By 1999, Amos Sabach, a ranger for Israel's Nature and Parks Authority, was sounding a battle cry against them: "There is no doubt that it is incumbent upon us, the guardians of nature, to fight an all-out war on species such as *Acacia saligna*, and in any case, we must eradicate the Acacia in nature reserves and open spaces."[100] The reason for the change of heart was its phenomenal success. The blue-leaf waddle had emerged as a classic "invasive species."

Israel's Mediterranean seaboard was once lined with rich sand dunes. As the population grew, beach sands were often paved over or mined—transformed into new neighborhoods. Only scattered remnants of sand ridges remained, with one significant undisturbed system of dunes surviving along the Nitzanim coast. Located in the south, between the cities of Ashdod and Ashkelon, Nitzanim provides a home for a variety of animals as well as humans seeking a noncommercial beach experience. Even if visitors do not catch a glimpse of the nocturnal creatures, footprints on the dunes reveal the rich nightlife enjoyed by surviving snakes, gazelles, foxes, hares, and hedgehogs.

With time it became hard to find open stretches of sand in the coastal park. They were hidden by the thriving blue acacia trees. When researchers tracked the state of the dunes in 2004, they found that over a thirty-four-year period, the trees' range had expanded by 267 percent. And the rate of proliferation was steadily increasing.[101] Ecologists became concerned that the indigenous flora that originally traversed the dunes would not survive the assault.

Buxton's jird (*meryon holot* in Hebrew or *Meriones sacramenti* for the taxonomically inclined) a little-known rodent, suddenly was brought out of obscurity. This adorable brown gerbil has large, dark eyes, and a tail as long as its body. It is also the only mammal that is fully endemic to Israel. First identified in 1922, its population is dropping fast enough to earn it a place on the IUCN international Red List of threatened species.[102] The jird lives and builds its burrows in the soft soils of the country's sandy deserts and clay soils of the southlands. and nowhere else. There it can eat the ample supply of seeds, roots, bulbs, insects, and whatever green vegetation and fruit it finds. Until recently, there probably wasn't a forester in Israel who had heard of the critter —much less be able to identify it. And yet, for it to survive the vigorous Australian acacia trees may need to be eradicated.

David Lehrer, director of the Arava Institute for Environmental Studies, did his masters research on the Nitzanim situation. He explains that once introduced to Israel, the *Acacia saligna* began to act as an eco-engineer, adapting the level of nutrients in the normally nutrient poor soil to its own advantage. In stabilizing the shifting sand dunes, the trees actually created an environment hostile to native flora and fauna. His solution is also uncompromising: "The only way to save the shifting sand dune ecosystem of Nitzanim is to eliminate *Acacia saligna*."

It turns out that this is easier said than done. Like many instances of invasive species, the persistence of a seed bank makes the task of eliminating the tenacious trees very trying. This is not only a common problem in ecological restoration, but a frustrating phenomenon with which any gardener will identify. Chopping down or uprooting the exotic species is not enough. The protracted dormancy of seeds, protected by a coat that keeps water and soil microbes out, makes the species almost invincible. Recently, field trials showed that soil solarization techniques might eliminate acacia seeds buried under the sands.[103] By mulching moist soil with a polyethylene sheet during warm seasons, soil temperatures can be raised as high as fifty-five degrees Celsius. This not only eradicates pests and pathogens, but it can bring weeds and acacia seeds to "sublethal" temperatures and reduce the viability of seed banks. Another strategy, reportedly successful in South Africa, involves utilizing "biological controls" via a rust fungus, *Uromycladium tepperianium,* that stunts its growth. Alternatively, the seed feeding weevil *Melanterius compactus* enthusiastically feeds on acacia seeds.[104] Needless to say, eradication will not be easy or cheap. Nor will it happen overnight.

Lehrer's recent economic assessment confirmed that with a projected willingness to pay of 8.6 dollars per citizen, the removal of the *Acacia saligna* would yield a net economic benefit.[105] Pua Bar Kutiel, an ecologist and head of the geography department at Ben-Gurion University, has spent more time researching the phenomenon than any other Israeli academic. While she is grateful that the planting of blue-leaf waddle was stopped early in the 1980s, she believes that the problem needs to be addressed at the national rather than the local level. Based on the phenomenal success in addressing a similar *saligna* problem in South Africa, Bar Kutiel has some basis for optimism. But, to her mind, the solution needs to involve a general biological pest-management strategy for open spaces. This will require considerable site-specific research.

In the interim, ad hoc measures like felling and even burning trees, along with reliance on herbivores, will have to do.[106]

In the past it could take a while for the JNF to internalize dangers associated with invasive tree species. Over time, awareness has improved. Zvika Avni was head of the Forestry Department in the year 2000, when complaints reached him about the *Prosopis*. The "white carob" (*Prosopis alba*) is a drought-resistant, South American tree whose high-energy fruits and seedpods are excellent fodder. Avni funded research that quickly confirmed concerns that the tree was becoming invasive. All planting was immediately stopped. With the exception of familiar eucalypts in the driest parts of the south, slowly but surely tree planting in Israel became limited to indigenous species. The evergreen *Tetraclinis articulate* is a Moroccan member of the cypress family which has proven to be very successful (80 percent survival rates), even in Israel's dryer regions, and could be an excellent tree for lining the boulevards of southern cities. Meticulous studies of the "invasion risk" were conducted before granting it a "kosher" status among the select nonnative species that are utilized in Israel's arid regions.[107]

This position by now is anchored in official policy. When the international board of directors of the JNF decided to adopt sustainability strategies in 2005, the issue was on the table and their position was clear: "The JNF strives to adjust plantings to the habitats and diversity of trees growing and being renewed naturally at every site, and particularly avoid planting invasive species of trees such as the Blue Leaf Wattle / Blue Acacia, Chinese Tree of Heaven, and Mesquite. . . . In recent years, the main task involved with afforestation is the replacement of the first generation of plantings that included the 'pioneer trees' with forests based on natural renewal and that include indigenous copse species and seeds whose origins are based in planted forests."[108]

This, of course, does not solve problems of the past. Foresters have a lot on their plate, and only in two cases thus far was the problem prioritized, with *Acacia saligna*'s seed bases eradicated: in the Balfour and Gvaram Forests. This accomplishment was due to conscientious foresters who were particularly committed to the task. Avni remains pessimistic that an eradication plan for the coastal region will be adopted, and even less certain that it will be implemented, as it involves a long process, requiring institutional stamina and a commitment by local land managers.[109] Like many unpleasant tasks, the longer a real response waits, the more daunting the job becomes.

Optimal Biodiversity

During the autumn of 2011, a heated argument broke out between two schools of ecologists on the pages of *Science,* the world's most prestigious scientific journal. The question involved what to do about the lands required to feed the ten billion people projected to be alive soon on the planet?[110] One group argued that the best way to protect global biodiversity was through "land sparing." This meant dividing up lands into two distinct classes: *agricultural lands,* which needed to be farmed as intensely as possible; and *protected lands,* which would be set aside for nature and for biodiversity preservation.[111] The opposing view prefers a strategy of "land sharing." This position seeks to improve yields on existing farmlands while simultaneously preserving biodiversity preservation through promotion of both objectives on the same lands.[112]

The *land sparing* group based its conclusions on research models that estimated crop yields and densities of bird and tree species in Ghana and northern India. The members of the group found that more species were harmed by agriculture than benefited from it, particularly among species with small global ranges. Ultimately, "sparers" did not believe farmlands would ever contribute meaningfully to preservation goals. It made more sense to focus efforts on maximizing harvest yields on certain plots, assuming this could be done through higher inputs of knowledge and labor rather than by means of chemicals. At the same time, they also assumed that mechanisms could be found to ensure the integrity of the remaining, sensitive protected areas.[113]

In their response, the *land sharing* school made several points: First, many countries lack the capacity to effectively protect biodiversity significant areas but have a tradition of environmentally sensitive farming. Second, there are many places where both agricultural yields and biodiversity are high and where biodiversity has even become dependent on agriculture. Areas with shallow soils or low rainfall are only suitable for nonintensive use anyway, so projections of intensive farming on much of the world's lands are unfounded. Finally, if increased yields in the land-sparing scenario are attained without chemicals, in practice the agricultural fields there would actually be very similar in function to those in the land-sharing scenario. Implicitly, they rejected the assumptions behind the land-sparing dichotomy, which divides the planet into "wild" and "cultivated" lands.

In effect, the land-sparing advocates propose a form of ecological "triage." When resources are insufficient for all to be treated, *medical triage* sacrifices some patients for others who can better benefit from intervention. Land sparing is rooted in pessimism: farming is inherently bad for nature. If future food production means losing lands, best to sacrifice intelligently and get the best deal you can for nature with what is left. To some extent the debate over the legitimacy of arid-land forestry in Israel resembles this heated argument about reconciling farming with biodiversity conservation.

Rothschild and the biodiversity-preservation perspective he represents are heavily influenced by the Millennium Ecosystem Assessment.[114] This vast 2005 report brought together a thousand of the world's top biologists to assess the state of the planet. Its authors considered the full array of natural systems that provide ecosystem services—the so-called "life-support system" for humans. Of the twenty-four ecosystem services evaluated globally, fifteen were declining. Beyond establishing a scientific consensus about a range of conservation issues, the report reframed the discussion about conservation. It shone a light on the many ways that human survival is dependent on nature and the critical need to protect ecosystem services.

While all agree that Israel's new dryland forests provide a host of such services, the nature-protection school fears that they come at the expense of local biological diversity. "The Millennium Ecosystem Assessment confirmed that diversity is the basis for Ecosystem Services—and not the other way around," Rothschild posits. "Diversity provides for system resilience that allows ecosystems to remain stable and secure under conditions of environmental change. Accordingly, any measure (like forestry) that increases ecosystem services at the price of diversity is conceptually wrong and ultimately unacceptable."[115]

This position can be criticized on both theoretical and empirical grounds. To begin with, a closer read of the Millennium Assessment's special volume on biodiversity does not actually find such a stringent position.[116] Biodiversity is important—very important. But along with its *intrinsic* value is its *instrumental* value: ensuring the provision of ecosystem systems and their services.[117] Recently, ecologists have begun to emphasize functional rather than purely taxonomic diversity.[118] A major review of ecological restoration in tropical rain forests published in *Science* reached the conclusion that a successful strategy requires an optimal *mix* of biodiversity and a steady supply of ecological goods and services.[119] The present dynamics of Israel's southern forests may involve a tradeoff whereby original species richness is slightly reduced while other services like soil conservation and carbon sequestration are greatly amplified.

Retired professor Yossi Riov, representing traditional thinking in Israeli forestry, takes this view to an extreme. Riov rejects the underlying assumptions associated with JNF's recent "infatuation" with biodiversity. In his view, biodiversity is always a means to an end, never an end in itself. Israel's nature reserves make up 25 percent of the land designations. Along with the thirty thousand hectares of natural woodlands. Riov believes that it is the reserves' job to protect local species—not forest plantations: "For twenty years, intensive research has been conducted on the subject of 'Biodiversity and Ecosystem Function' and the conclusion of many researchers is that diversity does not necessarily improve ecological functioning. Some even argue that in planted forests, it is preferable to rely on one or two species most appropriate for the habitat."[120] It is true that agriculture is far more widespread and has a far more significant ecological impact in the northern Negev than JNF forests, which only cover about 6 percent of lands.

Stanford professor Hal Mooney was one of the core authors of the Millennium Assessment's *Biodiversity Synthesis*. He offers an intermediate view which is closer to today's thinking in afforestation: "Some people argue: every genotype is crucial due to its 'option value,' so that future generation might benefit from it. Accordingly we have to protect every little thing. But there is another school that understands that certain species contribute more than others and those are the things that we have to protect first. Depending on the wealth of a society or its resources, you may not be able to protect everything. And of course, I don't think we're sad about stamping out small pox. But characterizing 'the relative contribution' is very tricky as it is driven by values and culture. Something may not do much in terms of regulating water or pollinating, but it may provide an important cultural service and need to be protected."[121]

During the 1990s, Mooney's Stanford colleague Professor Paul Ehrlich and his wife, Anne, developed an approach to the issue known as the "rivet-popping" hypothesis: Like an airplane, nature has some redundancy in its design. If you remove a few of the rivets of an airplane, it will probably still be able to fly. But there is also a threshold; at some point, if you continue and take one too many rivets out, the aircraft will surely crash.[122] The trouble is that nobody really knows the precise "threshold" rivet whose removal will lead to disaster. Similarly, it is exceedingly difficult to know which species is critical for the integrity of a given ecosystem. This leads to a conservative bias that assigns each species the same contribution to ecosystem functioning, fully cognizant that this is not the case. Lacking perfect information, it is well to preserve as much as possible.

The Millennium Assessment is of course very concerned about extinctions and processes that lead to the homogenization of life on earth: "The differences between the set of species at one location and the set of species at another location are, on average, diminishing." Land conversions and invasive species are driving a great deal of these trends and need to be addressed.[123] It is wrong to dismiss out of hand concerns that ecosystem services provided by afforestation will not only trump biological authenticity but will come at the expense of overall biological diversity. Eucalyptus forests around the world may sequester a great deal of carbon very quickly, but they can supplant local species and alter soil characteristics, undermining restoration efforts.[124] In South Africa, where eucalypt and pine stands consume 9.95 percent of available surface runoff, experts regret earlier alacrity for nonnative species and tree plantations.[125] In many areas of South Africa, pines have also become an invasive species no less vexing than Israel's blue waddles, with serious impacts for the hydrological balance,

At the same time, the assumption that Israel's planted forests are inherently hostile to indigenous wildlife has little empirical basis. It is true that they can pose dilemmas. For instance, the study in Yatir suggests that when vegetative cover increases, the number of species decreases. Maximizing plant diversity can come at the expense of soil conservation. Yet, when forest managers see themselves as full partners in biodiversity preservation efforts and are open to suggestions, such dynamics can be overcome. For instance, the research team studying the Yatir Forest, which included Moshe, suggested that a "spatial patchwork of forest stands in a matrix of rangeland may increase the overall regional biodiversity. Moreover, increasing the distance between trenches would not affect the anti-erosion benefits, while allowing for better regeneration by the endemic plants."[126] This is practical, valuable advice that can easily be implemented as part of a land-sharing strategic perspective.

The SPNI report highlighting dryland afforestion's dangers for biodiversity conveniently chose to ignore the final paragraph in the *Conservation Biology* article that postulated the ecological trap for lizards. It describes an adaptive management approach: "We used the results of our experiment to convince the JNF and other land management agencies to abandon the Savannazation plans for the remaining *A. beershebensis* habitats and to support establishment of a large sanctuary surrounded by sufficient buffer zones to protect the lizard and other species specialized in loess scrublands."[127] Indeed, even as the national master plan called for intensive plantings in the study area, forestry officials decided to leave the remaining habitat areas undisturbed as a sanctuary, concentrating additional tree planting around historic water holes and remains of ancient human settlements where mature acacia trees were already standing.[128] The encounter underscores the potential benefits of dialogue and a cooperative spirit. Conservation advocates and foresters are quite capable of working together, to produce optimal results where biodiversity and dryland afforestation are compatible.

Increasingly the world has come to recognize that the dichotomy between "wild" and "developed" systems is simplistic. A rich continuum exists between "natural" and urban or intensively managed lands. The fact is that human activities—intentional and inadvertent—have changed many ecosystems beyond recognition, creating new dynamics, combinations, and sometimes beauty. The new configurations are known as "novel ecosystems" (or "emerging ecosystems").[129] These areas are fully functioning ecosystems but the present species composition and relative abundances did not occur in previous biomes. The new systems may be quite stable or they may need management, but all are the result of inadvertent human activities or intentional interventions. Astonishingly, recent assessments suggest that only about a quarter of original wild ranges still remain on the planet. Novel ecosystems make up a full 37 percent of terrestrial open spaces.[130]

Ecologists and conservation advocates have come to recognize that it is very difficult and sometimes impossible to restore altered ecosystems to their original state. Even when it is possible, it may be prohibitively costly. The salient question is: "Can humanity move beyond the illusion of pristine authenticity in these areas and design management schemes that maximize beneficial changes and minimize damage to the ecosystem?"

Dr. Marilyn Jordan, a conservation biologist who works for the Nature Conservancy organization, takes a balanced view. On the one hand, she is vigilant in efforts to prevent introduction of invasive species. Sometimes it is possible to restore native species. But for the most part, there is no choice but to accept novel ecosystems and work with them to maximize the conservation value and ecosystem services.[131] She has compared conservation to working through the five stages of grief, arguing that more and more ecologists are reluctantly accepting that we live in a human dominated world. And some are discovering that patchwork ecosystems might even rival their pristine counterparts. The notion that humans might be able to actually improve on nature is not a new one. Indeed the environmental "pioneer," René Dubos—the same man who coined the phrase "think globally—work locally"—based an entire book, *The Wooing of Earth,* on this theme. In this book Dubos argues that the human ecological footprint not need be a heavy destructive one. On the contrary, he offers dozens of cases from around the world where human activities have had a beneficial effect, turning hostile natural terrain into magnificent gardens.[132]

Recently, the JNF adopted a new compendium of forest practices and principles called the *Bible of Forestry.* One of the main objectives it proposes for Israeli forestry is "improving the landscape." But Rothschild does not accept this as a touchstone for progress in forestry. "Who said the landscape needs improving?" he asks, demanding that the KKL better characterize an operational definition for "degraded lands" before it begins planting trees to rehabilitate the soil. Most Israelis do not agree. After thousands of years of human settlement, labels like "wilderness" or of "pristine nature" are misnomers. If humans act with caution and humility, degraded unfertile lands can be coaxed back to life and produce a variety of ecosystem services. The results can be fetching.

Professor Moshe Shachak spent forty years as a researcher of Israel's desert ecosystems. Before long he was an unofficial scientific advisor for Itzik Moshe and the JNF forestry program, designing and monitoring its savannization projects. Shachak has long since held the view that there is nothing natural about Israel's southlands. At the same time, there is a great deal to preserve. Shachak explains: "It's fine to argue, but first we all should share common assumptions. That is, that all of the ecosystems in Israel are the result of human activities. One can look at the interaction of species and their abundance, and try to preserve them. But this is not preservation of wild nature—only the preservation of species, whose assemblage is anything but natural. So today humans have the responsibility of deciding within this novel ecosystem what should we preserve and what should we create."[133]

Biodiversity preservation, of course, must be a critical threshold consideration for all land-use decisions. The genetic richness of the planet needs to be saved so that future generations will enjoy its many potential benefits. The planet is changing and it is important to leave the best possible genetic base for future evolution and adaptation. There are those like Rothschild who worry that if we focus too much on the general appearance of the "stage" without focusing on the details, we'll lose the "actors." This is a completely legitimate concen.[134] At the same time it is well to remember that many of Israel's natural habitats should not be glorified for their authenticity. The present distribution of the Be'er Sheva fringe-fingered lizard is the direct result of centuries of ecological degradation, with an opportunistic organism taking advantage of a landscape dramatically altered by grazing and human activities.

Undoubtedly, many species will not survive if they find their habitat completely transformed and can no longer rely on its natural support system. Butterflies, for example, tend to be particularly dependent on a specific, larval host-plant species.[135] Should a specific plant type be lost, the butterfly species will soon be gone as well. Even so-called generalist insect species, which are not as fussy, have their limits and usually can only eat a few different kinds of leaves. In its natural ecosystem, a plant can support 170 species, but outside its normal range may only support 5 species. Even if a forestry strategy is not driven by individual species conservation considerations, it has to be exceedingly cognizant of ecosystem stability and long-term productivity and soil fertility. But passive "preservation" approaches in a country like Israel frequently don't make sense. Restoration ecology is full of practical suggestions for actions that can return original biodiversity from seeding to wildlife repatriation.[136]

The challenge of creating the full range of ecological dividends in afforestation programs is of course universal. China during recent decades has invested the equivalent of a hundred billion dollars to increase its tree cover from 12 to 18 percent.[137] Today new forests fill half a million square kilometers of China's land.[138] More trees are planted there than in the rest of the world combined.[139] And yet, reports suggest that the trees may not always be a blessing for the land. Biodiversity often suffers. Birds and other creatures find the rows and rows of fast-growing, monoculture tree farms inhospitable. In others cases, plantations cause serious disruption to the hydrological balance. One meta-analysis there suggests that when pine plantations trees are planted on grasslands, they reduce water yield on average by 40% — eucalyptus by as much as 70 percent.[140] The heroic reforestation efforts have had little apparent effect on desertification rates or even the ferocity of dust storms.[141]

Moreover, newly planted Chinese saplings are not doing well. At least a quarter of the trees die during the first year and subsequent mortality continues to be high. Even the World Bank, never accused of excessive ecological sentimentality, advised China to focus on quality rather than quantity of saplings. After decimating so much of its natural woodlands during the past fifty years, China's decision to pursue a policy of afforestation and restoration is impressive and the scope of its efforts is unparalleled. But the resulting ecological benefits should be much greater.

Today about 4 percent of the world's forests are plantations.[142] In 2000, afforestation provided jobs for some eleven million people. Like the newly created woodlands, the number will continue to increase.[143] About a third of the world's timber is from planted forests and this number should reach 50 percent before long.[144] This is a very good thing, as demand is also growing rapidly. If nothing else, by planting new forests on degraded lands, more ancient woodlands, along with the biodiversity they contain, will remain untouched.

At the same time, the billions of new trees that will be introduced can either be planted with forethought and concern about the planet's ecological health or be an exercise in yield and profit maximization. Israel's experience suggests that there is a continuum between old-growth ancient forests and tree farms. When done properly, silviculture offers proof that "land sharing" is possible. Of course it is unrealistic to have one's proverbial cake and to eat it completely. When trying to generate ecosystem services, there may be cases where some trade-offs will be made between productivity of ecological services (including timber) and biodiversity.[145] But newly planted woodlands can offer a "middle ground" where nonnatural forests, even on very dry lands, improve quality of life and ecosystem functioning. Just as humans are capable of making enchanting gardens, they can create forests that offer magical experiences and homes for countless other creatures and plants. With sensitivity and resourcefulness, afforestation, even in the desert, can *directly* — and *indirectly* — be reconciled with biodiversity.

7

Of Fires and Foraging

Behold I will kindle a fire in you, and it shall devour every green tree in you and every dry tree, the flaming fire shall not be quenched.

—Ezekiel 21:3

The Mother of All Fires

Like all Israeli high school students, Elad Riven's high school requirements included volunteering for community service for at least sixty hours as part of a national "personal commitment" program. Elad, a student at Haifa's prestigious, science-based Reali School, was captivated by the Fire Scouts. The scouts are a local search-and-rescue initiative that monitors the nearby Carmel woodlands to facilitate early detection of fires. The Reali School is near the top of the Carmel summit, minutes away from the fire that broke out there around noon on December 2, 2010. Elad was among the first to see the smoke. This was the third major conflagration in the Carmel in the past two decades and would soon become the most catastrophic.

During recess Elad called his mother and asked her to pick him up and bring his fire-fighting gear so that he might head directly to the makeshift command post. She did not like the idea at all, and told him she was scared. Elad was an only child and was a particularly talented one. Popular among his friends, he was an outstanding athlete and nature enthusiast; he dreamed of being a pilot and had already flown light aircraft. Like many spirited Israeli adolescents, he was not particularly influenced by his parents' anxieties. He told his mother that if she didn't come to drive him, then he would take a bus. A few hours later his charred and lifeless body was found among the trees he had tried to save.

His mother eulogized Elad at his graveside: "He was close to the bus with the prison guards that went up in flames, ran to try and save lives and was trapped by the fire. That's how his life was cut short. A heroic boy who ran into the fire instead of running away from it and saving his life."[1]

The Carmel is one of two woodlands that miraculously survived the extirpation of most trees in Palestine during the Turkish rule. An average annual rainfall of seven hundred millimeters makes it one of the greenest corners of the country. Israel's quintessential Mediterranean forest, it offers a sprawling network of craggy evergreen hillsides sloping down to the Mediterranean Sea. Neanderthals roamed the Carmel aeons ago.[2] It was there also that Elijah the prophet reportedly faced off against 450 false prophets of Baal and 400 prophets of Ashera to see whose sacrifice might break a disastrous drought.[3] The Carmel is dotted with caves in which Elijah purportedly slept; the most famous is part of the Stella Maris Monastery, perched on a forested cliff overlooking the Mediterranean. The British army chose the area for staging a crushing final blow against the vanquished Ottoman forces to take control of Palestine at the twilight of World War I. The trees survived all this, as well as intermittent forest fires that struck the Carmel over the millennia. Indeed, fire was a critical component of the local forest ecology that evolved. But the blaze of 2010 was qualitatively different in its ferocity.

Elad Riven was the youngest but sadly not the only victim this fire claimed. The Damon Prison, nestled into the Carmel hills, was in danger of being engulfed by flames. By 12:38 P.M. the prisoners had been evacuated. But a bus carrying about fifty cadets in the Prison Service's officers course had been summoned and hastily driven to assist in the complicated transfer of prisoners. By the time they arrived on the scene the fire was burning out of control. The bus driver found the road impassable with a fallen tree blocking his way. He attempted to beat a hasty retreat, but was injured attempting a U-turn when the flames roared across the tree canopies and enveloped the bus. The rear exit door did not function due to the extreme heat. Eventually it was forced open and a few prison guards managed to escape, all sustaining some level of burns.[4] Most of the passengers did not have time to get out. Thirty-seven of them would die a horrible death.[5] Other heroes, like Ahuva Tomer, Israel's only female city police chief from the nearby Haifa station, would die that first day, trying to help amid the chaos.

For eighty-two hours, the flames raged in what was immediately recognized as an unusually violent conflagration. During the previous thirty years, along with dozens of minor fires that were controlled, there were at least nine significant wildfires that consumed 80 to 530 hectares of forests, damaging half of the upper-Carmel park.[6] But the 2010 blaze was by far the longest and most ferocious forest fire in the country's history. For four days the entire country's attention, including that of the political leadership, focused on the Carmel Forest and the communities that it encompasses. The tally of damage may not have been significant by international standards, but by Israeli standards it was unprecedented: over five million trees were lost on three thousand hectares of scorched lands; some 250 homes were destroyed in three communities; close to twenty thousand people were evacuated from their homes; dozens of people were injured from smoke and trauma, three of them with severe burns; and there were forty-four fatalities.[7] For four days Israelis were reminded that forest fires are a deadly serious business and trees are not just pretty scenery, but fuel.

Unlike most other major forest fires in the past that were traced to arson, the massive 2010 Carmel blaze can be attributed to carelessness. A fourteen-year-old Druze boy from Usifiyah was soon taken into custody. He admitted that he was "playing hooky" from school on a Thursday morning, toking on a hookah pipe with his friend. By around 11:00 A.M., the two had finished smoking and he casually tossed the smoldering coal piece from the pipe into the vacant lot behind their neighborhood, adjacent to the forest. When the parched woodlands burst into flames he was so overwhelmed and frightened that he ran back to school, hoping that the problem would go away. It didn't.[8]

The devastation that ensued to the Carmel can be attributed to several factors. Strong easterly winds and a topographical makeup of valleys and hills running from west to east created wind tunnels and very strong gusts that rapidly spread the fire. Because of a protracted drought, the flames were particularly powerful.

Traditionally, Jewish communities around the world begin praying for rain during the Sukkot festival in October, when the first showers break the long-summer dry season. But it was already December and there had not been a serious rainfall since spring! In retrospect, 2010 turned out to be the hottest year on record for Israel, a full 1.5 degree Celsius or 10 percent higher on average than any other recorded year.[9] Much of the Carmel is dominated by conifers, a throwback from the days of monoculture plantings. Pine trees are especially combustible because of their sap, whose flammability has been compared to that of "napalm." The desiccated trunks and branches provided the perfect kindling and trees turned into sizzling torches. The initial wave of firefighters, who usually had no problems snuffing out brushfires, was totally unprepared for the ferocity of the inferno. The flames quickly reached the pine canopies and began to leap over roads and fuel breaks.

Firemen on the ground characterized the blaze as "atomic" in its power and dimension, and they called in reinforcement fire and rescue services from Haifa. First on the scene were the venerable JNF foresters. Although they are formally not authorized powers or responsibilities for combating fires, the men who plant trees have always been zealous in their efforts to protect them. After so many years in the field, they are invariably the most experienced and skilled teams in forest terrain. They too had never seen such a mighty blaze and in retrospect called it "the mother of all fires." Professional firefighters from the country's different regions soon answered the call, showing up in their fire trucks. The Israel Defense Forces sent two battalions along with bulldozers, trucks, and water tankers. The military meteorological service provided up-to-the-minute forecasts. Even though internal policies prohibited providing any aircraft to fight the flames, the Air Force sent an unmanned drone to provide real-time reconnaissance photos of the fire's movement.[10] For an entire day, Israel threw everything it had at the fire, but it continued to spread.

Clearly outgunned, after the first day of the fire, Israeli Prime Minister Netanyahu turned to the world for assistance—which was quick to respond. Turkey and Greece, not on the warmest of terms with Israel, sent special aircraft and teams of seasoned firemen. The Netherlands sent four planes designed for firefighting. The Swiss air force sent helicopters, as did Cyprus.[11] And then the superpowers showed up: The Russian Ilyushin Il-76, a plane designed to carry forty tons of water, landed and was immediately pressed into service. By and by, the American "Evergreen Supertanker" arrived and began sorties over the forest, dropping its colossal eighty-thousand-liter loads.

One of the few bright spots of the four days was the participation of Palestinian and Jordanian firefighters.[12] Even as Israeli and Palestinian negotiations had stalled during the Netanyahu regime, fire trucks from Palestine were sent to Arab Israeli villages where other fires had broken out. After finishing these off, the Palestinian firefighters joined the Carmel front. Facing such a concerted international effort, the flames were bound to relent. By late afternoon on Sunday, December 5, after more than four days, the fire was finally extinguished.[13]

Of course the political fallout, finger-pointing, and recriminations were just beginning. While the official review of the firefighting service's "failed performance" was undertaken by Israel's State Comptroller, which focused on the institutional anarchy, unheeded warnings, and poor funding,[14] a separate "ecological" review took place. This review had little to add to the conclusions formulated some twenty years earlier, in 1989,[15] after the first major Carmel fire, or to the specific recommendations of five professional committees set up after massive fires broke out during the shelling of the Galilee during the 2006 Second Lebanon War. In fact, many of the new policies were already in various stages of implementation. But were they sufficient?

Perhaps as a result of climate change, forest fires around the world have reached hitherto unknown dimensions. Recently, four thousand square kilometers of forest were erased in Russia, in what may be the country's most massive conflagration.[16] In Australia, fires burned for a month, destroying 5,000 square kilometers at an estimated cost of four billion dollars.[17] This chapter considers Israel's recent forest fires and what might be done to limit their impact in the future. One of the central recommendations for preemptive management involves grazing. Once disparaged as the greatest of threats to successful local afforestation, expanded foraging by sheep and goats is now understood to be a critical component in a comprehensive strategy for preventing mega–forest fires

Wartime Firefighters

Along with the 1989 and 2010 fires in the Carmel woodlands, a month-long litany of conflagrations during the 2006 summer war in Lebanon heavily influenced local forest fire policies. It would fall on Omri Bonneh's shoulders to ensure their implementation.

Bonneh was raised in the 1960s in the western Carmel neighborhoods of Haifa, where he first developed his love for nature and birds. As a child he was enamored of farm life, spending his summer vacations in Moshav Merhavia, where his mother was born and raised, amidst the livestock and fields of the bucolic Jezreel Valley. He was also a very good student. And as he reached his full two-meter height, he enjoyed much success on the basketball court. For his high school applied-research project Bonneh went out to the Carmel Forest and studied the birds in the Kelach stream.[18] After completing his military service, he pursued these interests in academia.

In 1982, Haim Zaban, then head of JNF's Land Development Authority, began his discreet campaign to attract academics onto the JNF staff. Bonneh was among his first recruits, finding time for studying while enjoying a career as a professional basketball player on local teams. Although his formal doc-

Early aerial intervention: the 1989 Carmel Forest fire. (Chaya Schwartz, KKL-JNF Photo Archives)

toral research focused on empirical evaluation of afforestation practices, one could argue his real "postdoctoral" training involved advanced methods in firefighting. Within two years of starting at the JNF he was given what he still calls his "dream job": overseeing the Carmel woodlands that cross the hills overlooking the sea. (He still managed to moonlight, holding down the forward position on the nearby Hapoel Nahariah team.)

Bonneh first joined the JNF firefighters during a 1983 blaze outside of Kibbutz Beit Oren. In the subsequent postmortem he was given the job of assessing the damage and characteristics of some of the older arboreal casualties. Core samples that he took from fallen trees revealed indigenous Aleppo pines that were 105 and 110 years old. Bonneh returned to combat major conflagrations on the Carmel in 1989, 1998, 1999, 2005, and in 2010 for "the mother of all fires." With hundreds of other minor fire events every year, he became proficient at overseeing firefighting logistics, rising quickly in the ranks to become the director of the JNF's northern region and one of the country's top forestry experts.

The many blazes that Bonneh encountered in the Carmel, however, could not prepare him for the challenge of Israel's 2006 Second Lebanon War, when Hezbollah's militia barraged the Galilee with unyielding volleys of rockets. At times, it seemed as if the Lebanese irregulars were deliberately directing their fire at the vulnerable trees. Unlike the citizens in Israel's northern provinces, nature was unable to escape to bomb shelters. The thirty-four-day conflict took place between July and August, the peak of the dry season, when forests are most flammable. The attacks were relentless, with eight hundred forest fires started by an assortment of rockets, four hundred of which were serious enough to require intervention.

Given the intensity of the attacks, the resulting damage was relatively modest: The twelve hundred hectares of largely coniferous forests that burned was a tiny percentage of the woodlands threatened by the attacks. The Naphtali Ridge, above the city of Kiryat Shmona, was hit especially hard, with of 70 percent of afforested areas burned. Substantial damage occurred in the Biriya Forest, which lies below the city of Safed and the Israel Defense Force's Northern Command.[19] An additional seventy-one hundred hectares of pasture land and sixty-six hundred of nature reserve lands were also charred. The limited size of most forest fires was a testimony to remarkable human effort and a month filled with daily acts of heroism.

Once the war had begun, Israel's fire chief brought in reinforcements from the country's municipal fire departments so that 540 firefighters could be on duty round the clock, during three daily shifts.[20] His positioning of personnel highlighted the dichotomy between Israel's "professional firefighters," with their spanking new red fire trucks and the KKL's fleet of somewhat dilapidated old vehicles. It also defined their respective areas of operation. The red trucks restricted their areas of involvement to roads, factories, and urban communities where people lived. Preventing the fires in open spaces and in the forests was left to JNF's foresters. By then, Bonneh had been planting and tending to trees in the Galilee for twenty-five years, and he took charge of efforts to save them. Looking back, he calls that summer: JNF's "finest hour."

Initially, the foresters were caught by surprise. A few small local planes were recruited to drop fire suppressants on the landing sites of Katyusha rockets. These aerial forays bought foresters the time they needed to get their equipment together, reach the site, and set about extinguishing the flames. Planes make a huge difference in controlling the spread of fires, but ultimately only people on the ground can "finish the job."[21] With so many rockets falling, the teams had to prioritize which fires required immediate attention. Typically, they focused on fires that posed the greatest danger, such as those that started in the foot hills or in young plantations. It could take as long as twenty-four hours to put out an entire cluster of burnings ignited by a barrage of Katyusha rockets. By then there were any number of other fires to address.

The trouble was that the JNF fleet of fire trucks was really old and somewhat decrepit. Its newer models included separate tanks for water and fire

Firefighting on the ground in the Biriya Forest during the Second Lebanon War (2006). (Izik Zaburov, KKL-JNF Photo Archives)

retarding foam; the older, green Mercedes Benz and Abir models had water tankers with capacities of only a thousand to fifteen hundred liters.[22] This necessitated a great deal of shuttling back and forth to filling points. As the war settled into a routine, foresters from around the country left their duties and joined their northern colleagues. Within days Bonneh set up an, ad hoc logistical system, using a military-style division of forces. Each operated independently, with its own staff, communication network, and planes. Foresters in the north were accustomed to dealing with three to five hundred fire events a year. This incessant firefighting imparted a high level of familiarity with the area's woodlands and the uneven network of dirt roads and trails.

Firefighting is an inherently stressful affair. When coupled with missile warfare, it is a truly frightening prospect. Fired from multiple rocket launchers, Katyusha rockets tend to fall in clusters. Within ten seconds, ten missiles will land within a three-hundred-meter radius.[23] Sometimes during the 2006 fire, bombardments would hit in rapid succession and there was no place for firefighters to hide. Michael Weinberger is a veteran JNF forester and was one of the key area "commanders" in the hastily organized defense system. One day, as he moved his team to take on a nearby fire, a Katyusha rocket fell precisely where he had been standing only thirty seconds earlier. When Efi Stenzler, chairman of the JNF international board, came on a site visit to the Mount Naphtali Forest to view the damage around Kiryat Shmona and see the firefighting up close, a barrage of rockets missed him and his entourage by only a hundred meters.

Of course, the mission was even more trying for the foresters because the war was not limited to trees; the foresters were also worried about their families. Weinberger's house in the northern border town of Nahariah suffered a direct hit by a rocket. The explosion totaled the entire second floor but, miraculously, did not harm Weinberger's family, which had congregated on the first floor. Weinberger drove from the forest to survey the damage, turned around, and returned to save more trees.

Like any war scene, some level of personal trauma was unavoidable. Paul Ginsberg is particularly honest on this topic.

> I remember hearing that the war broke out when while I was representing Israel at a UNESCO World Heritage meeting in Vilnius, Lithuania. Hearing that my kibbutz, Sasa, was one of the numerous settlements under fire, I immediately returned home; the kibbutz soon emptied out with only a skeleton "crew" remaining. Naturally, I stayed and we [the JNF] immediately gathered for meetings and got organized at the Mahanayim Junction, near Rosh Pina, every morning. At first, there were only two of us there, with maybe a fire truck and a shaded gazebo for cover.
>
> I had done some firefighting before, but nothing out of the ordinary. Trying to put out fires when you're under attack during a war is entirely different. It's a very conflicting set of emotions. You feel a gut desire to protect the trees that you have spent so much time planting. You have a commitment to your job, and this ideal to create forests and make things beautiful for your country. But there's also a commitment to self-preservation and to your family. After a group of reserve soldiers were killed by a single Katyusha rocket while sitting not far from where we were working, it brought the conflict home. I had no desire to be a hero so that they could build a memorial to twelve foresters that were killed in the line of duty.
>
> You know, when Efi Stenzler and members of the JNF administration from Jerusalem were almost hit, we were taking them around in cars and it really was dumb luck. The entire leadership of the JNF could have been wiped out in a second. But you don't really think about it at the moment. In that case, I collected the motor from the fallen Katyusha rocket, threw it in my truck, and later gave it to him as a gift.[24]

Bonneh was aware of the stress his workers faced and soon after the war enlisted the support of psychologists to help the foresters work through their experience. But the problem of wartime firefighting raises a fundamental dilemma. Ginsberg explains:

> Foresters legally are not responsible for firefighting in Israel. Our involvement might be termed a 'luxury,' and we do it as volunteers, without formal training. Of course I care a great deal about the forests, but ultimately, I don't consider it a critical national asset. I mean if people didn't risk their lives, what's the worst that could happen? A million dunams would burn? You can always replant. And forests do grow back. When the big fire in the Carmel took place, I was already retired, but all the photos brought me back and actually took me to a place where I didn't want to be.[25]

One night during the last week of the war, Bonneh met in the Mount Naphtali Forest with Dani Hananya, chief of the Upper Galilee Region Fire Brigades. Dejected, Hananya shared his frustration that despite weeks of efforts to suppress the forest fires, piece by piece the forest was disappearing and by the end of the war would probably all be burned. Bonneh replied that he and his team would continue to fight the forest fires until the last tree. Eventually, while 70 percent of the Mount Naphtali Forest area was damaged, due to the stubborn firefighting efforts, flame intensity was reduced and many of the trees survived.[26] Driving along the scenic road of the forest six years after the war, a vigorous green tree line testifies to this battle's ultimate success.

As long as Israel is attacked by its neighbors, forest fires are likely to remain a permanent part of the wartime experience. This means that from time to time, the profession of Israeli forester becomes a perilous one. But there is great symbolic significance in these efforts. For a frightened and frustrated home front, saving the trees provides important validation and reassurance. Would that postwar recovery were as easy for the human casualties as it is for Israel's trees.

Death and Renewal

The fires of summer 2006 surely exacerbated a collective national wartime trauma. The 2010 Carmel fire was a heartbreaking, civilian tragedy for Israel. It can also be considered a temporary blow to the aesthetics of the Carmel landscape and a source, albeit ephemeral, of massive soil erosion.[27] But unlike the media spin and misguided press releases of many environmental groups, it would be wrong to characterize these events as ecological catastrophes. Like all forests in the Mediterranean region, there have always been fires in the woods that cover the Galilee and the Carmel. Tree species in planted conifer groves as well as in natural stands contain high oil, wax, and terpene contents, which are all highly flammable.[28] Without these disturbances and the renewal that forest fires bring, regeneration of diverse, healthy forests would not occur. After forest fires, soils are enriched with nutrients from the ashes and regrowth begins.[29] Many of the trees were planted before there was a clear prioritization of indigenous and diversified plantings. The forest that emerges after the flames subside is often completely different and far more diverse botanically. In narrow ecological terms, the 2010 blaze can actually be seen as enhancing successional integrity.

Twenty years earlier, for the first time, a diverse group of Israeli experts agreed that natural processes should drive restoration strategies after forest fires. In the aftermath of the 1989 Carmel fire, an interministerial committee was convened and decidedly opposed anything beyond symbolic replanting. Its position was part of an emerging international consensus. Professor Uriel Safriel, chief scientist of the Israel Nature Reserve Authority at the time, contrasts the Israeli committee's decision with that of a comparable American committee that considered an appropriate response after the massive 1988 fire in Yellowstone National Park:

> Both committees decided against reforestation, but for different reasons. The massive germination of pines, immediately following the Carmel fire, made it clear that reforestation was simply unnecessary and a complete waste of resources. The reasoning against reforestation in Yellowstone was more interesting. Prior to the 1988 fire, the distribution of tree species was patchy. Depending on this pre-fire patchiness, as well as on environmental factors prevailing during the fire, a new post-fire patchiness was created. Reforestation would then be an intervention in the natural patchiness, as well as in the genetic structure of stands."[30]

The formal recommendations at the time held that, for four years, after fires no management activities should be performed (including thinning and pruning) beyond monitoring and research.[31] This created an awkward situation, as the Israeli public had already been asked to contribute money for restoration activities.[32] In the end, most of the money raised in a high-profile 1989 telethon to replant the Carmel Forest was redirected to research. After years of monitoring by ecological experts, it appeared that "forest regeneration was almost entirely influenced by natural processes. Intervention had only a marginal effect."[33] Similarly, the return of animals to the forest was understood as being a function of progress in vegetative structure and regeneration.[34] Reintroduction efforts were deemed unnecessary. In the absence of human mediation, the salient question soon became: "What will nature do?"

Yohay Halpren followed fires in the Carmel forests closely as a teenager growing up in the region—so closely, that he would later adopt his mother's first name, "Carmel," as a new family surname to express this connection. Today he is an ecology professor at the Technion, the Israel institute of technology, in Haifa. He likens the dynamics after forest fires in the Carmel to a spirited race between the two key tree species that constitute the local ecology's greatest botanical rivalry. No sooner had the last plumes of smoke blown away than the competition between Israel's two tree prototypes—the Aleppo pine (*Pinus halepensis*) and the local common Oak (*Quercus calliprinos*)—took center stage.[35]

An initial appraisal suggests that pine trees have a tremendous natural advantage and should be the odds-on favorite. This was the conclusion of Haifa University ecologist Gidi Ne'eman after the 1989 fire.[36] Pines regenerate naturally by releasing seeds from their cones. Once established, few species grow faster than the *Pinus halepensis*. For the most part, pines are what botanists call "monoecious"—sporting both female and male cones on the same tree. (For Aleppo pines, the female cones tend to congregate on the upper branches and the males are relegated to lower shoots.)[37] It takes at least ten years for Aleppo pines to reach sexual maturity. After they do, male cones discharge their pollen. It wafts through the air to the waiting female cones. Having served their biological purpose, the male cones quickly fall off the tree and begin to decay. By contrast, female cones are in no hurry. Once pollinated, they take over a year to mature and produce viable seeds that can fall on the ground and germinate. Under normal, steady-state conditions, the seeds stay locked up and dormant on the tree in the fertilized cones. Getting them out and dispersing them effectively is the key to successful renewal of conifer forests. In a process botanists call "serotiny," pinecones tend to give up their seeds when stressed or "sensing" that the end may be near.

Some pine species have evolved to take advantage of mutualistic interventions by birds, goats, squirrels, or butterflies that forage, open the cones, and unwittingly spread the pine seeds. Other pines are considered to be "xeriscence." Their seed release is induced by naturally hot and dry conditions. And then there are the "obligate seeders"—pines whose seeds are only set free by fires. Aleppo pines, it turns out, are versatile and able to reproduce both ways. Massive quantities of seeds are released following fires, but also 60 percent of the "crop" release has been measured during periodic pulses of the dry heat waves (*sharavs*) that strike Israel.[38] When seeds are discharged under these fire-free conditions, forest shade does not provide hospitable conditions for young seedlings It turns out that the further the seeds are blown from the adult pine trees, the better the likelihood that the "escaped" seedling will survive.[39] The dynamic is not altogether different than some human adolescents who feel that they need to put some distance between themselves and their parents before they can flourish.

Fires, however, create entirely different conditions, essentially reshuffling the forests' botanical deck. The high heat melts the resin that fuses the cone's scales together. Once they are exposed to such high temperatures, pinecones open like flowers. Having been tested by millennia of inferno conditions, pine seeds evolved to be very hardy indeed, able to withstand unspeakably high temperatures of nine hundred degrees Celsius.[40]

Released by the fire, scores of seeds are carried by winds until they rest on the hot soil. Regeneration begins when the rains come, during the winter season that follows the fires.[41] The infant pines sprout from the soil, taking advantage of the new solar access that the freshly burned, open spaces provide. Once established on the charred soil, the seedlings begin to seek light. Indeed the very Hebrew name for pine, *oren,* possibly comes from the Hebrew word for light, *or.*[42]

Under ideal conditions, the life expectancy of Aleppo pines in the wild can reach 150 years. By conifer standards, this is not a particularly long lifespan for the trees. There is a conifer species in California that survives for thousands of years.[43] But many Aleppo pine trees never get the chance to live out their days. Their roots are shallow and generally don't go much deeper than two meters underground. If burning is severe, fires frequently prove to be fatal. When winds begin to blow, the weakened tree is knocked over. Alternatively, damaged pines simply dry up and die in the hot Middle Eastern sun.

In short, the pine's survival strategy "takes into account" its vulnerability and the likely fatal consequences of periodic conflagrations. Pine trees may succumb quickly to fires, but their evolutionary design guarantees a sufficient number of descendants to compete for a place in Israel's postfire forest community. If subsequent weather conditions are right, within a few months, innumerable tiny green seedlings sprout, spreading a lush green carpet over the blackened land at densities that may reach hundreds of seedlings per square meter. Only a tiny fraction of these sprouts, on average, survive through the second summer after the fire, and even then they are highly susceptible to pests.[44] But many, many trees endure.

After the flames subside and the ashes settle, the lifeless, charred pine trees stand like emaciated gravestones on the forest floor. Scorched oaks may stand alongside them looking just as dead. In fact, beneath the ground, the oaks are very much alive. For they have evolved to adopt a completely different survival strategy.

One of the reasons that oaks grow so slowly at the start of their lives is that during those early years they invest most of their resources in developing a deep root system and a nutrient reservoir.[45] In contrast to the pines, oaks are supported by roots that can reach down as far as thirty meters; they will not blow over. The aerial components of the tree — its leaves, branches, and trunks — may be entirely singed. But the thick bark of larger oaks can often protect them and their root systems are largely insulated, as the soil does not transfer heat.[46] Even before the ashes have settled, the root system can still perform its primary functions of stabilizing, absorbing water and minerals, and transporting them to the tree's photosynthetic mechanisms. Moreover, acorns that are buried by rodents are also quite capable of sprouting. In short, the oak tree has not lost its food bank. If pine trees do not overwhelm them and seize all available sunlight, oak shoots will slowly sprout again from the adventive root system, with stubborn coppices emerging from root crowns. Together with other marquis species, they can quickly form an impenetrable, chaparral-like shrub canopy that prevents erosion and retains water. Indeed, oak trees' survival after fires historically has depended on their ability to withstand the pressures of grazing sheep and goats. The same capabilities serve it well as a fire-resistance strategy. As the bitter oak leaves are generally distasteful to foraging animals, presumably the newly regenerated shoots have a better chance of being left alone.[47]

The obvious sports metaphor is the "sprinter" versus the "marathon runner." While Aleppo pines shoot up impatiently, oaks take a slower and steadier pace. Despite years of observation, Professor Yohay Carmel explains that predicting the winner in the ongoing competition for post–forest fire dominance is still extremely difficult, as it depends on numerous subtle factors. For instance, soil type is a well-known determinant of pine establishment. In a recent study, Carmel found that in addition to soil type, the season when fires take place may be a crucial factor affecting postfire vegetation at the landscape scale.[48] The massive 1989 Carmel blaze occurred in the autumn. It was followed by extremely rich and dense regeneration of pine trees that sprouted with the rains.

Carmel postulates that residual moisture from the winter rains might be sufficient to keep temperatures slightly lower during a fire — enabling pines to hold on and compete. But that is not always the case. There are also instances when fires break out in the spring and pines cannot make a go of it, because most of their seeds are eaten or collected by ants and other foraging organisms. In such cases, oaks hold a clear advantage. In some stands, for example, barely a single pine could be found twenty years after fires that occurred during the spring months.

Yet another hypothesis that Carmel and his team are checking is whether pine success might be linked to the effect of later fires on mycorrhiza. These soil microorganisms colonize on the roots of trees and play a central *mutualistic* role by facilitating better access to carbohydrates like glucose and sucrose. The impact of forest fires on mycorrhiza is not well characterized under Mediterranean conditions and needs to be further researched.

The sequence of fire and rebirth played out in the Carmel and Galilee fires as it has in Mediterranean woodlands from time immemorial.[49] With a clearer understanding of the ecology of regeneration, postfire management strategies have matured in Israel, now allowing natural processes to dominate recovery cycles. But this does not mean that firefighting practices should be laissez-faire. Foresters can and should do a great deal to prevent conflagrations and limit forest fire damage in afforested woodlands. Today's dynamics are different than those of the past; the frequency of fires appears to be growing. For some species this poses an existential problem. Pine trees, for example, are sexually mature only at age ten. If a forest burns too often, the conifers cannot regenerate. The pinecones will simply be "prepubescent," with no fertile seeds to deposit. The ferocity of the fires is also increasing, while the number of Israelis living in and around forests rises. The stage is set for greater and greater property damage and human suffering. Luckily, a suite of fire-prevention practices can help.

Protective Silviculture

Most of Israel's newly planted forests are designed to minimize damage from fires. While implementation is never perfect, the committees that met after the 1989 Carmel fire and those of the Lebanon War spoke in one voice. The JNF planners and foresters in the field, for the most part, adopted the recommendations. An entire system of "protective silviculture" has been developed that relies heavily on "presuppression operations" and "fuel management."[50] There are some common practices that appear to significantly reduce the risk of fire. These include elimination of the understory vegetation by prescribed burning and herbicides, as well as pruning of the lower branches two to three meters from the ground, to prevent the fire from climbing up the fuel ladder and reaching the canopy.[51] These practices also facilitate easier access to the forests. At the same time, "planted" or "shaded" fuel breaks have become a centerpiece of Israel's forest fire mitigation design.[52]

These open corridors that crisscross Israel's woodlands cannot stop large fires, but they can effectively slow the advance of small and moderate ones. The raging conflagrations that erupted in 2006 and 2010 easily hopped over the breaks, igniting trees on the other side of the lines. Even in these extreme cases, the breaks slowed the flames down by truncating fuel continuity. More importantly, they give firefighters a base line from which to launch a counterattack.

Maintaining fuel breaks over the years is challenging, especially when reduction of herbicide use is an important management priority. Nature is resilient, and when the sun shines on the open fuel breaks, vegetation of all sorts quickly returns. Keeping them clear is a constant struggle. Access roads in forests serve as de facto fuel breaks and contribute to the matrix. They too need to be maintained. But fire after fire confirms that it is worth the effort.

Establishing a matrix of bald corridors has downsides. For many, the straight, sterile, tic-tac-toe tracks belie any sense of "naturalness" in the forest. To respond to aesthetic deficiencies, *planted* and *shaded fuel breaks* have emerged as alternatives. These can still slow a fire's advance, and retain the feeling of authentic woodlands. *Planted fuel breaks* are defined in professional literature as highly maintained greenbelt strips of land converted to a nonflammable cover-vegetation through interventions that include irrigation and mechanical treatment.[53]

Fire-resistant trees such as a low density of oaks and carobs are planted between the more flammable stands or alongside roads and settlements. In the not-so-distant past, the venerable Australian Casuarina was popular among original Israeli foresters in this role. Cypress species' proven ability to stop the spread of fires is well recognized. JNF forest planners have begun to introduce belts of Cypress trees along major forest routes and ridge tops. Planted fuel breaks can also take the form of groves of fruit trees—from olives and figs to almonds and pomegranates.

An alternative type of fire-prevention planting involves shaded fuel breaks. These are breaks where "surface fuel"—the vegetative understory—is simply removed, and the distance to the base of tree crown is raised. In addition, the canopy is opened, as the distance between individual trees is increased through intensive thinning operations.[54]

Because arson tends to be a predominant cause of forest fires in Israel, it influences the utility of roads in fire-prevention activities. While providing fire trucks with better access, roads also make it easier for vandals to start a major conflagration. There were deep divisions between the members of the 1989 committee reviewing the Carmel fire about the merits of adding roads to the forest. Eventually the dispute was resolved by agreeing on criteria. These included the quality of existing roads, distance from roads to forests, range of fire hoses, road steepness, forest flammability, land gradient (susceptibility to erosion), and, of course, general landscape considerations.

A consulting landscape architect who applied these principles to the existing road network in the Carmel found that there were already too many roads in the forest and that any new roads were superfluous.[55] Notwithstanding these recommendations, the National Parks Authority reached its own conclusions, and paved a road precisely where the committee recommended not doing so. Still, a major road-paving initiative was prevented.

There is nothing new about the concept of separating residential areas from trees. Indeed, the Mishnah, the first section of the Talmud, compiled two thousand years ago, held that "a tree may not be grown within a distance of twenty-five cubits from the town or fifty cubits if it is a carob tree or a sycamore tree."[56] Despite international recognition of fuel breaks' value,[57] not all Israeli land-management agencies have shared enthusiasm for them.

When it was decided to zone the Carmel as open space rather than putting residential housing in the mid-1960s, there was a debate about whether these highlands should be forested as a JNF park or left as a sanctuary for nature. The JNF, along with the government Forestry Department, had just completed planting sixteen hundred hectares of new trees there, primarily conifers,[58] and fully anticipated jurisdiction. But a newly created Nature Reserve Authority also weighed in with expectations. In the end, there was a compromise. The director of that agency at the time, former General Avram Yoffe, was a powerful figure. In the ensuing negotiations he struck the better deal: 80 percent of the land was declared a nature reserve.[59]

The Parks and Nature Authority in Israel maintains a strict "take only your memories—leave only your footprint" ethos in lands that it controls. Although this has great visceral appeal, the policy has significant consequences. For the Carmel it meant that understory in the reserve's forests grew thickly alongside the highly flammable Aleppo pines. The network of fuel breaks was not maintained; thinning and pruning activities in the national park were scanty. The system of lookout towers for early warnings of forest fires was unmanned, a victim of budget cuts and indifference.[60] A "hands-off" policy might make sense for remote nature reserves in North America or Russia, but in a crowded country with lives in the balance, it is irresponsible.

Historically, there are four times as many fires in planted tracts as in comparable natural forests. The Aleppo pine's continued prevalence and extreme incendiary qualities contribute to this phenomenon. The increased incidence of fires might also be attributable to easier access and higher utilization by the public. With lightning fires almost unknown in Israel, it is people who cause almost all Israeli fires. It is therefore remarkable that the total area damaged in a typical fire event is far *smaller* in planted areas than in natural ones. On average, a fire event will damage 7 hectares in a natural stand and only 1.6 hectares in a planted forest.[61] Given the relatively modest losses in managed forests, and the public's hue and cry after the painful fatalities in the 2010 fire, policy makers in the Nature and Parks Authority began to reconsider their orientation.

A key management tool that has *not* been sufficiently explored by any of Israel's institutional land managers is prescribed burning. Throughout the world, forestry agencies rely on carefully monitored programs that initiate periodic fires to remove understory.[62] Assuming that fires are inevitable, this affords foresters the opportunity to choose the time, place, and dimensions, preventing the "megafires" that blaze out of control. In Israel, for instance, in the limited trials thus far, planned fires have been conducted during the winter months, when wood was wet and fires more easily controlled.

When René Karschon moved to Israel from Switzerland in 1949 to oversee research in the government forest service, his first assignment was to conduct a trial burn.[63] But the practice was left undeveloped. In 1987, the U.S. Forest Service sent experts to Israel to teach and demonstrate prescribed burning techniques. Delegations of Israeli foresters were then sent to the United States for more advanced training. Trials were conducted on-site in Israel to assess whether a prescribed burning program would reduce the risk of major future conflagrations,[64] but the practice soon ran into problems associated with Israel's high population density and the proximity of forests to homes. The associated smoke was unpleasant, and many residents grew frightened that fires might rage out of control. The few times that Israel applied controlled burning techniques, it was limited to remote military firing zones, garbage dumps, or site preparation for replanting.[65] The visiting experts shared an American adage, "Pay now or pay later," to explain the advantage of well-known and controlled costs in terms of smoke and discomfort, when compared to the dangers associated with the unpredictable, but ultimately ineluctable major conflagration.[66]

Public opposition to prescribed burnings highlights the human component in implementing fire-management practices. It is human nature to prefer avoiding a certain but relatively minor discomfort when an uncertain but severe risk remains poorly characterized. Moreover, people become attached to their neighborhood forest. The main reason many choose to live in faraway locations is a personal connection to the soothing presence of the woods. They certainly do not see the trees as "fuel" that needs to be suppressed or eliminated. When JNF foresters set out to apply the "lessons" of the 2010 Carmel fire, a noted Israeli environmental artist, Shai Zakai, led a local campaign against the introduction of JNF fuel breaks. Her home village of Srigim is nestled in the thick forest surrounding the Jerusalem hills. Zakai felt that the clear-cutting around the edge of the town was at best exaggerated and — at worst — superfluous. Appeals to logic or even self-preservation often are ineffective. Even though during the same Carmel fire twenty houses burned down in the artist village of Ein Hod, many residents were adamantly opposed to thinning efforts and the clearing of contiguous woodlands.[67] The JNF doggedly pursued its work plan over the following year and at a cost of over six million shekels, eventually trimmed back 831 hectares of forests contiguous to sixty-two communities.[68] Successful fire management clearly needs to entail "dialogue" with the affected public as much as it needs to control vegetation.

In addition to the natural understory that must be cleared, a perennial fire-prevention challenge involves removal of wood residues from pruning, thinning, and clear-cutting operations, along with dead trees. Fire strategies have changed dramatically since the early operations overseen by Yosef Weitz. Most challenges, however, remain the same. Weitz wrote emphatically about the dangers of leaving woody debris in situ in forests. After it dries, the organic material is easily set alight. It also obstructs subsequent efforts to clear out the understory.[69]

As labor costs rose, pulling the trees out manually and burning them became prohibitively expensive. Today, moreover, the goal of foresters should be to sequester carbon rather than release it. Eventually, a practice known as "whole tree logging" was developed for thinning and clear-cutting operations. Forest branches are removed and trees treated on-site or at a nearby roadside, where they can be chipped, and given away as fire wood.[70] After the 1989 Carmel fire, the management review team found that clear-cutting dead, burned trees held no negative impact on regeneration. "Salvage cutting" of burned trees, especially in areas of intense public activities where risk of recurrent fire is high, was recommended.[71] But this is easier said than done. Foresters often prefer a path of less resistance and simply leave discarded and dead wood in the forest, letting nature take its course. There are surely ecological benefits to this approach. Leaving some burned trees in place contributes to a spatially diverse mosaic that supports a higher-diversity habitat for bird populations in regenerating forests.[72]

Fuel breaks can slow down the flames, but will never replace efforts on the ground to extinguish them. It is critical that firefighters get there quickly. To successfully stop a blaze from reaching uncontrollable levels, it is generally accepted that firefighting teams need to arrive at the site of a fire within twenty minutes of an outbreak. This requires extremely high detection capabilities and accessibility.[73] To this end, Israel created a national network of fire lookout towers in the 1950s. Forty towers were built in strategic locations around major forest clusters. JNF specifications stipulate that they be at least twelve meters in height and located in areas free of trees or other obstructions. During the daylight hours, fire "scouts" need to have high-power binoculars, precise maps, and solid cell phone coverage or another reliable form of communication.[74] The manpower required to run this national grid is considerable.

Costs can be moderated by a prioritization process; for instance, monitoring capacity peaks during the driest summer "fire season" with clearly identified, high risk — that is, "red flag" days. JNF claims that when its system is fully operational, early detection is largely achieved: 43 percent of fires were immediately identified from towers, with foresters on the ground sighting an additional 16 percent. Presumably, in a country as small and closely observed as Israel, remote sensing and aerial observations should provide real-time information about fires.[75]

A critical postfire management decision that has evolved is the level of active reforestation to undertake after major fires. Following the 2006 Lebanon War, the JNF convened a complex review process that included five internal subcommittees and an informal council of ecological experts who were to consider forestry strategy and research needs in the wake of the fire.[76] In the past, major fires were a rallying cry for fundraising drives. Millions of dollars had already been raised for restoration efforts. Jewish communities around the world watched with dismay images of Israeli families crowding into bomb shelters and burning forests. They were eager to express solidarity by donating trees. But the clear consensus among ecologists is that the best thing to do, in fact, is to do nothing, at least until two rainy seasons transpire. A new message to the public was adopted that emphasized specific projects for physical infrastructure in forest rehabilitation, leaving space for natural regeneration. For instance, ancient terraces are reconstructed with picnic sites and roads restored.

A three-pronged strategy for reforestation emerged after analyzing the causes and outcomes of both the 2006 and 2010 fires. Roughly a quarter of the land was deemed able to regenerate naturally; half was to receive minimal supplementary plantings on isolated, scorched patches where natural regeneration was unlikely; and a quarter of the most damaged stands, particularly young plantation with limited capacity for natural regeneration, was replanted. In retrospect, actual practices were probably a little more aggressive than the ecological recommendations that almost completely favored natural regrowth. Yet it is still far from the default "replanting" campaigns of bygone eras.

Seedlings introduced in postfire restoration were almost exclusively broadleafed, indigenous species. These not only enhance forest diversity but make returning forests more fire resistant. Survival rates among cypress, oak, and other hardwood species are far higher than pine trees, which consistently prove to be highly vulnerable to fire. And even the pines are far from uniform in their flammability. Research shows that Aleppo pines, with their relatively thin bark, are particularly susceptible to fire. The average *brutia* pine bark is two to ten times thicker, leading to lower fire-related damage and mortality.[77] Indeed, expert recommendations included planting fire-resistant shrubs throughout the forests as well as slow-burning tree species (e.g., cypress and tamarisks) in rows along the network of forest roads to slow the coming fires.[78]

With relatively few replacement trees to plant, donations received after recent fires were often directed to upgrading the fleet of fire trucks so that future responses would be more effective. This is not an inexpensive venture. The price of a single "off-road" fire truck, capable of navigating the bumpy dirt roads in Israel's forests, is currently three hundred thousand dollars, about three times the cost of a conventional fire engine. JNF fundraisers had an easier time marketing the cheaper urban trucks, which were affordable, to a broader range of donors. Ironically, Israel's neglected *city* fire departments were the primary beneficiaries of a campaign whose original intention was *forest* protection. Nonetheless, the JNF made upgrading its forest fleet a budgetary priority, and a new generation of trucks, specially designed for rugged terrain, was acquired over a three-year period.

The new JNF four-by-four fire fleet is specially designed for Israeli forests. The prototype required a multinational "mishmash" of parts: a German "MAN" truck frame, with a 3.6-cubic-meter Dutch fiberglass tank (along with a separate foam tank), using Turkish pumps and sundry other parts from Austria—all assembled in Israel. Lighter than earlier models, the new fire engine is maneuverable. The truck has hand extinguishers for a team of twelve, with a hose that can reach for twelve hundred meters and a cannon capability that reaches tree canopies. The introduction of the new fire engines was impeded by several engineering and manufacturing kinks, typically associated with the adoption of new technologies. Among the problems that had to be solved were sloppy assembly, faulty Turkish pumps, and defective tires. But after adjustments, the fleet emerged as "state of the art" by international standards, providing the best possible equipment for firefighters in the field.[79] In 2012, seven new and even more nimble engines were ordered, replete with cannon fire controlled from inside the truck cabin. This brought the JNF fire fleet size to twenty-three.[80]

Following a government decision, in 2011 the Israeli air force was given responsibility for fighting forest fires from the sky. (Before this, private planes were paid by the JNF.) Within a year a new "air tractor" was purchased, with a capacity to hold thirty-one hundred liters of foam or fire retardants. The number of practice flight hours doubled to six hundred. Planes can now be in the air within fifteen minutes of notification. In its first year, the new air force unit undertook 1,176 sorties in response to 54 fire events. The unit was even discretely called to lend a hand when a forest fire broke out in Jordan. While the unit estimates that 30 percent of the fires could probably have been extinguished without its help, in 40 percent of the cases the speedy, precise interventions successfully prevented major damage.[81]

The new mechanical measures are expensive. Between the years 1994 and 2002, as much as 14 percent of the JNF Forestry Department's operating budget was dedicated to fire-suppression measures.[82] Especially following conflagrations, natural regeneration is usually very dense, requiring repeated thinning. Financial contributions to "rehabilitation work" after major blazes typically do not so much go for replanting as for shaping and thinning a diverse and well-spaced forest.[83]

Following the fire of 2010, special government funding of fifty-five million shekels was allocated to expand fire breaks and increase thinning in high-priority forests near human settlements on Mount Carmel and beyond. This will help remedy years of backlogged work in forest maintenance, but is still insufficient. It will do little to ensure that forests crowded with biomass don't become future firebombs, waiting for a match.

Ultimately, there is a conflict between competing management objectives. Professor Yossi Riov calls JNF's present strategy, which prioritizes natural ecological regeneration and the resulting profusion of plants and shrubs, "the antithesis of fire prevention."[84] This view conveniently ignores the dramatic long-term advantage of natural forest stands over first generation plantations in withstanding fire events. Replanting and buying new fire engines ultimately address symptoms and at best, mitigate damage. The real management challenge remains prevention. Fortunately, there are less costly ways to remove this "fuel" from the forest floor. Israel's forests can offer a "free lunch" to their traditional firefighting partners: grazing livestock.

The Fall and Rise of Grazing

Aware of how quickly animal herds could destroy the painstaking labor associated with new afforestation projects, British land managers were obsessive in their opposition to grazing, especially when goats were involved. In 1928, Colonel E. R. Sawer, then director of agriculture and forests for the British Mandate government, explained: "There remains the outstanding and distasteful problem of the goat—the alleged evil genius of the Mediterranean, against whom has been directed a larger body of legislation than has honoured, or otherwise, any domestic animal."[85]

The public-policy response was draconian, involving grazing bans in all forest reserves. The position was summed up in a 1946 report to the British government:

> The practice of extensive grazing, a tolerable and even sound practice in temperate regions, is in the Palestine climate and conditions the greatest single bar to rural prosperity. In the time of Abraham a few pastoral nomads roamed through great areas of forests and scrub and found an easy livelihood. Since then the population has vastly increased, the area and volume of vegetation has correspondingly dwindled, and it is now an inescapable fact that the destruction of vegetation by the grazing of animals is severely damaging the economy of the plains and bringing ruin to the hill country. . . . The remedy lies in the curtailment of the numbers of animals to be grazed and in radical change of the present regime, familiar to scores of past generations.[86]

Initially, Israel adopted this antigrazing bias. In 1950, in one of its first legislative initiatives, Israel's Knesset passed the Plant Protection (Damage by Goats) Law, which prohibited grazing of goats on public lands—forested and rangelands. The law even imposed grazing standards for private lands to ensure that carrying capacity was not exceeded. A two-tiered standard for different soil conditions was established. Specifically, one goat was allowed for four hectares of rain-fed lands. If lands were irrigated, the standard was relaxed to one goat per hectare.[87] The minister of agriculture (or his representative) was empowered to create a permitted-grazing program, but the overall message was clear: goats are an environmental hazard, no less pernicious than toxic wastes, that need to be excluded from the public domain.[88]

The Israeli scientific community, however, was not convinced. Over time, the decidedly hostile attitude toward goats and grazing in general underwent a dramatic reversal, especially in forestlands. During the 1970s, ecologists around the world, began to challenge the notion that biodiversity was highest and ecosystems most productive when they went undisturbed.[89] Rather, an "intermediate disturbance hypothesis" was proposed. It posited that when disturbances are rare and of low intensity, dominating organisms slowly push weaker species toward extinction. At the other extreme, when disturbances are relentless and severe (e.g., frequent fires, acute overgrazing) any number of species are wiped out. But ecosystems that experience periodic disturbances of moderate intensities (not too regularly and not too infrequently) attain maximum species richness.[90]

After looking at the impact of different grazing regimes on local biodiversity, Technion University ecologist, Ze'ev Naveh argued that this insight had enormous implications for Israel's forests and rangelands policy.[91] Empirical studies confirmed that even heavy grazing did not eliminate woody species.[92] Rather, foraging herds can create disturbances that maintain open patches and increase overall habitat diversity.[93] In perhaps a watershed moment, Naveh went public with his conclusions and was joined by other leading local ecologists. Their letter to the prestigious Israeli daily *Haaretz,* on June 15, 1978, argued that goats and grazing had a positive effect on the ecosystem and that the black goat law should be overturned.

Subsequently, Avi Perevolotsky and No'am Seligman, two researchers at the Ministry of Agriculture's Volcani Institute of Agriculture Research, conducted experiments that proved grazing to be a critical management tool for Israel's woodlands. They explain that when grazing stops, the dense oak thickets that quickly materialize become impassable and unpleasant.[94] Not only do such lands prove inhospitable for human recreation or visual pleasure, but biodiversity quickly plummets. This was borne out in Ze'ev Naveh's original research, which showed a 75 percent drop in species richness in ungrazed oak forest stands. Birds, rodents, reptiles, and insects were much harder to find there than in woods with opened areas.[95]

Grazing breaks up "homogenous carpets" into a more "heterogeneous mosaic," which can accommodate a wider range of plants and animals. Otherwise, trees and taller bushes in Mediterranean forests will steadily block the sunlight, a prerequisite for a rich variety of ground level plants to emerge. Grazing preserves the openings, facilitating an optimum shade-light balance. This also enables Israel's lovely geophytes to flourish. Every year, the striking red blossoms of anemones (*kalaniot*) brighten Israel's forest as a harbinger of spring.[96] In a simple controlled experiment recently conducted, a small section of Negev woodlands where anemones thrive come February was cordoned off and grazing animals kept out. Within a few years the flowers disappeared, overwhelmed by the tall understory, while just over the fence where foraging was uninterrupted, the anemones continued to prosper. In short, without grazing, species richness drops. When it is not cut back, the number of *overstory* species is far lower than the diverse range of plants and flowers in the *understory.*

Of course, the difficult question is one of calibration: when is a grazing disturbance "appropriate" and when does it become excessive? At the theoretical level, "overgrazing" is easily defined as a drop in primary and secondary production (animal foraging) due to excessive exploitation. Perevolotsky and Seligman represent the growing school of Israeli ecologists who argue that ever since the British brought their antigoat bias to Palestine, people have confused "heavy grazing" with "overgrazing." Assuming the land *can* support "heavy grazing," setting that optimum level is not simple and is surely a site- and season-specific exercise. Mediterranean ecosystems, especially woodlands, have evolved to be highly resilient. Some scientists argue that grazing on lands that have weathered intensive grazing over the years contributes to only trivial soil-erosion rates.[97] They believe that land managers should not fear extensive foraging in the forests. Other studies in Cyprus show, however, that goats still cause considerable erosion in historically grazed rangelands.[98] All agree that when controlled, grazing on forestlands can produce enormous benefits.

Local experience also confirms that even when grazing is excessive, damage is *not* irreversible. Israel's border with Lebanon offers a remarkable, almost controlled study in which Israel's forestlands emerge as robust and resilient. A 1946 aerial photograph shows a single, consistently degraded, landscape unit spanning the border. The same remote image taken fifty years after a fence divided the two countries and their very different grazing and deforestation policies reveals a striking contrast. In Lebanon, where intensive grazing and deforestation continued, major woodland species are still extinct. In Israel, which reined in grazing and wood removal, rich arboreal vegetation returned to fill the hillsides and valleys.

Grazing patterns also help explain the striking difference in species diversity between similar climatic regions in the "New World" versus the "Old World." For example, Hebrew University botanist Avi Shmida estimates that the number of plant taxa in Israel is four times greater than in a region with comparable climatic conditions in California, even though the California site was an order of magnitude larger.[99] Not only do grazing animals provide a wonderful mechanism for spreading seed and enriching soils with nutrients; long-term grazing can improve the competitive ability of other plant species less palatable to animals.

Grazing also brings with it unexpected aesthetic benefits. In one recent survey, Israelis were asked to assess different grazing levels in a park in the north of Israel. The majority preferred the look of lands that had been shared with grazing animals. The symbiosis of moderately grazed lands with shady trees and diverse brush that do *not* impair visibility holds visual appeal. The common metaphor used for these types of woodlands is "mosaic"; and indeed, the authors of the survey concluded that the preferred area is "complex, which makes it more interesting."[100]

Even if grazing did not constitute a blessing for biodiversity and landscapes —and it surely does—its utility as a fire-management tool that reduces fuel availability is beyond question. The frequency of fires around the Mediterranean Basin between the 1960s and 1980s was compared in a study which showed that bans on grazing in forest areas in the Mediterranean apparently tripled the number of fires.[101] Because forest fires begin in this lower level of stems, leaves, branches, grasses, and shrubs, the flames' ability to reach the higher canopy is dependent on igniting the rich understory that serves as a "proverbial" ladder, to fuel the climb. When grazing is managed properly, the ladder can be removed and the intensity of fires reduced or eliminated altogether, without having to resort to toxic and expensive chemicals.[102] Cattle grazing, for instance, has proven to be an extremely efficient way to keep fuel breaks clear.[103] Granted, cows can change the feel of woodlands from "natural" to "bucolic." Reintroducing natural herbivores to the forest has intuitive appeal, but was dismissed as an ineffective grazing alternative, due to insufficient numbers and low genetic variability.[104]

When the people of Israel consider their historic occupational inclinations, shepherding emerges as a livelihood that stretches back to the nascent days of civilization. Abraham, Isaac, and Jacob wandered across Canaan tending their flocks. During his period as an Egyptian political refugee in Midian, Moses married Jethro's daughter and tended to his sheep. David perfected his slingshot skills while watching over his herds. This heritage is something which Jews and Arabs share. Yet very little is done on either side to perpetuate the culture of shepherding.[105] Modern regulations and paperwork make pastoral occupations increasingly bureaucratic and less profitable.[106] The present incentive structure needs to be reexamined. Forest management offers an opportunity to revitalize grazing as a societally valuable occupation. If private shepherds do not answer the call, it would be well for Israel's foresters to establish a livestock division and take matters into their own hands.

From Ecological Grazing Theories to Forest Management Practices

While grazing, at least theoretically, has been fully rehabilitated, its reintroduction to Israel's woodlands has not been without problems. When the 2010 Carmel fire is evaluated, a gap emerges between official support for grazing regimes and the rehabilitated status of goats. Omri Bonneh, whose experience with land management and fires is unsurpassed, is convinced that part of the reason why the 2010 fire was so intense involves the steady disappearance of grazing.[107] When Bonneh first oversaw the forest, the Carmel supported some nine thousand black goats; today the number has dropped to about five thousand. In the distant past, goats were a part of the natural balance that existed across the Carmel. There was a period during which Israel pushed for a shift to sheep who are not considered as severe a soil-erosion agent. (More recently cows pasturing is on the rise.) In retrospect, it was the exclusion of goats that made the forest *more* susceptible to fire. Sheep are not as aggressive browsers and do not clear out the forest understory as thoroughly. Once gone, it proved impossible to simply push a button and bring the goats back.

Maximum fire suppression also requires an optimal balance. Consistent

Natural forest firefighters: sheep outside the Yatir Forest (2006). (Nira Zadok, KKL-JNF Photo Archives)

with intermediate disturbance theory, low-level grazing may not be sufficiently comprehensive to eliminate available fuel. On the other hand, excessively aggressive and prolonged grazing can destroy the forest. As it turns out, animals—like people—can be picky eaters. During the initial stages of grazing they will leave vegetation less appetizing and eat only the tastiest fodder. Just like fussy children, they must be pushed to "clear their plates." Once the first round of grazing has wiped out the more delicious and nutritious understory, a second round is needed to complete the clearing, even if this course is less tasty for the animals.

A grazing strategy to prevent fires should also be seasonal, with intensity and frequency driven by precipitation and the thickness of vegetation on the forest floor and breaks. A one-size-fits-all approach by definition will not be efficient. Nonetheless management procedures should not be too complicated, with default standards the easiest to implement. Typically, three months is considered an optimal duration for a grazing regime. During this time, grounds can be cleared without causing undue damage to the trees and other plant life. Present estimates are that one thousand kilograms of dry, flammable, organic material can be removed from a single hectare each cycle as part of a controlled-grazing regime.[108]

During the 1950s and 1960s it was laws that kept shepherds out of forests.

Today it is economics and cultural inclinations. As Israel's standard of living moves into the OECD comfort zone, the subsistence agricultural and pastoral lifestyles of the past cannot compete. The profession of "shepherd" is a not a lucrative one, and nomadic grazers have difficulty competing with feedlots. The drop in livestock in the Carmel translates into understory which becomes fuel and a fire hazard.

The JNF foresters have come full circle in their attitudes toward shepherding. Managers are now willing to "subsidize" herders to bring their flocks into the forests and provide low-cost fire-risk reduction.[109] In France, this has been the policy for some time.[110] Subsidies can take a variety of forms — from provision of fencing and water connections to allowing herders to set up temporary homes and tents in the forest. In either case, the JNF has begun to sign contracts with livestock owners to ensure that forests are systematically cleared of flammable understory.[111] But finding forest-based shepherds is not always easy.

In one interesting instance, a Bedouin faced certain death due to a blood vendetta declared against him by a rival Negev tribe. He managed to escape to the center of the country, and found refuge in one of the largest forests with his small flock of sheep, protected by the trees and JNF rangers. The lifestyle agreed with him, and as the numbers of his sheep and goats grew, the number of fires in the forest dropped. Policy implementation, however, must be carefully monitored. In this particular case, the rangers and the shepherd enjoy a warm relationship; yet the forest rangers still make sure that the shepherd and his many children do not build permanent residential structures that could lead to land claims. Some things never change.

Among the key challenges is determining which forestlands are appropriate for a given herd of animals. The available biomass waxes and wanes, and often there are simply not enough animals to cover every corner of the forests. Picking stands that are the most vulnerable to fires is important. Internal JNF policy correctly stipulates that understory on land which hosts human activities (settlement, forest roads, and picnic sites) should be cleared first.[112]

One of the most difficult dilemmas for foresters in implementing grazing policies involves the massive damage that animals cause newly planted trees. This, after all, was the reason why goats were banned by the British foresters in the first place. On the other hand, the sooner the animals can be utilized to reduce vegetation, the better the chance that trees will avoid extreme flames and survive to maturity. Here again, generic benchmarks for grazing intensity need to be utilized, even as site-specific circumstances should trump such standards when there is a clear, on-site logic to depart from them. For instance, scientists monitoring the Carmel rejuvenation after the 1989 fire recommended delaying heavy grazing until trees reach a height of three meters.

This can take as long as ten to fifteen years.[113] Present JNF guidelines call for a total ban on grazing until a forest stand is three to five years old. Depending on the condition of trees in the spring, some flocks can be allowed into "kindergarten" conifer forests. Slower-growing broadleaf woodlands may require a longer period for seclusion.[114]

Yet another dilemma faced by foresters is appropriate management practices for newly planted, "young" forests, which are particularly susceptible to fires. These trees and their low canopies are not tall enough to survive even modest combustions; typically, a conflagration will mean total loss if a forest that is less than six years old. Furthermore, it takes many years for the bark on oaks to grow thick enough to be an effective fire retardant.[115] Because grazing will destroy young saplings, chemical spraying has been recommended for these stands during their early years, provided that it is limited to relatively nonpersistent treatments.[116] Herbicides have both ecological and economic drawbacks and should be a last resort, subject to clear stipulations that biodegradable chemicals are used that will not contaminate water or harm nontarget organisms. Considering the alternatives, the sooner grazing can be introduced the better.

Symptoms and Causes

The number of forest fires in Israel has increased dramatically over the past twenty years, despite increased sophistication in prevention and response. A recent study summarized the data over a fifteen-year period and found that the number of annual fire events ranged from five hundred to eleven hundred per year.[117] This is an extraordinary figure in light of the country's six-month rainy season, during which forest fires are almost unknown. In the peak summer months, there can be five to ten significant fire events a day scattered across a very small country. When one considers this relentless litany of conflagrations and the desiccated state of the forests for much of the year, it is amazing that anything is left at all.

The story of Israel's fight to save its forests from fires, therefore, is an impressive one. While there is no single magic bullet, the suite of structural management practices, a grazing program to control understory fuels and quick response to conflagrations largely works. Still some people argue that present policies have not gone far enough. Numerous experts and JNF board members have called for an outright ban on lighting any fire at all in the proximity of forests during the dry half of the year, even proscribing Israel's sacrosanct familial grilling rituals. Technically, the old British prohibition on smoking in the forest is still "on the books." Even if such draconian measures were adopted, it would be impossible to prevent all major fires. So it is also important to recall that Mediterranean forest ecosystems have evolved to survive and even thrive after fires. It can even be argued that forest fires in Israel, at a reasonable level, have made a positive contribution to replacing the "pioneer" pine monocultures with biologically richer, more indigenous second-generation stands.

Implementation of Israeli forestry fire policy will have to continue to improve since climate change does not appear to be abating. This makes the forests more dehydrated than ever during the country's long summer months. The number of people living next to woodlands will surely continue to grow, raising the stakes during major fire events.

The real question, however, does not receive enough attention—perhaps because, politically, it is considered to be a "hot potato": what is *causing* this epidemic of combustion? Fires tend to break out between May and Octobers, the dry season when Israel has neither rain nor lightning. In other words, people cause Israel's forest fires. Very often it is not a mistake. A review of fire statistics over a fifteen-year period indicates that roughly half of the thousands of fire events in Israel's forests remain unexplained. For those conflagrations where subsequent investigations *do* find clues, slightly more than half are caused by negligence. The other half are the result of arson. During years when tensions are running high between Israeli Jews and Arabs, politically motivated arson becomes the predominant cause of fires, with hundreds of torchings occurring every year.[118]

Israel's forests are filled with the usual Smokey the Bear messages that remind visitors about the grave consequences of a single careless match. But this hardly engages that part of the public that *intentionally* endeavors to burn the country's forests down. Admittedly the actual number of people who lift a hand against the forest is extremely small. But the motive is nationalistic. The few arsonists who have been caught are adolescent Arab boys. Presumably, they are operating in an environment that either condones or encourages this behavior against Israel. There is no similar arson phenomenon in the forests of Jordan, Syria, Turkey, or Iraq, which all have dissatisfied minorities. Even the defiant Basque minority in Spain or the militant Chechnyans find avenues, other than razing forests, to express their displeasure.

Beyond improving enforcement, therefore, forestry policies must place greater emphasis on the public's relationship to Israel's forests. At the same time, more information needs to be provided to the public about necessary prevention practices. For example, prescribed burning needs to be integrated into forestry management protocols. These programs tend to be unpopular around the world,[119] and politically it will be difficult to expand them without outreach to neighboring communities. Such efforts need to include Israel's minority and disenfranchised communities. There is no reason why a national consensus cannot be forged that values open spaces and expresses solidarity with local forests. It is time to better understand and address any underlying dissatisfaction with the country's forests. Better ways are needed to engage members of a diverse society as committed and equal partners in the planning and re-creating of Israel's woodlands.

8

People and Trees

She is a tree of life to those who embrace her; and happy are those that support her.

—Proverbs 3:18

Forest Feuds

Evegeny Podolsky was one of many immigrants from the former Soviet Union who made Israel his home when the Iron Curtain collapsed and Perestroika opened the gates to over a million Russian-speaking Jews. Like the majority of these émigrés, Podolsky arrived with a valuable skill set, including formal training and experience in forest ecology. Working in the eastern Ukraine he once mapped an eighteen-thousand-square-kilometer forest, an area almost as large as all of Israel. After an initial absorption period, learning Hebrew in his new country, he landed a forester position in the JNF's southern region. Even if the reduced dimensions and dry climatic conditions of Israel's woodlands were somewhat new, the professional impulse was the same.[1]

Work in the field immediately brought him in touch with the area's Bedouin citizens. This collection of nomadic Arab clans and tribes first wandered into Ottoman Palestine centuries ago.[2] Best estimates suggest that in 1948 there were some sixty-five thousand Bedouins who were either settled or maintaining a traditional, pastoral lifestyle around the northern Negev region. Having sided with Arab forces during the war, they were justifiably wary of the new Jewish regime. Most thought it advisable to move on and join clan members in Jordan or Egypt. When the dust settled after the war, only eleven thousand remained in Israel's southlands.[3]

Those Bedouin who stayed were viewed with great suspicion by the Israeli security establishment and cordoned off into a small area of the Negev until emergency restrictions were finally rescinded in 1966. By then the Bedouin had begun to integrate reasonably into Israeli society: learning Hebrew, serving in the military and eventually electing candidates into the Knesset. With the benefit of modern Israeli medicine, better nutrition, and the revival of traditional polygamous practices, their birth rate skyrocketed, to an average of eight to ten people per family. Today the number of Bedouins living in the Negev southlands has swelled to two hundred thousand, almost a third of the region's population, 54 percent of whom are under the age of fourteen![4]

Just as Caine clashed with Abel, nomads have quarreled with their sedentary neighbors from time immemorial.[5] In modern Israel, the itinerant lifestyle led to innumerable conflicts involving damages caused by trampling herds or simply discomfort with trespassing. There were also occasional accusations and indictments for theft.[6]

Decision makers in Jerusalem did not take long to decide that the unregulated nomadic lifestyle could not continue. The government decided that it was time to institute a policy to "settle" Israel's Bedouin. Seven cities were planned for the northern Negev region, ostensibly to offer the largely indigent and underprivileged Bedouin the trappings of civilization and economic opportunity. In 1967 the city of Tel Sheva opened its doors to its new residents, with living quarters that were quite spacious by Israeli standards of the time.[7] It seemed like a more than fair deal. But about half of Israel's Bedouin citizens were disinclined to embrace modernity and the constraints of urban living.[8] Some claimed rights to alternative "ancestral" lands; some refused to share neighborhoods with members of rival tribes; and some were simply not interested in a modern lifestyle, preferring the traditional pastoral ways.[9]

Those who resisted remained in what became known as "unrecognized villages"—a euphemism for sprawling, tenement settlements of dilapidated, corrugated iron shacks on lands that are not zoned as residential. Many continued to raise livestock, with varying degrees of nomadic management. In many cases, individual Bedouin submitted claims of ownership to these or

Evegeny Podolsky: planting trees from the icy Ukraine to the hot Middle East. (Photograph courtesy Evegeny Podolsky)

other lands, citing historic use of the lands by their parents or grandparents. In most cases, the state attorney's office denied these claims. In any event, long before the legal papers were filed, laws had already been enacted nationalizing most of the disputed lands. If evidence could be provided to the courts that supported claims of historic ownership, then at most, Israel would pay the Bedouin plaintiffs compensation.

The government's position was grounded in the perspective that in a crowded country, with very little private ownership of private estates, open spaces were needed to serve the general public.[10] And there were demographic concerns that the Negev region remain under Jewish control. Much of the contested lands were designated as firing zones for the military; others were zoned to be forests; a large chunk was slated to be nature reserves; agriculture claimed a growing portion of the countryside. Only 2 percent was designated for the Bedouin community. And so the seeds of conflict were sewn.

For eighteen years, Podolsky worked closely with the Bedouin. Having had experience with Muslim foresters in Kyrgyzstan, he instantly felt that he "understood the mentality"; the situation required toughness: "When someone is weak, they give them trouble. If you are strong, they respect you." But at the same time, Podolsky believed that if he was completely honest and transparent in his dealings with Bedouin and made every effort to be fair, he could get along with them. Usually he did. For example, when there were disagreements and protestations over land-right violations, he immediately stopped all forestry work to better assess the legitimacy of claims and attempted to reach a consensual resolution.

Podolsky was particularly frustrated when a 350-hectare forest that he had started planting in 1994 in Dudaim, outside of the city of Be'er Sheva, became the target of vandalism by Bedouin families who wished to settle there. He was especially proud of the 98 percent success rate for planted seedlings that was achieved by his team of workers whom he accompanied in the field. "I felt as if I knew every tree we planted there personally!" he boasts.

The area was clearly zoned to be a forested region, under National Master Plan 22, to provide a greenbelt for Be'er Sheva, the largest city in Israel's southlands. Because of their proximity to the Negev's "capital city" the woodlands also offered a prime location for Bedouin families who sought to acquire lucrative real estate by establishing "facts" on the ground. An increasing number of incidents occurred where Bedouins uprooted trees while setting up tents or huts and "squatting" alongside the newly forested land. Podolsky tried to impose some sense of order but realized that he was coming up against a more violent, hostile kind of Bedouin than he had encountered in the past. "There's going to be a disaster here," he reported to his boss.[11] His dire prediction was soon confirmed.

On February 22, 2009, Podolsky set out to the field to review the forest's progress, meeting his assistant forester at the site early in the morning. Immediately, his vehicle was accosted by a young Bedouin man who began yelling at him. His assistant, a Moroccan Jew whose Arabic is better than the Ukrainian Podolsky, yelled over to him: "Evegeny—you're in for a tough day." Podolsky immediately responded: "Call the police." No sooner had their two cars parked and Podolsky opened the door to alight, than four vehicles filled with Bedouin pulled up while a Subaru pickup truck drove straight into his car almost knocking his door off. Podolsky hurried out of the car to take stock of the situation, but the Subaru did an immediate U-turn and tried to run him over. At this point he leaped on top of his car to avoid being crushed—and then quickly jumped back inside to avoid injury.

Before he could turn on the ignition, the Bedouin poured out of their trucks, covering their faces with their keffiyehs to protect their anonymity, and began to scream in Arabic: "Death to the Jews." Growing increasingly bold, they pulled Podolsky from the car while they pummeled him with stones and sticks. Then they hurled a Molotov cocktail at his car, which burst into flames.

Podolsky had not only studied forestry in Russia; as a youth he had also learned karate, and was an accomplished martial artist who had competed in Krav Maga competitions. A physically powerful man, he instinctively began to hit back. Surprised at his resistance, the Bedouin assailants stepped back for a second, giving him an opening. He bolted into his assistant's car, which had been attacked but was not burning. The assistant hit the gas and they made a narrow escape.

Looking back at the incident, Podolsky is most frustrated by the warped dynamics that the incident embodied and the way it was framed. It typified much of the recent violence associated with Israel's Negev Bedouin population: *hooligans* immediately attained "victim" status. He explains:

> Who was the first to show up at the site? Not the police. Rather, it was a representative of a human rights NGO group that came accompanied by a journalist. And they surely didn't come to help us—but only to take photographs for Bedouin propaganda. When Taleh Bassana, the Bedouin parliamentarian, showed up next, I realized that the ambush had been well planned and coordinated.
>
> To be honest I was just grateful that God didn't let me bring my pistol to the field that day. It had jammed in a recent target practice and so I didn't have it on me. As a result, I was the only one hurt that day. (Just between the two of us—those Bedouin really didn't know how to hit very hard. Now in the Ukraine, they know how to hit Jews hard!) I was hospitalized with fractures, but after two days sitting at home in bandages, I decided that I had to go back to the field and show them that I was not afraid. And of course I went back without my gun.[12]

After twenty years of relatively harmonious work in the region, Podolsky found considerable consolation in the number of phone calls that he received from many Bedouins across the Negev with whom he'd worked in the past. They all apologized and assured him that he and the JNF forestry projects did not deserve this sort of abuse. Podolsky was grateful for the expressions of solidarity. But when his Bedouin friends asked if he wanted them to set up a *sulcha*—a traditional Arab ceremony to reconcile adversaries—he emphatically declined: "Tell them that my respect has been violated and that I am insulted. If I see any of them around Be'er Sheva, I'll personally finish them off—and I don't care if I go to jail as a result. This isn't a personal grudge. It was the JNF that was attacked. And I am not in a position to forgive on behalf of the trees."

For a year, Podolsky and his foresters were provided with police protection. This couldn't stop the torching of several bulldozers or—on one occasion—open sniper fire from M-16 rifles that took aim at foresters planting trees. The problem, in Podolsky's view, is linked to the general lawlessness that has inundated the Negev region, making afforestation a dangerous business. Most Israeli landscape contractors are no longer willing to sign a contract to plant trees in the southern regions. One young Druze contractor from the Galilee who had just gotten out of the Israeli army agreed to help out. He had two tractors and told Podolsky that he feared no one. But then his brother, who was working with him, was attacked and almost killed when he was thrown off his tractor. The Druze manager explained why he was quitting: "You know—I pity you. When I was in the army, I knew I had someone watching my back. But here, you are completely on your own. How are you going to survive with only that toy pistol?" Podolsky, however, is not intimidated. He sees himself on a mission. "I think that forestry is the most interesting profession in the world. Some people might say that that money is the most important thing; others will say food; but I say it's forests."[13]

There is nothing new about the violence against the trees in the land of Israel. Indeed, as detailed in chapter 3, since the British Mandate almost a century ago, it has been an unfortunate part of the country's afforestation experience. For many disgruntled minority citizens, sabotaging woodlands is seen as a legitimate expression of Arab civil disobedience. And although nationalism is usually is not the paramount consideration, in the south of Israel the trees continue to have strong political implications. Professor Yossi Riov, from Hebrew University, is one of Israel's distinguished and senior forestry researchers. He acknowledges that, given the demographic dynamics of the Negev and the profusion of illegal squatters, staking claims on open spaces in the desert has come to affect his own perspective on the issue of arid-land forestry. "I, personally, particularly like desert landscapes. Therefore my own opinion is that in principle, in order to preserve the landscape, there shouldn't be any planting in this region. At the same time, there are exigencies that have led me to recognize the importance of planting in the Negev. The planted forests constitute an important factor in the conservation of national lands. I have followed activities in the Negev for thirty-five years and can attest from experience that I have never seen Bedouin settlements in planted areas. Recently, JNF has begun to plant trees outside of National Master Plan 22 at the request of the Israel Land Authority that finances the planting. This testifies to the importance of planting for protecting national lands."[14]

Israel's forests remain a controversial political matter. This is ironic because when trees are not planted for timber profits — but only for public recreational and ecological objectives—garnering support from society for forests and forestry should not constitute a "hard sell." Unrestricted, shaded parks and leafy canopies ought to be well received by all of the local inhabitants. In fact, amid Israel's complex political reality and ever critical environmental movement, many forests are resented.

The aggressive nationalistic opposition of a small minority of Bedouin to the planting of trees is a particularly extreme example of the dissonance between the country's environmental and leisure-time agendas and the outlook of an embittered faction within Israel's diverse society. It demonstrates the kinds of passions that forests can unleash among the country's citizens. Bedouin "antagonists" and vandalism in the south are not the only forces in Israeli society that pose an existential threat to Israel's woods. Seemingly more respectable adversaries may in fact pose more ominous dangers. Many of the forests' "enemies" simply see trees as an impediment to the expansion of city borders or high return real estate ventures. Others commandeer the trees when they can for personal economic gain.

When such conflicts result in flagrant defiance of the law, they are best resolved by law enforcement agencies. But this ultimately constitutes a response to "symptoms" rather than a cure for "causes." In other cases, clashes over the forest policy spill into the political arena. With all its imperfections, Israel is still a robust democracy: public opinion matters. Votes matter. A critical part of a forest conservation strategy, therefore, must involve engaging the public as partners in efforts to preserve and manage the woodlands. Many Israelis are indifferent to the country's forests or unaware of the edifying experiences and the diverse ecosystem services that they offer. Getting these citizens to appreciate the value of Israel's woodlands constitutes a paramount long-term policy objective.

Without a strong consensus across society backing forestry programs, trees will not be safe. Modern environmental history teaches us that the greatest threats to forests come from humans: the harm from *natural* fires, weather, and pests combined produce but a fraction of the biological damage caused by people, intentionally and unintentionally, to forests. For instance, there have been cases in Israel where herbicides used by neighboring farmers to stop pest infestation inadvertently decimated adjacent forest stands.[15]

The tensions documented in this chapter are not unique to Israel, even as the cast of characters and institutions is. Many other societies share a similar constellation of circumstances: indigenous groups with nomadic lifestyles; growing populations in need of housing and services; unscrupulous opportunists who appropriate profits from the commons; and the never-ending search for "suitable" lands by developers with powerful profit motives. It is well, therefore, to consider the interplay between the people in Israel and the country's forestry program.

A New Precedent for Public Participation

Public participation for planning in general and for forestry in particular has become a procedural cornerstone of successful public policy.[16] At a minimum, it requires two things: first, a legal framework that grants the public access to decisions while balancing openness, transparency, and efficiency; second, a *knowledgeable* public with strong convictions who can make a meaningful contribution to the policy discourse. Of course "the public" is something of an abstraction. Defining and facilitating authentic representation in a multicultural society like Israel is particularly difficult.[17] At best, the public is a disparate collection of individuals. Like many revolutions, the dramatically expanded public involvement in the planning of Israel's forests can be traced to *one* very passionate and very knowledgeable woman.

Aviva Rabinovitch was a particularly colorful figure in the lively cast of characters who pioneered Israel's environmental movement. As a small child during the British Mandate, she would disappear from her parent's home in Rehovoth and return with plants, stray animals, and birds. At age seventeen, she ran away again, but this time she didn't come back so quickly. Lying about her age, she managed to get inducted as a combat soldier in the incipient Jewish army that would soon wage a War of Independence. When she returned home this time it was on a stretcher, after being wounded during an attack of her crack "Palmach" unit in the Jerusalem corridor. After the war she settled down to a more predictable life as a biology teacher and mother on a Galilee kibbutz. No one expected such a conventional lifestyle to last for someone as fearless and adventuresome. It didn't. When her children were no longer small, Rabinovitch decided to acquire "credentials." Based on her encyclopedic knowledge of nature and abiding interest in the interplay between geology and vegetation, she received what may be the fastest-attained biology Ph.D. in the history of Hebrew University.[18] Rabinovitch then took a position in Israel's fledgling Nature Reserve Authority, and for most of her twenty-seven years there served as its chief scientist. She stirred up enough dust to win a UN Ecology Award in 1997 for her outstanding contribution toward a better planet.

In this capacity Rabinovitch fomented institutional frictions with the rival forestry agency that to this day have not entirely dissipated.[19] She blamed the JNF for much of the country's ecological woes. When the foresters held public hearings or conferences, it seemed like she was always in the audience to express her revulsion at the pine monocultures of the period. She lobbied her old comrades-in-arms, many of whom had become senior military officers and politicians, like future Prime Minister Ariel Sharon and right-wing party leader Rafael Eitan. In retrospect, many foresters admit to enjoying the periodic skirmishes she initiated that forced them to reconsider many assumptions.[20] At the time, Rabinovitch never believed that her ecological critique enjoyed any real traction among the macho forestry community.[21] She was wrong: when Rabinovitch died in 2008, the JNF forestry journal published a warm obituary, praising her stormy but extraordinary contribution to Israel's new generation of forests.

In 1998, long since retired but still very frustrated at local forestry practices, she decided to change her tactics and take on the role of legal advocate. Rabinovitch took the bus from her home at Kibbutz Kabri to the offices of Adam Teva

Aviva Rabinovitch (ca. 1964), just before her long tenure as chief scientist at Israel's Nature Reserve Authority, where she launched an unrelenting ecological critique of JNF forestry practices. Her position was ultimately vindicated by the Supreme Court's 2001 decision. (Buxi Arbel, Kibbutz Kabri Archives)

V'din, the Israel Union for Environmental Defense, a public-interest law organization in Tel Aviv. The organization initially did not know what to make of her passionate claims. But Rabinovitch was extremely persuasive about the need for a total reform of forestry practices and it took little time to recruit the young attorneys to join her battle cry.

Later a law professor, Irus Braverman was a young staff attorney when the assignment fell in her lap: "It was literally my first month on the job. I remember when a more senior staff attorney Elli Ben Ari came to my office and said: 'There's this women coming here today who has been harassing me about JNF and forests. I can't get her off my back. Just sit with her, hear her out, and get it over with.' I was used to getting passed junk cases as the junior team member from my days as a prosecutor. So this very skinny woman with white hair shows up, but she had an extremely powerful personality and it didn't take long for me to get engaged. I agreed to go up with her to the Galilee so she could walk me around and show me the destructive practices of the JNF. And though I was getting into her head and understood her concerns, I also knew that we were going to have trouble making this legal case based on environmental concerns, which was the usual path for the organization. If we went that route it would turn into a scientific debate about the merits of pine trees that the court wouldn't hear."[22]

To be sure, by then forestry theories had already changed in Jerusalem. But in the periphery of Israel, where much of the planting was still being done, Rabinovitch saw countless examples of the old industrial, ecologically obtuse practices. And she didn't mince words. When the upstart environmental lawyers fired out their letters detailing the legal basis behind Rabinovitch 's ecological complaint, there was indignation in JNF's Jerusalem headquarters at the audacity of the challenge and little willingness to engage. So the lawyers decided to let the Supreme Court decide who was right. Rabinovitch convinced veteran ecology professor Ze'ev Naveh, as well as retired director of the Nature Reserve Authority Dan Perry, to join her as co-petitioners and a suit was filed in Israel's High Court of Justice.

In her sworn affidavit, Rabinovitch took aim at the pervasive burning, spraying, and bulldozing of lands and the pine monocultures that followed:

> During the last decades the JNF has adopted a much more aggressive approach, utilizing heavy mechanical methods. . . . Afforestation practices are almost uniform, resulting in heavy damages to natural habitats, the systematic destruction of Israel's Mediterranean ecosystems, and the biodiversity and species they contain. . . . The JNF acts within its forested lands as if it is a private holding. Its approach is flawed and narrow, including spraying (chemicals) that are hazardous to humans, flora, and fauna, brutal mechanical activity, as well as a limited focus on the trees—with a preference for planting and cultivation of a certain type of tree (pines, with all their implications). . . . The JNF's attitude to these open spaces is no different than that practiced in agricultural areas.
>
> Each natural habitat in Israel contains hundreds of species. By the time the JNF completes its work preparing the ground for the planting, there is usually only one species left; after a few years, as few as thirty species might survive instead of hundreds that would be there naturally (among them, many that humans could utilize in the future as spices, medicinal herbs, or essential progenitors). All other creatures are destroyed: plants such as algae, lichen, moss, and ferns; also insects—among them pollinating bees that are essential to agriculture—rodents, birds, lizards, skinks, and in fact everything that lives in the area.[23]

While Rabinovitch's motivation was ecological, the attorneys understood that they would only be successful if they could show procedural delinquency. The petition's legal arguments were based on the right of the public to participate in the planning of forests. The Planning and Building Law offered a recognized framework for promoting Rabinovitch's environmental agenda of forestry reform.

Under the existing system at the time, forest plans were extremely vague. The only review they underwent was internal, by means of a "Grazing and Forest" committee that enjoyed no formal legal standing. The attorneys argued that the 1995 National Master Plan, promulgated under the Planning and Building Law, changed that and fashioned a transparent process of notice and comment. Accordingly, the petitioners demanded that, as with any other development project, prior to tree plantings a detailed plan be prepared for the entire forest area and reviewed by the public. The detailed plan had to be consistent with the type of forest zoned and the associated parameters listed in the National Master Plan. For instance, a detailed plan should include land designation, anticipated uses, planned forestry activities, access routes to the forest, roads within the forest, and a precise demarcation of boundaries.

The petition was filed in 1998, requesting an injunction that would impose fundamental changes in the way Israel's forests were planned and approved.[24] The spectacle of one "green" organization suing another in the highest tribunal of the land was picked up by the press, and in the environmental world was something of a cause celebre.[25] The court was clearly uncomfortable with resolving the dispute and causing damage to a venerated public agency. The judges encouraged the sides to "work things out." But the JNF stonewalled the environmentalists. After two years Rabinovitch and her ecological dream team returned to the court and demanded a decision. Left no choice, a lengthy judgment was rendered that quoted extensively from Rabinovitch's disparaging description of forestry practices. Then it went on to the legal questions:

> Here's the crux of the matter: the JNF has busied itself for many years across the entire country in sundry afforestation activities, and the question that is asked is: what is the law regarding these activities after 1995, when National Master Plan 22 came into force? More specifically, while section 5(b) of the Master Plan speaks of preparation of detailed plans for a variety of programs, not a single detailed plan has been submitted in the matter of land zoning, the approved uses and their division according to Master Plan 22, or about the anticipated forestry activities (with the exception of construction and road building, etc.).... Nor has a single program passed the series of processes that a proposed program is supposed to pass through in order for it to become an approved plan according to law.[26]

Mischa Heshin was the senior presiding Supreme Court justice in the three-judge panel that heard the case. Heshin had been a brilliant law professor and attorney, and was considered a particularly opinionated and flamboyant judge for Israel's usually subdued Supreme Court. Once the sides failed to reach an agreement, he did not hesitate to decide who was right.

Now that Israel's forests were regulated by a national master plan, Heshin held that any and all plantings of forestlands were illegal unless they received prior approval from the relevant regional planning commission. For this to happen, forests needed to be based on very detailed plans that specify what trees will be planted, how the forests will look, and what activities they support: "Our conclusion, therefore, is that for purposes of National Master Plan 22, the law applies to every person and body in the state, including the JNF. And what forestry activities are prohibited to all persons or bodies is also the law for the JNF. The respondent's position, if we were to accept it, would lead us to recognizing the JNF as a 'state within a state.' Rather, the JNF is subject to the laws and the statutory master plans, just like any person or body in the state—including National Master Plan Number 22."[27]

After pursuing a thirty-five-year quixotic crusade for ecologically friendly forests in Israel, Aviva Rabinovitch was at long last victorious. In boxing terms it was a technical knockout! In fact, however, only then did the real work begin.

When Pinchas Kahana took over planning at the JNF in 2000 he had no idea that he would end up with the tedious job of turning the high-minded language of the court into an entirely unprecedented planning system for Israel's forests. With training in economics and regional planning from the Technion University in Haifa, he had already enjoyed a fruitful career in the Jewish Agency, designing new farming communities. But it turned out that opening up forest planning as a participatory process was a lot tougher and more expensive than anyone had imagined. In retrospect, it was an important phase in the maturation of national afforestation efforts.[28]

Even before Kahana's arrival, preparation of plans had already begun for two of the largest forests in the center of the country: Britania and Ben Shemen. Kahana saw the process as an opportunity to establish a model system that would familiarize regional and local planners with national standards and conditions. In fact, these two forest development projects constituted the largest physical planning initiative going on in the country at the time. Together, they covered eight thousand hectares in the country's heartland.

Kahana convened a steering committee that decided that the forest blueprints should conform to the emerging national principles that call for maximizing conservation of open spaces. Physical development was held to a minimum and concentrated in a single cluster. For example, several buildings that had been proposed to provide guest services to the scores of anticipated visitors from Jerusalem and Tel Aviv were abandoned, and restaurants scattered throughout the woods were rejected.[29] As some of the forests were already relatively mature, there was great sensitivity to preserving existing growth. Under these detailed plans, for example, no mechanical machinery was allowed for digging holes. Of course, the kinds of tree species to be planted were specified up front in order to ensure the indigenousness of the forest. Preserving a natural ambience was also considered to be of critical significance. In some woodlands, in order to maintain the local character, even information centers were considered superfluous physical development and not worth the aesthetic price tag. Once revised and trimmed back, the draft forest-development plans were then reviewed by the various stakeholders.

As a result of this approach, by the time forest plans are submitted to regional planning commissions, their provisions contain the outcomes of animated (and sometimes angry) debates, negotiations, and compromises. Some experts warn that forestry is a science and that excessive influence by an unprofessional public can lead to foolish and even irreversible mistakes. For example, in the United States, public discomfort with thinning practices led to massive pest outbreaks and the loss of millions of trees.[30] But, more commonly, these procedural "speed bumps" can lead to meaningful improvements. Making the most of public participation has been compared to a tapestry. You can bring all the threads in the room, but a professional weaver is needed to create a product that allows each strand to make a contribution.[31]

Three examples are instructive in this regard. When the Erez Park, near Israel's border with Gaza, was in the planning stages, discussions with the elders on the neighboring kibbutz revealed the existence of a historic road lined with sycamores. Its preservation was built into the program. The second example concerns a forest in the semi-arid region of the Negev. After a public hearing with farmers near the Lahav-Shomriyah forest, the pine plantation was modified to spare historic agricultural areas. Today these area offer a protected sanctuary for rare flora in the heart of the woods. The third example involves another southern stand, the Be'eri Forest. Locals urged that an old road and quarrying site from the British Mandate be set aside as a tourist attraction within the forest—and so it was.[32] Substantively, Israel's forests are richer because of the procedural requirement to "stop and listen."

That said, detailed plans can be wearisome, and the deliberations with an ever-opinionated Israeli public a "pain in the neck" for foresters who in the past could simply "go with the flow" and their on-site intuition. Once approved, however, detailed plans offer critical benefits in the ongoing preservation skirmishes. When forest dimensions are articulated in a statutorily approved blueprint, it is much harder to change them. For instance, the very lovely Beit Keshet Forest is home each spring to some of the country's most astonishing blossoming wildflowers, providing free picnic and natural sites for celebrations to the Arab and Jewish populations of the Galilee. But when a new development strategy was proposed (Master Plan 35), the mayor of Nazareth wanted to include the forest inside the greater municipal boundaries, and the Planning Administration at the Interior Ministry was inclined to accede to his request. Anticipating eventual urban development at the expense of the forest, Kahana refused to have the approved plan revised in the National Planning Council meeting. He threatened to stall the approval of a new national development strategy until the forest was ultimately returned to its place as protected woodlands in the map of National Master Plan 22.

It was during this time when that a new internal forestry policy was put in place by then JNF chairman Yehiel Leket that provided an additional backup for a more conservative, planning policy: Leket decided that any concessions that involved subtracting lands from an approved forest plan required his personal approval, after recommendations from top-level JNF management in Jerusalem. In the past, the default dynamic had been just the opposite: a forester who wanted to *oppose* development needed special approval from Jerusalem.[33] When supported by this sort of a conservation bias, detailed plans offer forests and foresters a modicum of certainty in a rapidly developing country where trees are constantly under attack.

Defending existing forests from the challenge of legitimate interest groups is a taxing, relentless task. It can only be done by getting planning commissions to set boundaries and then lobbying politicians to respect them. Protecting forests from individual Israelis who would assail the forests for their own personal gain involves another kind of challenge altogether.

A Sherriff for Israel's Forest

Amikam Riklin is a large man. This is a good thing, since he has a very big job. It requires obstinacy, an ability to intimidate, and an uncompromising sense of purpose. To succeed, he needs to create meaningful deterrence—no small order when his formal authorities in some areas are rather anemic. For many years Riklin enjoyed the tranquil existence of a geography teacher and informal educator of high school hiking groups. But in a midlife decision, he embraced a new career overseeing efforts to enforce the laws that protect Israel's forests. This puts him in conflict with hundreds of citizens from all walks of life. Yet he seems to enjoy the rough-and-tumble that goes along with standing up for the trees, or at least the sense that he is starting to win this unremitting battle.

Riklin brings an unusual passion for the land of Israel to the position of JNF's chief enforcer. As a personal hobby he began to chronicle "charismatic" individual trees that adorn Israel's countryside. The photographs and stories of these grand arboreal specimens were ultimately published in a fascinating book: *101 Special and Amazing Trees in Israel.*[34] For his day job, Riklin has served as the trees' bodyguard. "When I meet violators I tell them that there are four things people need to know about me," he explains: (1) I was here before you: my family has lived in this land for seven generations; (2) I speak Arabic; (3) I speak Yiddish; and (4) I'm the son of a judge. I hate criminals.[35]

In 2002 the JNF decided to create its own independent "inspection" unit to

"The Enforcer": Amikam Riklin, head of Israel's forestry Inspection Unit. (Photograph courtesy Amikam Riklin)

curb the countless violations against trees. In part, the restructuring was done to comply with demands imposed by the Supreme Court case.[36] In addition to their environmental critique and demand for public review of detailed forest plans, the petitioners also challenged an inherent institutional "conflict of interest." The day-to-day afforestation conducted in the JNF regions, it was argued, might undermine the organization's Inspection Department, that needs to operate in a strictly regulatory oversight capacity. Rather than transfer "enforcement" back to the Ministry of Agriculture, the JNF management negotiated permission to create an autonomous unit. It then gave the job of running it to the most autonomous member on its staff.

Riklin confirms that conflicted dynamics characterized past enforcement efforts.[37] Once the authorities were transferred to an independent "national" unit, institutional resolve and efficiency immediately improved. Prior to that, forest inspectors were overseen by regional managers, whose top priority was planting, tending to trees, or taking care of visitors. In those days there were sixteen forest inspectors, spread across the different regions of the country. In a good year, an inspector might file ten reports of violation. Their bosses were not obsessive about protecting every inch of woodlands and imposing the letter of the forest protection laws that safeguard the forests. All that changed when Riklin took control. By 2012, there were only eight inspectors—half of the previous staff. Operating in concert as a national enforcement unit, they were filing hundreds of citations annually and addressing significantly more problems.

Every year, the Forest Inspection Unit publishes an annual report that offers a snapshot of its activities. Slow but steady progress in addressing the pathology of noncompliance in Israel's forests emerges from the statistics. The annual violation count ranges between five hundred and seven hundred. Typically, illegal felling of trees is the most common offense identified, sometimes reaching as high as 40 percent of total violations. As it became clearer that infractions were being vigorously prosecuted, deterrence set in and the numbers of violations began to drop. Table 8.1 (reproduced from the 2010 report) offers representative examples of the nature and scope of the unit's work.

In 2003, a year after its formation the Inspection Unit issued its *Work Procedures*.[38] A self-proclaimed "Bible for inspectors," the handbook offers a step-by-step manual for the unit's personnel, describing the measures to be taken when people break forestry law. Violations include the illegal felling and transport of trees or trespassing in the forest. Once they spot the illegal activity, JNF inspectors are expected to document it, beginning with a thorough filming of the violation site. Then they identify the precise location on their GPS and compare it to the zoning in existing planning maps to be completely certain about the nature of the violation. This is crucial because not all forests in Israel are the same. Different categories of forests enjoy different levels of protection: the old Mandatory forest reserves have one set of rules and procedures for addressing trespass, while nondeclared, open spaces have another. Forests planted according to National Master Plan 22 are another normative "kettle of fish" altogether. Each has its own laws and enforcement procedures. To be effective, inspectors rely on relatively simple equipment. A recent shopping list requesting new requisitions for the unit is instructive: it includes basic monitoring paraphernalia such as long-distance cameras and MP9 video recorders, high-power binoculars, night-vision glasses, a variety of vehicular accessories such as towing capabilities, and hydraulic lifts for lifting illegally logged timber.

Because of its long-term implications, trespass is probably the single most problematic forest-related infraction. This can take a variety of forms, from a cavalier landowner who borders the forest deciding to extend her backyard, to would-be homesteaders. Because of the legal directives, violations are divided into two categories: "fresh trespassers" and "veteran trespassers." A fresh violation involves someone who has encroached on forest lands for thirty days or less. Such cases are relatively easy to determine, with the help of aerial photographs and satellite images. Moving trespassers out of the forest is of course a much harder task.

Notification and implicit threats of coercion frequently are not enough, and removal of illegal forest trespassers requires physical force. The problem is that forestry inspectors are not government employees with full police powers. After documenting violations and conducting the basic interrogations, an appropriate "enforcement partner" needs to be called in. This is typically the nearest police station or sometimes the "Green Patrol." The latter is a special government enforcement unit established in 1977 on behalf of Israel's Land

Table 8.1. Activities of Forest Inspection Unit, 2012

Type of Violation	Number of Violations	Number of Cases in Court
Forest Ordinance—Felling	191	33
Trespass	131	44
Evacuating trespassers	112	
Grazing without licenses	79	
Violation of Litter Law (dumping)	175	
Quarrying Law violations	5	
Planning and Building Law violations	2	
Vandalism	18	
Other	28	
Inspection of contractors	143	
Total	879	77

Source: Jewish National Fund, Inspection Branch, *Summary Report of Activities of the Inspection Branch for 2012* (Jerusalem: JNF, 2013)

Authority and Department of Agriculture to combat trespassing on public lands.[39] In some cases, the forest actually lies inside municipal boundaries and the local police station is the appropriate "partner." Similarly, when trucks carrying wood are suspected of illegal logging, only the police have authority to intervene and detain vehicles.[40] In 2010, 107 fresh trespass infractions were identified and summarily evacuated thanks to backup from legally authorized enforcement units.

It is never simple, but expelling a fresh interloper from the forest is a practicable task. Israeli law makes it is much more difficult, however, to deal with a "veteran" trespasser. Should a protracted "veteran" trespasser manage to assert a reasonable claim to "squatters' rights," a longer and more tedious legal process ensues. There are considerable economic incentives for grabbing public lands, so trespassers frequently put up intense legal battles. As most forests are located on state lands, legal enforcement action requires the involvement of the attorney general's office in Jerusalem or one of its district representatives. But Israel's law enforcement authorities have a lot of other matters on their hands. The hundreds of violations, large and small, associated with forests cannot compete for attention with the organized crime, corruption, and violence that plague modern Israel. Because of the case backload and recognition of its limited resources, the state attorney's office finally agreed to "privatize" the legal prosecution of offenders.

In 2008 the JNF received permission to employ private attorneys to prosecute violators of the nation's forest laws. Lawyers with a background in criminal law were hired on a retainer basis. After receiving authorization from the attorney general to file cases on behalf of his office, they offer a far more nimble legal team. The private attorneys are "duty-bound" to the forestry service which can always "let them go" if their performance falls short of expectations. Technically, the private lawyers are bound by the directives of the state attorney's office. In practice they must rely heavily on the technical information and evidence collected by the Inspection Unit.[41] The result has been a dramatic increase in the number of cases filed against contraventions of forest law. For example, in 2009 sixty court verdicts were handed down in sundry legal actions initiated by the Inspection Unit with 120 cases still pending in court.[42]

Accompanying Riklin on an inspection-site visit is a captivating experience because he has a story for every corner of every forest and is fairly fearless in his willingness to confront violators. Every year there are about 150 documented instances involving Israelis who trespass and try to create facts on the ground at the expense of Israel's forests. While typically people tend to associate trespassing with Israel's rural Arab minorities, in fact, the interlopers come in all shapes and sizes. During rounds one August morning, Riklin visits a Jewish suburban community to follow up on a case in which—despite a series of notifications—a local family could not resist the lure of the forest that borders its home. When warnings went unheeded—given the "fresh violation" status of the encroachment—the unit simply sent in a tractor, confiscated the family's trampoline, horse carriage, and other outdoor furniture, and destroyed any paved trails made by the offender into the forest. Riklin is satisfied that the message was internalized and there are no signs of renewed trespassing.

But this is an easy "fresh" case. A far more acrimonious battle involves a Yeshiva—an adult center for studying Jewish religious texts and theology—which has sprawled out of control. The center is located near the site which some believe to be the grave of "Dan Ben Yaakov," the son of the patriarch Jacob, whose progeny formed the tribe of Dan. In the 1950s, Israel's Ministry of Religion decided to protect a network of graves in which Jewish ancient "saints" were purportedly buried, lest the graves be usurped by Muslim competitors, claiming that their hallowed dead were actually buried there. Dan's grave was officially recognized and given the status of a holy site. The tomb was even zoned as a religious area, with a 352-meter radius that is surrounded by a lovely young forest. It did not take much time for a Yeshiva study center to establish itself there under the charismatic leadership of one Lior Enriques.[43] Soon a whole network of illegal, ramshackle structures sprang up for use as dormitories and other facilities. Naturally, this came at the expense of the forest. As part of the new focus on "long-time violators," Riklin initiated legal proceedings to expedite the Yeshiva's expulsion.

Upon visiting the site, Riklin immediately was buttonholed by a rabbi from the Yeshiva staff. The pressure of the pending legal trial appeared to be taking its toll and the rabbi attempted to seize an opportunity to bypass the attorneys and try to reconcile directly. But Riklin was unmoved: "You are a criminal," he told him. Referring to the prayers of Kol Nidrei ("All My Vows") and the opening appeal before Judaism's most solemn day of repentance, he continued: "Only on Yom Kippur am I allowed to sit among criminals."

The Yeshiva is not without its political patrons. In the present case, Enriques enjoys a strong affiliation with the religious Sephardi political party, SHAS. Ironically, the party is headed by Eli Yishai, an intermittent minister of the interior, who technically oversees the integrity of Israel's planning system. The flagrant conflict of interests does not seem to bother the minister. The day before the visit, his office brought pressure, yet again, on the inspector responsible for the case, in an attempt to soften Riklin's obdurate stance. But this was to no avail, as Riklin explained: "I told them: we aren't going anywhere. There is a national forestry plan, and it says that the JNF owns this land. It is zoned to be forest—except for the 352 meters of grave site." Well aware that his conduct could be used against him in the inevitable court battle, Riklin interrupted the discussion: "Look: perhaps you will succeed in getting a change in the land ownership. But as long as this land is designated as forest, I am going to throw you out. You don't need to talk to me. Talk to your lawyer."

More recently, the JNF Inspection Unit began to make such long-standing taking of public lands its top priority. In 2010, for example, twenty-six legal cases were filed against squatters as opposed to only sixteen in 2009.[44]

Conflicts like these make forestry inspection a high-anxiety business. Inspectors are routinely threatened and occasionally beaten up. Because these encounters can quickly inflate into national controversies, the Inspection Unit coordinates not only with the police but also with the JNF spokesperson. Managing the public relations around enforcement is an important part of the process. Most of all, the work requires dispassionate, cool-headed personnel, and Riklin has very clear ideas about what this means in terms of "job description" and the kind of workers he is willing to hire:

> I don't want young people in my department. Things can heat up too quickly. A young person should be out doing other things, like chasing girls. An older, more mature inspector usually has the ability to be available for work at unconventional hours, which is when a lot of the violations are taking place.

> But these also must be very special people—people who are willing to leave a warm bed and spend the night in the elements outdoors. You know after a long struggle, the personnel department agreed to give forty hours of "overtime" that we might award workers each month. But typically we'll use that up in a week![45]

Keeping a cool head is particularly important when one is working with Israel's Arab minority. Riklin has improved his Arabic dramatically and takes enormous efforts to show respect to the Arab community. But he also keeps his distance, refusing to become engaged in the typical Middle Eastern rituals of coffee drinking and informal conciliatory consultations that violators so often offer. "This can infuriate them, but I tell them that I'm here because they did not behave properly and it would be hypocritical to get engaged in a friendly discussion. But I never raise my voice and I would never speak to an Arab the way I spoke to the rabbi there at Dan Ben Yaakov's grave."[46] Like many foresters, Riklin has also been assaulted, and his vehicle has been vandalized and burned—not by Arab Israelis, but rather by an angry Jewish hoodlum caught violating the rules of the forest.

Despite the urgings of the JNF security department, Riklin refuses to carry a gun with him in his work. Most of the inspectors who work for him, however, are not as confident and pack a pistol. Intuitively, he understands that making peace between Israel's Arab citizens and the country's forests is not going to be advanced by a gun-toting sheriff. At best, enforcement can control the most extreme expressions of antiforest sentiment, but it does little to reach the hearts and minds of Israel's Arab citizens. Their uneasy relationship with the nation's woodlands needs to be better understood if their perceptions are to change.

Between Israel's Arabs and Israel's Forests

Jews and Arabs rarely mix in Israel. One out of every five Israeli citizens is Arab, and the vast majority of them—over 70 percent—prefer to live in exclusively Arab villages and towns.[47] Only about 25 percent of Israel's 1.4 million Arabs live in "integrated cities" such as Haifa, Jaffa, Acre, and Ramla. Even here, they usually prefer to live in their own neighborhoods. Both Christian and Muslim Arabs in Israel choose to attend separate school systems, enjoy different music, go to different cultural events, root for different soccer teams, and speak a separate language than their Jewish countrymen. Israel's Christian and Muslim Arabs also tend to live separately from each other, reaching their own uneasy equilibrium.

With very few exceptions, the 114 Arab towns and cities in Israel emerged spontaneously, without the top-down, systematic urban planning that informs the symmetrical, but relatively sterile building in the Jewish sector. Arab communities often display an authentically local and unregimented aesthetic that is envied by many Jews. But it also means that there has been little or no land set aside for parks and open spaces. Historically, Israel's planning commissions were dominated by Jewish officials with clear ideas about maximizing Jewish control of national lands. So most Arab cities were never allowed to expand their municipal boundaries. Limited space and crowded conditions ensued, with precious few accessible recreational sites.

It would seem, therefore, that Arab Israelis would be among the biggest champions of the country's forests. Not only do forests provide breathing room and shaded venues for outings and celebrations, they are also an open link to the land about which Arab Israelis too are awfully passionate. And yet the attitude of many Arabs to the country's forests is profoundly ambivalent. They feel, with some justification, that over the years trees have been disingenuously used as a tool for Jewish nationalistic ends, in order to stymie Arab development and property claims.[48]

The confrontational atmosphere created by the Turkish land laws that played out in the war of attrition against colonial forestry during the British Mandate was exacerbated when abandoned Arabs villages were razed and forests planted over them after the establishment of the state. These dynamics continue to the present. In addition to the usual biological and physical "ecosystem services" provided by forests—such as pollination, soil conservation, and soil sequestration, Israel's land managers often remain fixated on a *political* "service": *tfisat karkah,* "holding onto the land." In the ongoing struggle for sovereignty, trees have played a significant role. Once established, forests keep out Arab settlements. Arab Israelis were very aware that the trees being planted around the edges of their villages were not necessarily there to improve the quality of their life, but to constrain them. The fact that to this day, the JNF charter prohibits any leasing of its lands to non-Jews sends a hostile and racist message that even the most loyal Arab Israeli citizens resent.[49]

Legal scholar Irus Braverman dedicated a book to the issue in connection with Israel's occupation of the West Bank. She likens the role of trees in the region to "planted flags," a phrase that she used for the title of her book: "For decades, Jewish people from around the world have been invested both economically and emotionally, in planting trees in Israel, while Palestinians have been cultivating and harvesting olive groves on what they consider to be Palestine's hills. It is difficult to imagine that these ostensibly apolitical acts could actually fuel a brutal yet clandestine war that has been going on in the area for more than a century: the war of the natural landscape."[50]

The easiest way for Arabs to express their displeasure with the trees they see as agents of oppression is to burn them down. Unfortunately, after almost a century of nationalistic arson, this is considered normative if not legitimate behavior. Although leaders in Israel's Arab community are quick to downplay the phenomenon and, of course, distance themselves from lawlessness — incendiarism is common. Forests are designed with fire breaks ready and waiting for the inevitable assaults.

It is a common enough phenomenon that in 1968 A. B. Yehoshua, one of Israel's leading authors, devoted a well-known short story to the topic, "Facing the Forest."[51] Thirty-one years later, the plot was relevant enough to be reworked and made into a major motion picture of the same name. In the story, a somewhat alienated and nihilistic Israeli graduate student who works as a summer watchman for early fire detection becomes strangely sympathetic to an Arab forestry worker. The worker's tongue symbolically has been cut out. Although never entirely clear, the worker appears to have lived in a village that has become the site of the watchman's forest. The mute Arab worker tries to burn the forest down with the quiet complicity of the Jewish watchman. Here is the final, climatic scene:

> He starts counting the flames. The Arab is setting the forest on fire at its four corners, then takes a firebrand and rushes through the trees like an evil spirit, setting fire to the rest. The thoroughness with which he goes about his task amazes the fire watcher. He ought to run and raise the alarm, call for help, but his movements are so tranquil, his limbs leaden. Flames surge as in a frenzy high over the trees, roar at the lighted sky. Pines split and crash. Wild excitement sweeps him, rapture. He is happy. Where is the Arab now? The Arab speaks to him out of the fire, wishes to say everything. everything and at once. . . . Will he understand?[52]

The truth of the matter is that Israelis may understand why some Arabs *once upon a time* wanted to burn down forests — but today, most have little empathy. After sixty-five years, it is time to accept the new geographical reality and move on. Until such time, for Israel's Jewish majority, the woodlands constitute another front in the never-ending battle for survival. In fact, the vast majority of arson cases remain unsolved. When arsonists are finally caught, their motivation typically appears to be adolescent restlessness and showing off, rather than a high-minded battle on behalf of Palestinian national aspirations.

While the vast majority of fires are started by Arabs, the vast majority of Arabs are not nationalistic arsonists. On the contrary: the violent opposition to forests today is the work of a tiny minority. Forests are ultimately pleasant places to spend time, which Arab Israelis have come to enjoy. With their houses crowded and with few local parks, where else might they go with their large families to refresh their spirits? When visiting forests in the Galilee or the Negev, where there is a large concentration of Arab citizens, one typically hears as much Arabic spoken as Hebrew among the picnickers and much more Arabic singing. To be sure, many of the forests were planted to assert Jewish sovereignty. But all Israeli citizens and visitors to the countryside have free access to them and seem to revel in them equally.

In recent years, the heart of the hostilities has moved to the south of Israel, where the conflict continues to simmer between a small group of Bedouins, who wish to claim public lands and foresters like Evegeny Podolsky. Notwithstanding the press's sensationalism, the majority of Bedouin citizens have come to appreciate the new forests in their native southlands. Abed Abulkian is the head forester for Yatir, Israel's largest forest, and a very proud Bedouin Israeli. He explains the growing popularity of forests in his Negev community:

> Today, the Bedouins have discovered the forest. Most Bedouins are now living in villages. The Bedouin accepts that he doesn't have his ancestral lands anymore. He has also come to understand that he now shares full access to these forests. That's where they go to have picnics, and often they'll spend entire days in nature. In Yatir you'll see the whole range of the Bedouin community discovering forests—starting with the schools, teachers on vacation, conferences. You know, at first they are amazed, because it's such a large forest and soon they start to take pride and clean up after themselves.[53]

It is little wonder that there has never been a major forest fire in the Yatir Forest. The local Bedouin from Abulkian's tribe are ever available to lend a hand and help squelch conflagrations before they get out of hand. They rely on the forest for wood, animal fodder—and now recreation.

Of course, such good news and the satisfaction of the silent majority are of little interest to the press, which focuses its reporting on isolated conflicts in order to sensationalize the news. No dispute has attracted more media attention than the controversy over the "Al-Arakib" area, located to the west of the Tel Aviv Be'er Sheva Highway just south of the southern Bedouin city of Rahat. Seven times now the Israeli police have enforced a series of court rulings there and destroyed the hastily constructed "shacks" and temporary structures.[54]

The history of Al-Arakib is not unlike many other parts of the Negev Desert. In 1953 a handful of Bedouin nomads were refused entry into the area as new military training grounds were to become operational there. Soon after, in 1954, the lands were officially annexed by the nascent state of Israel for the purported purposes of development, settlement, and security. By the time the

Israel Lands Authority had begun to manage the country's open spaces, the entire region was defined as "state lands." As Israel matured, it began to consider the optimal utilization of the desert region, which, it turned out, was hardly as vast and sprawling as people liked to think. The area of Al-Arakib was designated to become a green belt that would provide open spaces for the forty-five thousand residents of the adjacent Bedouin city of Rahat. The forest would also serve to truncate the creeping expansion of the city of Be'er Sheva in the south and of Rahat in the north, providing the residents of both cities access to high-quality shade, recreation, and open spaces.

The official plans for the region were of little interest to two Bedouin clans: the Abu Medium of Rahat and Abu Jabar of K'far Kasem in the center of Israel, who claimed that their grandparents had "owned" thousands of hectares of the land prior to the establishment of the State of Israel. Joining the wave of illegal Bedouin construction, they decided to create facts on the ground.

The basis for their ownership claims involves internal, handwritten documents that were exchanged between Bedouins. The problem is that at no point were any of these land rights recognized by the ruling governments in pre-State Palestine. Nor did they appear in the land registry of the time. Aerial photographs from before the state show occasional tents appearing there, and a very small family graveyard. But there was no meaningful settlement to speak of.

That began to change in 1998, when the clans decided to move in. They initially made their presence known through nonintensive agricultural cultivation. The first buildings they threw up a few years later were corrugated huts—but later these were upgraded to become structures with cement cinder blocks. As the illegal construction grew, so did the attacks on foresters and their operations. The Israel Land Authority decided it had had enough. The agency initiated a series of legal actions to remove the squatters, which consistently yielded the same result: the courts noted that the lands had long since been nationalized under the law of eminent domain and that the land belonged to the state. The Bedouin defendants could claim compensation for any lands that they could *prove* their families had owned, but they could not live there. The courts granted permission to destroy the illegal structures and to implement the zoning for trees.[55] This inherently unpleasant eviction procedure was done in the face of varying degrees of resistance, recorded and posted in real time on YouTube. Injunctions were then issued against the Bedouin clans, yet these were consistently violated. The Bedouin trespassers insisted that they had historic rights and kept coming back.

And so a cycle of hostility ensued. Time and again, the controversy was adjudicated, including one appeal that reached the Supreme Court, but the courts' decisions were always affirmed. No credible evidence existed for past ownership. Even if such ownership could be proven, the area had been public land for over half a century and compensation was the only realistic remedy.

This cycle of destruction and squatting in the Al-Arakib hamlet was the focus of protests and, later, a rallying cry for progressive organizations and an item that surfaced periodically in the Israeli and international press.[56] The JNF was the target of much criticism in anti-Israeli and human rights blogs because the proximate reason for the demolition was to enable the JNF to plant trees on these public lands.

Dr. Ouad Abu Fariach is a highly articulate leader of the Bedouin families contesting the ownership of Al-Arakib. He is also head of the biotechnology program at Sapir College in the Negev. Despite his scientific credentials, he has become well versed in Israel's land laws, filing innumerable legal actions. "I was born here," he explains. "There is a graveyard at Al-Arakib where my ancestors are buried. When the JNF plants, it is not just adding vegetation. It takes over the land. Its terraces change it. This creates a very bad atmosphere and ensures that there will continue to be trouble." He dismisses the more conciliatory positions of the mayors in the surrounding Bedouin cities as simple cases of politicians being bought off by trivial sums of money. And he rejects considering alternative lands—at least until all legal remedies are exhausted. "After sixty years, why the rush to plant trees?" he asks.[57]

Bedouins who claimed a historic relationship to Al-Arakib found that it was not hard to sell themselves as victims, and a long list of progressive Jewish organizations bought the disinformation.[58] These groups organized demonstrations in Jerusalem across the street from JNF forestry offices headquarters, protesting that "land in the Negev desert should be used for 'housing and education' and not for 'racist forestation.'"[59] Bedouin advocates charged that Israel was choosing trees over people, leaving some of the most destitute Israeli citizens homeless, a claim that was never properly refuted. The government eventually cobbled together a fairly feeble PR response that published photographs of the permanent homes owned by many of the Bedouin squatters. Some of them were palatial. The message was that the underdog Bedouin interlopers were anything but homeless victims, but in fact disingenuous opportunists. Bedouin families are large, so even these villas did not offer enough room for all the progeny. The residential opportunities offered by the Negev's open spaces beckoned. And there were also the on- and off-site benefits associated with erosion control and open-space preservation. The rebuttal did not gain much traction. As a gesture to the perceived "rule of law," the JNF froze the actual planting of trees until the court could offer a final ruling. But this was merely a temporary tactical move.

It is not clear that the voices of solidarity with Al-Arakib understood all the implications of their demands. The shabby and superficial media coverage did not address the most fundamental question: If the State acceded to these demands it could mean the freezing of all tree planting and development on lands where individual Bedouin have made claims. Did the sympathizers really want to bring all such activity to a halt in an area which so desperately needs investment? Clearly not. A case in point is the Bedouin city of Rahat, which has a yearly birth rate of almost 4 percent. Rahat urgently needs to expand its residential neighborhoods, but the entire city is built on lands that are being claimed by individual Bedouins as their private real estate. Had the court ruled otherwise, Rahat would not be allowed to develop and grow. The same is true for the industrial zone in the development town of Arad, as well as most any forestry project in the Negev. Even though the lands have long been state-owned, any new project would run up against claims that Bedouins had once grazed there. Does increasing the wealth of a tiny handful of Bedouins justify stymying the quality of life and economic progress of Israel's poorest and most neglected citizens?

The JNF had actually already begun to undertake a series of initiatives to provide support for the Bedouin communities in the Negev who live in legally recognized settlements. Beyond the traditional forestry work, funds were raised for major initiatives like the multimillion-dollar restoration of the Grar stream. This ambitious project will serve as a central park running through the heart of the city of Rahat, the largest Bedouin city in Israel and the second largest city in the south of Israel. Sadly, despite its size, there is not yet a single significant park in Rahat. In another project, the American offices of the JNF pledged three million dollars to the city of Hura's "Wadi Atir" initiative, which would establish a sustainability center focusing on traditional Bedouin agriculture and skills. The JNF has consistently provided trees and playgrounds to Bedouin cities, even when these were vandalized (as was the case in the city of Segev Shalom—where a hundred-thousand-dollar playground was destroyed twice).

On the whole, the Bedouin community understands what is at stake and that it needs to find a way to help plan and maintain these projects. Indeed, the mayor of Rahat has openly expressed his appreciation to the JNF as the only green organization in Israel that has ever helped his city. When an anti-Israel "Al-Arakib" symposium was held in Europe, the JNF sent the mayor of the Bedouin city Segev Shalom, Omar Abu Muamar, to speak on the organization's behalf. Abu Muamar chastised the critics for not understanding the real needs of the Bedouin, and the Israeli detractors as "unpatriotic." The mayor of Hura, Dr. Muhammad El-Nabari, has gone on numerous speaking tours throughout North America on behalf of the JNF. His purpose has been to advocate local sustainability as well as to promote the many projects that support and improve the quality of life in the Bedouin communities.

This is not to say that the JNF and the forestry program have not made mistakes in their dealing with the Bedouin community. But surrendering to the clamor of the press and human rights advocates may make matters worse, as it sends a message which rewards lawlessness. The recent conflicts do not mean that some conciliation is impossible. Many, if not most of the specific land claims may be inflated and imprecise. Bedouins were surely part of the Negev Desert geography before the establishment of the State of Israel. But reality has changed as have the collective needs of the Bedouin. Israel is a small country and cannot allow individuals to grab homesteads on public lands. Providing open spaces for all needs to be a priority and Bedouin need to be full partners in the design and implementation of such policies. Ultimately, the forests which are being planted in the southlands are critical if the Bedouin are to retain any semblance of their ancestral, pastoral heritage. Some 65 percent of Bedouin herds graze in JNF forests, which provide the shade and the biomass to make the economic equation come out positively.

The issue ultimately is philosophical: Should Israel adopt a "utilitarian" approach to land allocation and zoning to ensure that the general public — Jewish and Arab — enjoys forestlands? Or should they reward individual claims by recognizing individual rights and consenting to demands for private estates? Beginning with the British Mandate, the public's overall interests have always informed local land policies and Israel would do well to stick to this orientation.

Standing firm behind a democratically approved national forest plan may make sense from the perspective of physical planning, but it also heightens the importance of proactively engaging Arab citizens. It is not surprising that many Arab Israelis feel disenfranchised and angry due to their minority status, years of discriminatory policy, and no small amount of latent and overt racism among large parts of the Jewish public. Innocently or disingenuously, forests became part of the Zionist territorial strategy. But trend need not equal destiny. Israel has demonstrated that with the right approach it is possible for its forests to literally "turn over a new leaf." One such successful outreach effort involves the public schools of Daburiah, a Bedouin village that lies at the foot of historic Mount Tabor.

For years Daburiah residents neglected the lands surrounding the holy mountain where Deborah, the biblical prophetess, gave Barak God's military orders and Jesus' alleged transfiguration took place so that his countenance and clothes began to glisten. The mountain historically was heavily wooded,[60] but during the Ottoman rule, the trees suffered the total deforestation common to the period. The JNF replanted with its usual gusto, but maintenance was spotty and the woodlands leading up the hillside looked a mess, littered with debris and marked by signs of intermittent arson.

A far-thinking Daburiah principal challenged his middle school to formally "adopt" the forest around the mountain. With JNF and Ministry of Environment assistance, a curriculum was developed to create better understanding of the local ecology. It included frequent excursions into the forest. Hiking trails were cleared, cleanup campaigns conducted, celebrations held. The golden *halmaniot* (*Sternbergia clusiana*) wildflowers provided the gem in the crown —bright yellow blossoms that are eagerly awaited each year and can now be seen by all who come to visit on the new and well-maintained footpaths. Parents are pulled along by their enthusiastic children. The students have extended their commitment beyond their village and now pile onto buses to help plant trees and clear debris after forest fires around the country. There are no longer any fires in the Tabor Forest. Daburiah shows that a successful, bottom-up, outreach program, implemented through Arab public schools, can completely transform the face and fate of a community and its forest.[61]

This is but one example of reconciliation. There are others as well. Israel's forestry program was established by immigrants from Europe who often lacked a knowledge of local flora and aesthetics. There is much that they might have learned from the indigenous Arab communities. Suohil Zedan has been a forester at JNF for almost thirty years and is blessed with a surplus of indigenous knowledge. He is in fact a second-generation forester, as his father was a JNF field worker. Zedan studied agronomy and rose to senior management in the organization that was impressed by his exceptional horticulture talents. For many years he has served as the in-house expert responsible for the remarkably successful transplanting of mature trees that need to be uprooted due to development projects. His personal passion, however, is introducing *bustan*s, the traditional Middle Eastern fruit groves, into Israel's forests.[62]

Once considered a quaint idea, today more and more forests include groves with olives, carobs, almonds, dates, and grapes as part of the new diversity. This approach works because even forests planted in the driest of drylands will direct the drainage from flash floods into the base of the valleys, where Zedan's trees flourish. An additional bonus is that the fruits are free for the picking. The contrast provides an aesthetic break in the monotony of the usual tree lines, refreshing to the eye and palate.

Zedan is a professional who tries to keep away from politics or playing the role of token Arab. Nonetheless, his expertise, familiarity with local Arab culture, and unparalleled field skills proved invaluable when the JNF began to assist the Palestinian Authority in landscaping. The Palestinians have been in the process of building a new city, Ruwabi, in the West Bank, north of Ramalah. When interviewed during the planning stages, Zedan told the press: "There have been a number of meetings, both at the location where the city will be built and also at the JNF greenhouses. We have contributed with our know-how, by advising on how to prepare the ground for the planting and how public gardens should be planned, as well as the best times for planting, and what kinds of trees it is preferable to plant. We did not talk about politics and we shall not talk about it. We deal with trees and understand forestry, botany and greenhouses."[63]

This sort of outreach shows how forests could and should be a bridge between Arabs and Jews in the region, rather than a source of conflict. Israel's environmental movement has many active Arab leaders who are extremely pragmatic in this regard. They are willing to make a distinction between historic land claims and environmental cooperation and to compartmentalize. While they recognize that conflicts of perspective on zoning and equitable distribution of land rights are unlikely to be readily resolved, this does not mean that it is impossible to forge real partnerships to work together on issues ranging from forestry to water management.[64]

Changes in attitudes on both sides can be achieved. It will require a reassessment on the part of Arabs about where their real future interests lie. It is time that Arab Israeli leadership assert its full moral authority and join the fight against the ongoing assault on Israel's forests. It also means that the JNF may want to reconsider some basic assumptions, exhibit enhanced cultural sensitivity, and even alter semantics. For instance, it is time that some forest areas in Israel have Arabic names, honoring geographic locations or local heroes venerated by Arab Israelis. Such seemingly small shifts would send a significant message to Israel's Arabs that the forests also belong to them.

Other reforms are more fundamental and may take longer to attain. It is time to address the problematic dynamics of having a Jewish public corporation responsible for national forestry in a country where at least one-fifth of the citizens are not Jewish. There are several ideas about redesigning the forestry administrative framework so that policy decisions would be made by a representative national council, similar to that which oversees the Nature and Park Authority. There is a debate as to whether the minister of environment or the minister of agriculture should be the authority that appoints such a committee. But for Israel's Arabs, government oversight at the strategic level would signal that forestry is no longer a "Jewish" program but one that is linked to participatory national democracy and designed to serve all of Israel's citizens.

Even if such reform is adopted, the JNF—with its technical expertise, funds, and hundred-year history of success—should continue to play a central and influential role in afforestation policy discussions and their implementation. It would be foolish and ecologically imprudent for it to close up shop. But the organization should make extraordinary efforts, if not take affirmative action, to integrate Israel's Arab citizens into the country's forestry program. This starts with planning but continues on to governance. When the Meretz political party appointed Radi Safori, an Arab Israeli affiliated with the party, to be its representative to the JNF International Board, many eyebrows were raised. Safori's appointment was actually challenged in court by the National Union, a right-wing Israeli political party. Safori weathered the attack and stayed on the JNF Board of Directors as its first Arab member. This constituted a small step toward progress. It is critical that, parallel to uncompromising efforts to protect Israel's forests from sabotage, even more meaningful efforts must be made to give Arab Israelis a place at the table of forests and afforestation policy.

Forests and Development

Israel is a very crowded place and is rapidly growing more so. Population density for the entire country is roughly 320 people per square kilometer, ten times the average in OECD (Organization for Economic Cooperation and Development) countries.[65] In practice 92 percent of Israelis live in the more temperate center and north of the country, where population density reaches 710 people per square kilometer.[66] Present trends will push that number to 900 by 2020. By way of comparison, even a very crowded nation like the Netherlands has only 400 people per square kilometer; Belgium and Japan 354 and 337 respectively. This demographic swelling has not been good for Israel's trees. When a mayor starts thinking about new residential development for his town, frequently, he can't see the forest . . . for the expanded property tax base.

Zafrir Rinat, Israel's daily newspaper *Haaretz*'s esteemed environmental correspondent, considered the phenomenon in a column entitled "There Once Was a Forest":

> The immense amount of construction in the area (Israel's heartland) is gradually destroying the possibility of developing parks and preserving natural recreation areas, a goal fought for by so many planners. What is even more disturbing is the disappearance of areas that have been considered nature reserves, like the forests of Ben-Shemen which are being eroded from every possible direction. Parts of the new city of Modi'in itself will also be built at the expense of the forest. The experience from recent years has proven that environmental considerations are not enough to halt construction. It is the army, that is, for the moment, preserving the last remains of wilderness. But

> the land does not serve the public at large. Given the realities of planning in Israel, it is reasonable to assume that as soon as the army relinquishes the ground, industrial areas and roads will quickly spring up. Residents of the central region will have to spend their leisure time in another shopping mall, rather than in the great outdoors.[67]

The article was written in 1997, when there were less than six million people living in Israel and the city of Modi'in was but an idea on the maps of regional planners. In 2013, the population is eight million, Modi'in will soon have 150,000 residents, and the trees' situation is even graver.[68] There is no end in sight. If the wonderful forests surrounding Israel's urban centers are to have a chance of surviving, so they can provide recreational and spiritual services to the country's future citizens and visitors, societal dynamics need to change. There needs to be a much more profound political commitment to protect them.

Israelis care deeply about their woodlands, but few understand how vulnerable they are. Part of the problem is that, typically, development does not wipe forests out in one fell swoop; rather, wooded areas slowly slip away. The public does not grow alarmed until it wakes up one fine day to discover that the forests that it has taken for granted for so long is no longer there. Like the proverbial frog that cannot feel gradual temperature changes and slowly boils to death without sensing the danger,[69] people in Israel's cities are being slowly stripped of their woodlands without even noticing.

Nowhere is this phenomenon more pronounced than in Jerusalem. In 1949, David Ben-Gurion shared his dreams of lush forests transforming the bald hillsides encompassing Jerusalem and Israel's foresters heeded the call. The JNF quickly translated this grand vision into the mechanics of silviculture, within the constraints of the capital's zoning restrictions. Starting in the Judean Hills, forests were steadily planted on some 21 percent of the surrounding lands.[70] By 1958 the effort reached the city itself, with seven separate forests planned for the capital. Israel's president at the time, Yitzschak Ben Zvi, planted the first tree in what was to be the jewel in the crown:[71] "the Jerusalem Forest" located on the city's western hills covered 450 hectares by 1967. Here, residents could escape the bustling and ever intense urban Israeli experience and stroll along a quiet shaded trail that lines the forest's Ravida stream. Youth organizations and schools could introduce the children of the capital to nature. Picnickers from around the country could take pleasure in a scenic haven and fresh air, enjoying a diversion from the stone and asphalt scenery. Lovers, young and old, could have a little privacy.

Other forests planted in Jerusalem are even more remarkable. For instance, in the northern side of the city, between the Ramot and the Ramat Eshkol neighborhood lies the Mitzpeh Naftoah hillside. Miraculously passed over by past building booms, its terraced hills are a haven for the area's indigenous wildlife and flora. When Hebrew University botany professor extraordinaire Avi Shmida did a census of the plant life there, he was stunned to discover that the groves and surrounding open spaces were home to over five hundred plant species, many of them considered endangered and protected.[72] Here, the patient visitor is rewarded by daily sightings of the *Gazella gazella,* the local antelope species that jumps across the hills as if auditioning for a video clip of Solomon's Song of Songs. Over one hundred bird species have been counted.[73] For the secular and religious residents of the nearby neighborhoods, the ability to escape to such a magical ecological smorgasbord constitutes a cherished gift. But there has long been a sense that it is a gift on borrowed time. For over a decade representatives of the ultra-orthodox communities in Jerusalem have eyed the spot as an ideal location for a new neighborhood they would like the government to build to help accommodate the crowding caused by their high birth rate.

Indeed Jerusalem's population is growing very fast: in 1948 there were 165,000 Jewish, Muslim, and Christian residents in the city—today there are almost 800,000. The Jewish residents look at the demographic statistics that show the steady rise in the percentage of Arab residents and grow very nervous, looking for ways to expand the Jewish presence in the city. Whether or not Yasser Arafat's quip about "the Arab womb being his strongest weapon" is apocryphal, Palestinian birth rates are among the highest in the world. For the ancient city, the result is a demographic "race to the bottom." Rather than seeing the city's twelve hundred hectares of total forestland as the most precious of resources for the city's residents, Jerusalem's Jewish mayors and planning departments consistently view it as a reservoir of residential lands for development and a vehicle for ensuring Jewish demographic dominance.

Neighborhoods such as the Har Nof or Givat Shmuel quarters morphed onto lands that had once been lovely corners of the Jerusalem Forest. When a huge oil depot was needed to hold fuel reserves for the capital, not surprisingly the forest provided the natural solution. A new multistory municipal graveyard was placed on the edge of the forest, and has been expanding. There are, after all, a great number of Jews who yearn to be buried in the Holy City. Even the directors of Israel's iconic, national Holocaust museum, Yad Vashem, did not think twice about expanding the museum's grounds in the direction of the forest. The JNF website reports that today, only 125 hectares of the Jerusalem Forest are left—a quarter of the original stand! Mitzpeh Naftoah faces a similar fate as the city of Jerusalem has decided to back plans that remove the remarkably rich woodlands from the National Master Plan for forests.

Jerusalem residents have been quick to organize and fight for their open spaces. The Ramot Environmental Association has been spearheaded by three dedicated women who brought considerable pluck with them when they moved to Israel: Hilary Herzberger, a South African Israeli therapist; Rachelle Adam, an American-born attorney; and Dutch émigré Esther Reiss have galvanized their neighborhood on behalf of their little corner of heaven. The association sued on behalf of their neighbors, the *Gazella gazella* species, whose habitat would be destroyed, and surprisingly gained de facto standing in the Supreme Court. The court found procedural flaws with the plans to convert a recognized forest into residential zoning. As a procedural victory, it could do little more than buy time, so it hardly affected the Jerusalem city planners' long-term designs. Across town, the Coalition for the Jerusalem Forest brought together neighborhood organizations from the diverse populations of Israel's capital, whose qualify of life stood to be poorer without the forest. As almost entirely local initiatives with little support from national organizations, these campaigns have been very impressive. But prospects are not promising.

Washington, DC, has a resource similar to the Jerusalem Forest. Its 1,750-acre Rock Creek Park cuts through the heart of the American capital city. When it was established according to presidential decree by Benjamin Harrison in 1890, it was only slightly larger than the original Jerusalem Forest. Since the 1930s it has been faithfully administered by the National Park Service.[74] It is still 1,750 acres, now more than four times the Jerusalem Forest's shrunken dimensions. The city of Washington would not dream of cashing in this scenic vestige of American natural heritage for a few additional residential developments or to temporarily improve transportation conditions. Sadly, the political reality and the priorities of Jerusalem's local government are different. Rather than reconsidering its development policy, the City Council now wants to solve projected traffic jams by building a massive intersection off a new highway. This intersection will cut another piece out of the Jerusalem Forest, at the entrance to the Revida stream. As of this writing, the grassroots protest by a coalition of neighborhood groups has managed to attract only modest attention from Nir Barkat, the city's mayor.[75]

Barkat, a former high-tech entrepreneur, presumably should be a paragon of environmental sensitivity. He is a distance runner who likes the outdoors. When campaigning for mayor he was hosted by the Ramot Environmental Association and in a visit to the Mitzpeh Naftoah site joined the hundreds of citizens filing legal objections against the new neighborhoods. He even placed the Jerusalem director of the chapter of the Society for Protection of Nature in Israel, Naomi Tsur, on his list of city council candidates. After the elections, he appointed her to be his deputy. But once Barkat became mayor his priorities changed, and he quietly allowed his city engineer to move plans forward for new neighborhoods at the Mitzpeh Naftoah site.

The major environmental organizations and the JNF have made token protestations over the projected plan, but they never truly prioritized Jerusalem's urban forests. In private conversations, they acknowledge that they are more worried about saving the Judean Hills outside the city limits from expanding suburban development. The concern is that since construction is inevitable, if they try to stop development in the city's forests, the open spaces outside Jerusalem will pay the price.

Often, when local Jerusalem residents speak up and try to remonstrate, they are castigated as narrow-minded, unrealistic, antiprogressive and—worst of all—suffering from "NIMBY." The NIMBY ("Not in My Back Yard") syndrome, an American acronym based on the dynamics in many environmental conflicts, has quickly entered the Hebrew vernacular. It is intended as a pejorative term for citizens who put narrow personal interests ahead of broader civic values in national affairs. But in Israel, where the "micro" is often indistinguishable from the "macro," what could be more in the national interest than the forests that shield the nation's capital?

Jerusalem city's municipal boundaries include 125 square kilometers. Paris is slightly smaller—only 105 square kilometers—even as its population of 2.1 million is three times the size of the Israeli capital's.[76] If Jerusalem's city planners were a little more conscientious or courageous, they might insist on utilizing the tens of thousands of residential sites within the city that have already been approved for building and that are far from the city's forests. There are ample land reserves for the growing demand. Yet sometimes it seems that the only lands deemed available for development are the weary remnants of what was once a grand forestry project.

The sad fact is that those who love trees have not succeeded in engaging the public and drafting them in the battle for conservation of this resource. In a word, there are no political consequences for preferring property tax revenues or coalitional support in the City Council over trees. It is unwise to assume that the general public is like military reservists, waiting for a call to jump into battle. In a crowded land, where there are constant economic incentives to develop on wooded lands, any preservation strategy demands proactive vigilance that mobilizes the general public and specific users on behalf of their neighborhood forests. The best defense indeed is a good offense. Fortunately, the offensive has begun.

Community and Forests

Because Israeli cities are so dense and life is so "intense," it is little wonder that when Israelis have some free time they look for respite in their forests. During the course of a typical national holiday such as Passover or

Succoth (the Feast of Tabernacles), two million Israelis — or two citizens out of seven — make their way to the shelter of the trees.[77] It was not always like that. Israel's original forests were designed for timber production, erosion prevention, and perhaps an existential sense of collective actualization. Little thought, if any, was given to recreational facilities. It was not until the 1970s that forest managers began to think about the forest's potential as a tourist and leisure-time destination.[78] JNF chairman Moshe Rivlin gave his blessing to explore the new approach.[79] Twenty years later, the JNF's international board of directors formally approved what was already a fait accompli: a strategic decision to "open the forest up" to the public.

The motivation for the decision was part political and part philanthropic. It also is not clear whether or not, in approving the new approach, JNF leaders understood the exceptional cost implications. Crowds require infrastructure and innumerable services. Just picking up trash after visitors leave involves an annual cost of over four million dollars, which JNF pays to outsourced cleanup companies.[80] Millions of visitors need parking, playground facilities, and picnic tables, and they appreciate running water and clean bathrooms very much. The establishment of such facilities is often funded by donations. A picnic table costs six hundred dollars. A children's playground costs about four thousand dollars (but can of course be far more depending on how elaborate it is). And that is just the initial expense. Although the new generation of recreational amenities in Israel's forests is designed to withstand the elements, vandalism, and time, those amenities still have a lifespan. After about ten years the tables need to be replaced; recreational accessories need to be revamped after fifteen years or else the forest begins to take on a dilapidated look.[81] Constant maintenance is required, for which someone has to pay.

As the JNF acquired lands on behalf of the Jewish people, free entrance to its parklands became ideologically axiomatic. The forests literally belong to the people. So the ongoing expense of the public's recreational, tourist, and natural experiences are funded from the JNF coffers, which rely on donations and revenues from land leasing. All told, the total budget line comes to twenty million shekels (over five million dollars) a year, a full 80 percent of the funds allocated for planting new forests.[82] And this amount doesn't even include the four million shekels spent on maintaining roads and adding to the network of trails in the forests.

Keeping the forests open for the public may impose a massive expense. Tactically, however, it is the best imaginable investment in the forest's future. Although few Israelis have ever heard of cultural ecosystem services, most intuitively appreciate how much they gain from their newly planted woodlands and how much they stand to lose if they disappear.

In recent years forest infrastructure took on new challenges and functions which attract new clienteles. In 2008, the number of regular Israeli cyclists crossed the one million mark, and the figures have continued to grow 5 percent annually in Israel's Jewish sector.[83] Unlike countries such as the Netherlands, most Israelis do not yet see cycling as a feasible commuting option, but rather as an exhilarating recreational outlet. This creates many problems. Israeli highways and back roads can both be treacherous and unattractive. There are few bicycle lanes. As part of the community outreach program, the JNF cobbled together a seven-thousand kilometer network of designated trails through Israel's main forests. Dusty and bumpy, the hundreds of trails through the woods require a mountain bicycle. They need to be expanded, upgraded, and maintained—but they offer an excellent start. In 2008, JNF international chairman, Efi Stenzler, struck a deal with Israel's government. The government agreed to partner in the hundred-million-shekel multiyear funding of a "national bike trail" that will crisscross the country from north to south and rely heavily on bike paths in the forests.[84] As the years go by, Israelis will come to see forests as synonymous with off-road cycling—another reason for people to care about the forests.

For most of the country's history, Israel's community of people with disabilities saw the forest from a distance. If the general public had a hard time finding its way to the woods, it was completely inaccessible to the eighty thousand Israelis who suffer from paralysis or motor injuries.[85] In 2005, new national legislation called for improved accessibility in a range of public spaces.[86] In response, the JNF unveiled an infrastructure strategy for providing greater access to the forests. Not only are Israelis in wheelchairs increasingly able to enjoy the forests, but measures to accommodate the twenty thousand legally blind and eight thousand deaf citizens are being introduced.[87] This increased accessibility to forests is part of an international trend.[88]

The new accessibility strategy is ambitious. It states: "The JNF views the disabled community as an inseparable part of the public and the people on whose behalf it toils. Opening the forest to the public addresses this public in its totality, including people with disabilities. . . . The disabled community should not be separated from the general public, as many families have one member with disabilities who requires special adjustment and care relative to one's physical limitations. . . . KKL-JNF understands that a person with physical disabilities . . . wishes as well as deserves to spend time in the outdoors with one's family and the general public."

The strategy acknowledges that it is much easier to integrate elements that ensure accessibility in new forest developments than in existing forests, but it promises "to ensure that, even at an existing site, most of the designated and functional areas be accessible to the entire public, which includes people with disabilities."[89]

Dedication of wheelchair-accessible playground in the forest (2004). (KKL-JNF Photo Archives)

Some aspects of accessibility planning are fairly self-evident: Trails need to be widened, handicapped parking places designated, obstacles removed, ramps incorporated. But other aspects are trickier and may involve tradeoffs. Sometimes a form of accessibility for people with certain disabilities becomes a nuisance or problem for people with other disabilities. For instance a fifty-five-inch-wide ramp may be suitable for people in wheelchairs, but may be completely unwelcoming for people who need to grasp handrails on either side of a path, for whom trails cannot be wider than thirty-seven inches. Changing the surface texture of a walking area will help the visually impaired but make sites less comfortable for the physically impaired. And providing accessible toilets in remote locations, where temporary facilities are the rule, is anything but simple.[90]

Making forests accessible is a process that will take decades, and there will always be room for improvement. Ultimately, meeting *all* the needs of *all* disabled Israelis is prohibitively expensive and constitutes an ideal rather than a practical objective. But the very fact that Israel is seriously trying to make its forests available to all sends a message that the country's disabled community does not hear often enough. It also creates another partner for a forest coalition that will need to be actively engaged in future land conflicts.

Partnerships can be both private and public. Recently, for example, Israel's powerful (and profitable) Electric Corporation has adopted forests, paying

JNF to plant indigenous broadleaf species and pave bike trails near its facilities. While such financial support is helpful, galvanizing public support is even more important. Experience around the world confirms that getting the public involved will not happen without concerted endeavors to engage communities. Even then success is highly uncertain,[91] and often difficult to define.[92] As reflected in the Jerusalem dynamics, protecting forests in and around crowded urban areas constitutes a considerable political challenge. A dedicated constituency must be cultivated that will serve as a perpetually vigilant lobby on behalf of the forest. Some areas naturally select for forest advocates. In the rural region of Megiddo, when the local government surveyed the public about its priorities, local citizens ranked preservation of the open green landscape at the top of its list, ahead of education and even ahead of employment and economic development.[93]

In other areas, the public needs to be convinced—this is the model behind the JNF's Forest and Community project. Rosh HaAyin, a midsize town in the center of the country, serves as the "poster child" for the Forest and Community program. Israel's forests offer the only real recreational refuge for its residents. The town of Rosh HaAyin has grown rapidly and the 105-hectare forest that was first planted in the 1970s offers a hedge against excessive urbanization, a green belt for the city, and a destination for citizens.[94] When JNF set up a new Forest and Community unit, it found a truly enthusiastic mayor and proposed a collaboration that would serve as a pilot program. A "covenant" was prepared and signed between the Rosh HaAyin City Council and the JNF Forestry Department. It called for partnership in the management of "forest life" between forest managers, the city government, and the local community. Each party has its own responsibilities: the JNF pledges not only to plant the trees but to include the local community in the forest planning process. The city commits to enforce laws, help clean the forests, and expand accessibility. Citizen representatives, for their part, agree to be involved in physical and educational projects associated with the forest, respect the nearby homeowners, and treat the forest right.[95] Nurturing political support for such initiatives is critical. In the case of Rosh HaAyin, a City Council member was given the specific "portfolio" of community forestry, thereby demonstrating the mayor's personal commitment to the initiative.

The model was then copied in towns like Migdal HaEmeq, Shoham, and later, Meitar, Kochav Yair, and, ironically, Jerusalem.[96] Early assessments revealed that half of the residents of Rosh HaAyin and Shoham made their way to the forests at a surprisingly high frequency, averaging nineteen annual visits per person. When the other residents were asked why they did not visit the forest, the most common answer was "lack of free time." In Migdal HaEmeq, initial participation was lower—only 25 percent even came to the forest. There, the most typical reason given for not visiting was that "there was nothing to do in the forest."[97] Four years later the overall individual visitation rate had not changed appreciably. But the number of annual visits for those who did frequent Migdal HaEmeq's forests had shot up to twenty-six. Even most of those residents who were not making their way to the forests now knew of the Forest and Community initiative.[98] The program shows that loyalty can be developed: In Rosh HaAyin, 60 percent of the residents reported that the forest constitutes an important reason for their enjoying life in the city. This does not mean that their local forest no longer face the perils of development and proposed urban expansion. But it does means that attempts to take away the forest will be met with resistance.

Periodic surveys are conducted which suggest that the public is quite happy with what they find in Israel's forests, notwithstanding their imperfections. What is the profile of this forest bound public? Forest visitors are usually families with children under the age of eighteen (mostly with one or two children). They are twenty-five to forty-four years old, of Western origin, with at least a high school degree. Over 50 percent of visitors have academic backgrounds, are employed, and earn an average to above-average income. They are generally nonreligious, own private cars, live in apartments, and enjoy average to above-average socioeconomic status. The flip side of this cohort is the many Israelis who do not visit forests. They are comparatively less educated, poorer, more religious, and without cars.[99] These conclusions match earlier studies. which found that visiting the forest is primarily a recreational outlet for families who are from relatively advantaged socioeconomic backgrounds. A recent survey of twelve hundred cyclists in Israel reveals an even more homogeneous profile: male, well-educated, upper-income bracket, and a skilled rider.[100] These data suggest that it is time to consider programs that offer less privileged communities greater access to Israeli forests. Programs such as bus lines and shared ventures with community centers in disadvantaged neighborhoods could further this end.

The level of ecological commitment of forest visitors suggests moderate but clear environmental leanings. Less than 20 percent of the visitors are members of environmental organizations. But a high percentage of visitors claim to carry out environmental activities, such as buying environmentally friendly products, subscribing to a nature-environment magazine, notifying local authorities about nuisances, and signing environmental petitions. In addition, 50 percent of visitors correctly answered most questions regarding environmental issues and the vast majority expressed supportive attitudes regarding forest preservation. While it is hard to prove causality, the actual experience of spending time in a forest probably colors perspective. Some 92 percent of forest visitors object to considering forest lands for future development, while only 67 percent of the nonvisiting public share this conviction.[101]

Public backing already exists and can grow as the nations' forests become a more integral part of the public's lifestyle. Forestry advocates need to harness this support and magnify it, turning it into genuine patronage. It is a complicated mission in a nation whose population looks like an increasingly chaotic mosaic. But it can be done. Compared to other countries, Israel's forests are unusual because practically none of them would exist if *people* had not planted them. Clearly, it will also take *people* to preserve them.

9

Ecosystem Services and Israel's Forests

He plants a cedar and the rain nourishes it. Then it becomes fuel for man. He takes a part of it and warms himself; he kindles a fire and bakes bread.

—Isaiah 44:14–15

Undervalued Natural Capital

What are Israel's forests worth? This is not a question environmentalists rush to answer. There are some things in life that should not be subject to economic evaluation.[1] Yet such philosophical musings have not stopped Israeli economists from dispassionately proposing price tags for Israel's woodlands.

Professor Moti Shechter, along with colleagues and graduate students, was the first to undertake such a valuation. The researchers claimed empirical validation from the Israeli public's responses to the 1989 Carmel Forest fire as reflected in a national telethon whose proceeds went to support restoration efforts. Many of the pledges were made by passive users—people who were unlikely to visit the Carmel in the foreseeable future. Some had never been there at all. Nonetheless, for these donors, the very "existence" of the forest held value.[2] Shechter estimated that this group represented more than a third of Israelis—at the time nearly two million people. In 1996 w the researchers added up "passive" benefits or "existence value" along with the active utility the park offered. At the time, they calculated the total value of the forest to be six hundred million shekels, well over two hundred million dollars.[3] Later studies were more complex,[4] such as Nir Becker's calculations that relied on "contingent valuation" techniques. A representative sample of the Israeli public was asked what they would be willing to pay just to save Israel's old-growth trees. When the responses were aggregated, the amount ranged from 2.4 to 24 million dollars per year.[5]

These are not inconsequential sums. Yet, such figures constitute a gross underestimation. Most economists have little grasp of the many ways that forests contribute to human well-being or the quality of daily life. Forests contain "natural capital"; intuitively, people sense the manifold contributions of forests to society, even if they personally are not direct consumers of its many products and services. Recently, efforts have begun to better characterize precisely what these are, so that they won't be overlooked when the value of Israel's forests is decided.

The year 2000 not only brought an end to both a century and a millennium, but it also provided an opportunity for the world to take stock. After verifying that they had survived the numeric transitions of their computers and clocks, many people began to wonder whether it didn't make sense to think a little bit "bigger" and consider how their planet was doing. There was a sense that the degradation and disturbances caused by humans in their march through modernity was unprecedented, but no real scientific consensus existed about the degree of damage. So a consortium of UN programs launched a global "check-up" to methodically appraise conditions on earth. It would take 1,360 experts from ninety-five countries five years to assemble and analyze the data and then offer the prognosis. Before they could begin they faced a threshold question: "How do you define the health of a planet?" When the massive report was finally published in 2005 it was called the Millennium Ecosystem Assessment (MA) and its conclusions were generally disconcerting.[6] Yet, the MA signaled the coming of age of a new paradigm that held that the most logical criteria for a planetary evaluation were global ecosystem services, or the "benefits that people derive from ecosystems."[7]

The MA coincided with a range of books, articles, scientists, and activists who had reached the tactical conclusion that it was time to reframe ecological advocacy.[8] Beyond the inherent value of biological diversity, the earth's "natural capital" holds inestimable value for humans. It provides food, fiber, and pharmaceuticals, purifies runoff, maintains climate, reduces erosion, mitigates floods, pollinates plants, and of course offers wondrous places to play and find inspiration. Stanford ecologist Gretchen Daily, a pioneer researcher in this field writes: "Much of nature's labor has enormous and obvious value, which has failed to win respect in the marketplace." Nature could be seen as "capital assets, supplying human beings with a stream of services that sustain and enhance our lives."[9] In 1997 a group of experts even tried to put a rough dollar figure on the many services humans receive from nature. The calculations were extremely rough, with a broad range of estimates, but the sums were staggering. The average value of thirty-three trillion dollars enjoyed each year by humanity was twice the global GNP produced by humans.[10]

For too long, afforestation projects were only considered in the context of timber.[11] The "cost-benefit" equation was driven by narrow calculations of board feet per acre and potential lumber revenues. A tree's utility was gauged according to the number of logs it provided or its value as commercial wood. But the same tree stores carbon, protects soil, filters rainfall, and provides habitat, recreation, and beauty for humans to enjoy. The problem is that no market exists where these services can be sold. Indeed, some of them are hard to fully identify much less quantify, and some benefits have yet to be recognized.

In Israel, where low rainfall makes timber only a secondary benefit of forests, characterizing the value of these other ecosystem services should be a central mission of forestry researchers. Until recently, however, little systematic effort was made to this end. This is both a tactical and a strategic mistake. As real estate and other development interests compete with the country's forests for land, it is critical to understand the full range of benefits that forests provide in order to make a compelling case for preservation. Impartial decision makers, who are not inclined to automatically embrace the suppositions of "tree huggers," need reliable information to reach the right conclusions about optimal land use. Monitoring ecological services can help foresters measure their own progress in raising healthy forests of maximum assistance to society.

For many citizens, identifying with Israel's forests has always been linked to a broader, nationalistic narrative. Seemingly intangible impulses like collective Jewish rebirth, sovereignty, and patriotism drove perceptions about trees. There is nothing wrong with nonscientific motivations for supporting forestry. But they are only part of the picture, and sometimes, in the rough-and-tumble of planning debates, they are not persuasive enough. Nor do they allow for an objective characterization of the inherent and instrumental value of restored woodlands, so important in national or local land-management discussions.

For other countries around the world to seriously consider Israel's approach to dryland afforestation, it is important to enumerate the full menu of ecosystem services provided by its woodlands. This chapter summarizes what is known about the diverse benefits derived from Israeli forests. Unfortunately, many uncertainties remain. But present knowledge is still sufficient to suggest that the planting of 260 million trees has brought with it a range of blessings. Many of these remain wholly unknown both to the decision makers and to the public. It is important that they appreciate the fact that ecological restoration is not just about helping flora and fauna; humans are also real beneficiaries. Ecosystem services are typically divided into four categories: (1) provisioning services (e.g., providing food, fodder, and fiber); (2) regulating services (e.g., providing pollinators, controlling floods and pests); (3) cultural services (e.g., celebrating religious or historic sites as well as offering venues for hikes, recreation, and socializing); and (4) supporting services (e.g. formation of soil, cycling of nutrients, and photosynthesis).[12] All these and more are provided by Israel's forests.

Provisioning Services: Timber and Foraging Makes a Comeback

For the majority of forests in the world, the most lucrative ecosystem services fall in the "provisioning" category, measured as timber revenues. JNF's official "sustainability" policies formally downgrade wood production as an operational objective in forestry program.[13] Yet Israel has not "foregone" wood products completely, and today the country's forests yield between 30,000 and 120,000 tons of timber a year. This typically constitutes more than 15 percent of the country's total lumber consumption and the majority of firewood supply.[14]

Wood is largely used for medium-density fiberboard and shipping pallets.[15] While Israeli woodlands should not be designed for maximum wood harvests, practices that improve recreation, soil conservation, and landscape beautification frequently contribute to timber production. Because Israeli wood is not of the highest value and labor-saving clear-cuts are rare, rights to forest timber are often granted merely in return for clearing out woody debris. When lumber prices are depressed, sometimes there are no takers, even on these attractive terms.[16]

Table 9.1 shows recent Israeli timber-production patterns. The amount of overall wood taken from the forest has been increasing, especially in the south where younger forests have been maturing. The dip in production in 2007 is of anecdotal interest. The year was anomalous, because of the biblically man-

ble 9.1. Wood Production by Category in Israel, 2005–10

	Sanitation/pruning (tons)			Logging (tons)			Thinning (tons)			
ar	South	Center	North	South	Center	North	South	Center	North	Total
05	2,655			1,034	931	1,453	409	3,136	4,376	13,994
06	2,899	311	3,734	16	2,771	4,030	323	3,623	5,211	22,918
07	1,326		5,275	59	367	3,046	746	1,753	4,617	17,189
08	2,013		4,520	87		511	316	2,626	18,909	28,982
09			2,709	106	977	4,606	55	3,270	23,810	35,533
10	2,101	68	3,786		406	1,055	1,463	3,324	14,627	26,830
Total	10,994	379	20,024	1,302	5,452	14,701	3,312	17,732	71,550	145,556

ırce: Jewish National Fund, Forestry Department (2012)

dated "sabbatical" cycle; every seventh year, Jews are expected to forego working the soil and allow the land of Israel to rest. The JNF typically does not plant trees during that year and harvesting is curtailed.

The exigencies of land restoration have changed the dynamics and economics of Israel's timber production. In cases where soils are degraded, tree plantations may offer the most cost-effective alternative for agricultural operations. This became apparent after Israel's agricultural sector underwent an economic crisis during the 1980s. For the most part, Israel's farmers largely opted for sustainable water management, adopting drip-irrigation practices.[17] There were a few areas, like the Jezreel Valley, however, where lands historically cultivated under earlier flood-irrigation regimes suffered from water logging. Water collecting below the surface creates high water tables, saturating the soil, cutting off oxygen supply, rotting the roots, and preventing the leaching of salts from the plant's root zone. Such soil abuse is as old as the collapse of the ancient Fertile Crescent civilizations.[18] It still constitutes a global scourge that undermines agricultural productivity in developed and developing nations alike.[19] Today, roughly 17 percent of arable lands in Israel are considered abandoned.[20] Farmers whose lands were damaged sought a low-maintenance crop that could be grown without costly irrigation and still produce reasonable revenues on salinized lands.

Because they grow so quickly, eucalypts are raised commercially in developing countries for construction lumber, particleboard, and biomass for energy. Israeli farmers received a triple dividend by utilizing them for soil restoration: First, given the high costs of drainage (especially subsurface), eucalyptus trees can play the role of biological pump, drawing groundwater down so that salinization of soil is averted; second, land unfit for most crops can happily grow salt resistant eucalypts with reasonable yields and profits.[21] Finally, because eucalyptus flowers are on an Australian southern hemisphere timetable, they can produce pollen during a particularly important period when indigenous plants are dormant or withered.

It was in its capacity as an agricultural-aid agency with experience in drainage that the JNF decided to provide eucalyptus to private farms for commercial production. About a hundred thousand free eucalypt saplings are distributed annually in Israel to local bee operations and two hundred thousand more planted in tree farms. The trees can be harvested after a period of ten years and frequently earlier.[22]

Eucalyptus wood is dense and of high enough quality to meet European wood standards, which Israel has adopted. Israel's two major lumber-producing plants estimate their annual capacity for eucalypts at a hundred thousand tons each. A recent economic analysis estimated the commercial "net present value" of these "commercial" small-scale forests as between 19 and 190 dollars a year per hectare, depending on climatic conditions.[23] This corresponds to a potential of about ten thousand hectares of farmland. Until now, utilization of eucalypt wood in furniture production has been modest; given the steady rise in timber prices on the world market, this may change.[24]

Fruit production is another provisioning ecosystem service found in the forests that only recently has begun to reach a critical mass. Some fifteen thousand hectares of fruit trees have already been planted as groves (commonly known as *bustans*) embedded in the country's woods, typically in the valleys. The forests are also home to an additional two thousand hectares of olives and thirty-five hundred hectares of carob trees.[25] All told, there are at least two thousand tons of olives grown in Israel's forest each year, freely available for the public to pick. With present market prices set at four thousand shekels per ton, the forest's olive crop is worth some two million dollars.

The value of this service has little to do with market prices. The Hebrew language has a special word for "picking olives," *masik,* reflecting the importance of the activity in local heritage. Even with the increase in market prices for olives,[26] the total yields from Israel's forest do not begin to capture the cultural value to families who join a long chain of ancestors in this seasonal, autumn adventure.[27]

A relatively unheralded provisioning ecosystem service provided by Israel's forests is grazing. Most discussions on the subject emphasize the fire-prevention benefits that the grazing animals provide forests. Still, it is well to recall that the foraging livestock are nourished by the biomass of the woods, which soon translates into protein for humans. Officially, there are roughly three hundred thousand domestic animals in Israel that graze on public lands, although an increasing number are given supplementary feed or simply held year-round in indoor feedlots.[28] The vast majority are sheep, even as goat cheese grows increasingly popular. Two-thirds of total grazing livestock are in the nomadic Bedouin sector.[29] Forests provide shade and windbreaks for the animals. For half the year, the vegetation offers a proverbial "all you can eat buffet." Birthing ewes and goats use trees as a place to hide from the herd and bond with the newborn, which has important behavior significance. Recent research also points to medicinal benefits that goats derive from eating trees and shrubs, most notably from the mastic tree (*Pistacia lentiscus*) and the small evergreen *Phillyrea latifolia* (*bar-zayit*), which prevents worms, and other parasites. Druse shepherds in the Carmel traditionally feed oak branches to weaning lambs to prevent diarrhea.[30] The economic value of the pasture is hardly reflected in the symbolic licensing fees, which shepherds pay each year to the Ministry of Agriculture for access to the food source.

Estimating the precise pasture value of forests and their understory has been called an "elusive endeavor." This is because estimates vary so widely among forests, depending on the botanical composition of the vegetation, tree density, canopy cover, and site-specific habitat characteristics. There are also tremendous fluxes in vegetation between, and even within, the seasons. And the amount of forage ultimately depends on actual rainfall during a given year, the specific habitat, and the forest characteristics.[31]

Having some sense of the upper limits of allowable herd size is important because *overgrazing* can be detrimental to forest health. At the same time, *undergrazing* leaves forests more vulnerable to fires. The actual sustainable pasture or "effective carrying capacity" can be calculated and ultimately translated into "stocking rates" by regulators. Lacking site-specific data, generalizations were common in the past. For instance, according to the original Israeli regulations, one goat was allowed on four hectares of land. If, however, the land were irrigated, then one goat was permitted on a single hectare.[32] This one-size-fits-all approach by definition is imprecise and considered outdated, for it leaves many forests underutilized. The standard has long since ceased to be a factor in the setting of stocking rates.[33] There is a need to develop new, simple assessment methods to generate reliable, site-specific quotas.

In fact, there is a team of JNF foresters and agricultural experts doing just that. In a recent study, the available forage for goats in three predominantly pine forests located near Jerusalem (Sataf, Hamisha, and Adulam) was assessed. The forests are located in a dry, subhumid climatic region with average precipitation around five hundred millimeters per year. Between twenty and fifty forest stands were assessed in each of the three grazing locations. The understory available for grazing was comprised of herbaceous vegetation, including oak and pistacia shrubs. The researchers calculated that one hectare of woody vegetation in these forests can support 450 to 750 grazing days per year, which translates into a carrying capacity of 1.25 to 2.1 goats per year on a hectare of land. The total "browse" provided by the forest to the goats came to as much as 1,103 kilograms of woody vegetation per hectare—over a ton of available food! The conclusions were comparable to findings in Spain.[34]

In a subsequent study, the research team made a distinction between "grazers" and "browsers": grazers were *sheep and cows,* which are relatively less aggressive foragers, preferring herbaceous vegetation; "browsers" are *goats,* whose constitutions are more vigorous and who have no problem eating woody understory and even young trees. Indeed, goats typically eat ten different tree and shrub species every day.[35] Six forest agglomerations in Israel's south and central regions, covering an area of twenty-six thousand hectares, were assessed. The team estimated that the average carrying capacity for a 240-day grazing season was somewhat higher than in the earlier Israeli study; this figure indicated the capacity to support 2.4 to 3.4 goats per hectare. Because of their herbaceous dietary constraints, these forests could only support 1.3 to 1.8 sheep units (or cattle equivalents) per hectare for a grazing season of 120 days. When the calculations were aggregated, it turned out that forests, representative of most planted woodlands in the central region, could maintain sixty-three thousand goats for 240 days of the year or thirty-four thousand sheep for 120 days.[36]

The actual number of animals presently utilizing these forests is far lower, suggesting that this ecosystem service is not fully utilized. In Israel today, *extensive* pastoral grazing has difficulty competing with *intensive* animal husbandry. Shepherds must manage herds far away from their homes, feedlots and meat processing installations. Moreover, Bedouin herd owners are often limited by inadequate infrastructure. It is important to make livestock grazing in forests more convenient. The researchers recommend providing watering facilities and better shelter to attract more herd owners.[37]

Today, some 65 percent of the animals grazing on public lands find their food in woodlands. Herds are rotated according to a cycle designed to both utilize available forage and facilitate renewal of vegetation. This allows for respite during the driest, nonproductive months. Picnickers are growing used to sharing the forests with herds of domestic animals, which have been a part of the local landscape from time immemorial. More work needs to be done across the range of Israeli woodlands before pasturing ecosystem services are fully understood. The required methodology is well characterized and the work is under way.[38]

Regulating Services: Sequestering Carbon

With growing concern about global warming and atmospheric concentrations of greenhouse gases, carbon sequestration has emerged as one of the most prominent "regulating" ecosystem services that forests provide humanity. Every year the world's forests absorb roughly 2.4 billion tons of carbon from the atmosphere—about a third of the carbon emissions that humans release into the air.[39] Accordingly, if there were twice as many trees on the planet capturing carbon, they would absorb *two-thirds* of anthropogenic carbon emissions. Clearly, afforestation should be a key part of any future international climate change strategy.

An appropriate management practice is to maximize afforestation's and reforestation's contribution to reducing atmospheric concentrations of greenhouse gasses.[40] One way to increase sequestration levels is to extend the intervals between harvesting rotations in plantation forests.[41] In Israel, where forests have little to do with timber production, such delays are not a relevant factor, although optimizing thinning practices may be significant.[42] Reducing forest fires, forest pests, and diseases can also help.

Although primary forests typically hold higher carbon stocks than planted woodlands, properly managed, fast-growing plantations can outpace them in C sequestration.[43] Considerable quantities can be captured for relatively modest investments.[44] Harvard economist Robert Stavins calculates that one-third of U.S. carbon emissions (five hundred million tons) could be offset each year by planting trees at a cost of only thirty to ninety dollars a ton, a relatively inexpensive intervention given the enormous global returns.[45] Other estimates run somewhat higher.[46] In 2009, the Israeli government hired international management consulting giant McKinsey and Company to assess cost-effective measures for reducing greenhouse gases. The company reported that the price of reducing CO_2 through afforestation was roughly thirty dollars per ton of Carbon Equivalent, on the low end of the American assessment.[47]

Forest ecosystems actually contain five separate reservoirs that store carbon. The first pool provides much of the raw materials for trees' actual physical form: 50 percent of most trees is pure "C." Along with the trunk, roots, stems, branches, and the foliage of trees—living and dead—hold enormous quantities of carbon. Carbon also collects in the shrubs, bushes, herbs, and grasses that make up understory vegetation of a forest. The tree litter and humus, collectively referred to as *fine woody debris* on the forest floor, is the third type of stock. Next is dead wood—stumps and branches lying on the ground that have not yet given up their carbon to decay and decomposition. The last pool is the soil itself, which constitutes a relatively stable carbon stockpile.[48] To these five natural reserves is added the timber that is harvested and transformed into carbon containing furniture, paper, and buildings. Wood products can last for decades, centuries, or longer, although of course, eventually they too will decompose.

Most of the increase in ecosystem-scale carbon stocks from afforestation takes place in the buildup of tree biomass *above* ground.[49] This sequestration "service" is provided by the miraculous mechanisms of photosynthesis: Trees break down CO_2 from the atmosphere into basic amino acids or glucose.[50] (The "C" is fixed and the "O_2" released.) When trees are young, the net flux of carbon from the atmosphere is quite high. With the passage of time, the balance changes, and carbon loss through respiration begins to occur, as does net carbon loss from decomposition of plant material. Ultimately, trees die and give back their carbon to the troposphere.

On a global scale, the ongoing sequestration process is massive: all told, forests hold half of the world's organic soil and vegetation carbon. But these overall statistics have little to do with actual local conditions. Carbon lying in a given forest's biomass may be highly ephemeral. Tree clearing, fires, diseases, and any number of hazards that forests face can release carbon that has accumulated over eons. This instability makes carbon sequestration a less than certain strategy for a *local* climate-change mitigation effort.

In the past, strategies for sequestration services focused on carbon fixing in *trees,* though recently there is growing interest in the forest's role in facilitating sequestration in *soils.*[51] This is because policies seeking to reduce atmospheric greenhouse-gas concentrations prefer more lasting and reliable storage strategies. Once organic carbon reaches soils, it becomes encapsulated by stable micro-aggregates that trap it in the ground. Microbes that would otherwise break carbon down have no access to it, making the enrichment process more permanent. With the right design and management practices, atmospheric CO_2 can be transferred to soils with impressive results.

As Israel considers potential benefits from sequestration services, its dryland soils are germane. At first glance, it would seem to be a trivial factor. A recent overview of the field concludes: "The effects of silvicultural practices on soil carbon are comparably small given the large small-scale variability of soil carbon stocks."[52] One of the most comprehensive studies of forest soils ascertained an "age effect" in the soil carbon cycle: young trees in afforested stands actually deplete carbon in the soil. Only after thirty years does soil carbon begin to increase.[53] With much of the local soil impoverished after centuries of desertification and human abuse, expectations for Israel might be especially low. But it turns out that for carbon capture, the eroded conditions are advantageous.

As carbon content drops in soil, the potential for additional "sink capacity" grows. A study in Sudan demonstrated how arid agricultural areas served as atmospheric sinks for carbon after lands were restored to a natural, savannah state.[54] Increased soil-carbon content was also measured following vegetation and seed production in Chinese and Mongolian drylands. Soon after low-density afforestation had begun, 80 percent of original carbon levels were measured.[55] Organic carbon sequestration rates in *soils* tend to be far greater in drier regions (350 millimeters or less of rainfall per year) than when forests are established on grasslands with high precipitation levels (660 to 1,070 millimeters per year).[56] Israel's Yatir Forest has even more dramatic figures—carbon in the forest's soils doubled after only thirty-five years.[57] When lands are degraded, planting trees counteracts a default "carbon loss" that often occurs in degraded drylands, producing a "double carbon dividend."[58]

Many factors appear to influence the overall sequestration effectiveness of afforestation: these include previous land use, forest age, and environmental conditions.[59] For instance, climate and especially precipitation are known to play a major role in soil-sequestration rates. The key question in the Israeli context is: what kind of "carbon dividend" might forests offer a country that is almost entirely comprised of drylands? Conventional wisdom holds that because trees in dryland regions grow slowly, only modest CO_2 uptake should be anticipated. Surprisingly, Israel's experience calls the pessimistic conventional wisdom into question, offering a much stronger endorsement of woodlands' sequestration potential in semi-arid regions. Indeed, there has been massive underestimation of tree efficiency in drylands.[60] This was confirmed after years of research by Professor Dan Yakir and his colleagues at the Weizmann Institute, a premier scientific research center in Rehovoth.

Yakir understood that Israel's forests would never absorb globally meaningful quantities of carbon because of their small size. Nonetheless, the country had special significance, as it represented an entire geographic and climatic zone that was ignored in the new global-monitoring networks.[61] Yakir cobbled together the resources and equipment to make Yatir the driest permanent forest sequestration monitoring site in the world.[62] When his first findings were calculated, Yatir's sequestration performance was remarkably impressive. During an average year the forest sequesters between 2.5 to 2.6 tons of carbon per hectare. This is comparable to the 2.7 ton average uptake of European trees.[63] When Israel has a rainy year, carbon levels can reach 3.5 ton per hectare.

Why do the pine trees in Yatir sequester carbon so well? Like all forests, carbon content in the trees is positively associated with photosynthetically active radiation, vapor pressure deficit, air temperature, and soil water content.[64] But in the case of Yatir, the pine trees were also especially selected for eco-physiological hardiness, as "drought- and salt-resistant" varieties.[65] Their ability to absorb carbon dioxide without having to open their stomates excessively appears to be one explanation. (Stomata are the pores in leaves responsible for the gas exchange of carbon dioxide and oxygen.) This feature helps to maintain trees' water balance, in a climate with extremely high evapotranspiration rates.[66] Other studies in the semi-arid regions of Patagonia showed similar results in newly planted forests.[67]

While Yatir's high performance sequestration surely is good news, it probably will not affect Israel's climate-change strategy meaningfully. Israel is a small country; most of the lands that can be forested have already been planted. The McKinsey national assessment only projected an additional 0.30 megaton potential for sequestration of CO_2 equivalents from expanded forestry. By comparison, either a shift to more efficient lighting or retrofitting of buildings can produce 160 times greater greenhouse gas reductions.[68] As of the time of this writing, Israel's forests absorb about 250,000 tons of carbon a year.

In the global discussions of sequestration strategies, dryland regions are more or less written off, overshadowed by the rainforest "superstars." But this approach appears to be ill-advised. About 70 percent of Africa's greenhouse gas emissions are associated with deforestation, much of it in dryland regions.[69] Beyond the contribution to atmospheric carbon increases, loss of tree cover has accelerated the heartbreaking symptoms associated with the continent's desertification pathology: massive soil degradation, poor yields, famine, and refugees. Global funding for afforestation to sequester carbon can improve Africa's carbon balance as well as bolster many of its other beleaguered ecosystem services.

In *theory*, gaining international carbon credits for new forests in the international Clean Development Mechanism (CDM) framework was possible from the outset. It soon became possible in *practice* as well. China, also a developing country under the Kyoto Protocol, was the first country to receive approval for a major tree-planting project in the Pearl River Basin in Guangxi Province, meeting the somewhat convoluted documentation demands set forth by the United Nations. Moldova,[70] Bangladesh, India,[71] and Uganda were soon to follow,[72] receiving carbon credits for afforestation under the UN system, which they could then sell to a wealthy nation. Technically a developing country, Israel was also eligible to put together an afforestation package and trade on the world carbon market. But for numerous reasons involving difficulty meeting UN standards for "additionality," the small size of future Israeli stands, and the inability to assure long-term sequestration benefits, it was decided not to seek certification for Israeli afforestation projects.[73] Rather, an "offsetting" program was launched whereby the public could voluntarily plant trees in order to symbolically balance their personal emissions.

This institutional caution was soon validated by Dan Yakir's ongoing scientific research. After extremely optimistic initial publications, Yakir and associate Eyal Rotenberg published an article in *Science* in which they tempered their original estimates about forest potential in semi-arid regions to combat global warming.[74] Assessing the *total energy budget,* rather than only carbon sequestration levels, they reported that while the Yatir forest absorbs considerable carbon — it also absorbs substantial quantities of heat. The article posited that the natural cooling system of the trees in Yatir is influenced by the so-called *albedo effect.*

The albedo effect describes the way sunlight is reflected from the earth's surface. It became a household world — at least among climate-change wonks — when the feedback loops associated with melting ice emerged as a significant factor in global warming. Where the earth's surface has a light color (like glaciers in the polar regions) it reflects light and heat back into the atmosphere, contributing maximally to a cooling process. In areas where the earth is dark — such as oceans or forests — it absorbs some of this heat. The scientists argued that relatively pale colored desert lands reflect more light than the relatively dark forest canopy, which retains the heat and suppresses infrared (thermal) radiation. The high levels of carbon sequestration still make semi-arid forests a positive proposition for overall global balance. Yet, because they contribute to absorption of heat, the net climate "benefit" from forests planted on desert terrain will take longer to "kick in," and will ultimately provide only about half of earlier estimated benefits.

These findings were not universally well received, and in letters that appeared in subsequent editions of *Science* some of the study's pessimistic assumptions were criticized as exaggerated.[75] Yakir feels confident about the veracity of his conclusions and insists that scientific nuances not cloud the bigger picture:

> To put things more in context, I do believe climate change will be significant and that our CO_2 emissions are mostly responsible for this. But I am also part of the effort that shows things could be much worse: about one-third of the carbon we emit is reabsorbed forests. The ocean is warming, so its uptake will go down. This makes the terrestrial biosphere our best hope to slow atmospheric carbon increases until we wake up. . . . But we need to know how the system works, how to predict its response to change in order to maximize this "sink."[76]

Yakir does not believe that the albedo-related findings should affect dryland afforestation policies at all. To begin with, the Yatir Forest is an extreme case, with particularly arid conditions. The balance between the *carbon effects* and *radiation effects* begins to quickly shift as one moves up the local rain gradient where productivity increases and the albedo effect is far smaller. In lands with four hundred millimeters of rainfall per year, the sequestration services of forests are far more compelling. With regard to the drier, arid regions, his solution is simple: decisions makers need to consider planting twice as many trees.

Birds and Bees: Pollinators and Plants

In recent years increased attention around the world has focused on the fragility of crop pollination by bees and other animals. Fertilization can occur when pollen is successfully transferred to the stigma—the sticky surface at the top of the pistil—that traps and holds a plant's pollen.[77] If no pollen is delivered, there will be no reproduction. No process in nature better exemplifies interspecies coexistence and symbiosis. Only 10 percent of the plants on the planet can be pollinated and sexually reproduce abiotically—without assistance from insects or animals. The rest rely on pollinators, many of which make their homes in woodlands.

This makes pollination one of the forests' noteworthy ecosystem services, critical for high agricultural yields, quality produce, and dependable harvests.[78] One recent study suggests that 75 percent of all crops with global importance benefit from pollination services.[79] When quantified, the annual economic benefit reaches two hundred billion dollars.[80] Israel's farms are no different. As a country that prides itself on producing most of its own food,[81] and with agriculture providing 3.6 percent of total exports, Israel's foresters need to worry more about local pollinators.

There are an estimated two hundred thousand different species of organisms in nature that participate in pollination processes. Most of these, of course, are insects. While moths and bats sometimes play a role in pollination, bees have evolved to do the "heavy lifting" for nature. One estimate suggests that 84 percent of the world's *commercial* plants, including most fruits, vegetables, and nuts are pollinated by bees.[82] Many farmers rely on managed bee pollination for ensuring pollination in their fields and orchards. For certain crops, wild pollinators provide the most meaningful source of pollen.[83] For instance, a survey of almond orchards in Israel showed that natural habitats and lands on the edges of groves held significantly larger populations of bees than the orchards themselves.[84] For most of the world's subsistence farmers, domesticating bee populations is simply not an option.

In recent years, there has been an alarming decline in wild bee populations around the world.[85] In the United States, for example, during the three-year period between 2007 and 2009, consecutive drops in honeybee populations were measured at 32 percent, 36 percent, and 29 respectively. Precipitous reductions were measured across Europe as well.[86] The impact is manifested directly in crop-production levels. The U.S. Department of Agriculture estimated economic losses of fifteen billion dollars as a result of declines in honeybee pollination. One Australian study projected that the impact of a single invasive species (varroa mites) on feral honeybee pollination services cost twenty-one million dollars in lost yields.[87] Happily, no significant collapse of honey bee colonies in Israel has yet been identified.[88] Given global trends, however, a precautionary approach is prudent.

There still is no clear explanation for the precipitous decline in native bees and honeybee populations. Among the many reasons postulated are damage from parasites—in particular a deformed wing virus transmitted by mites,[89] insecticides, and malnutrition. Habitat loss is unquestionably a critical factor. Fragmentation and degradation of pristine and even seminatural habitats have decimated bee communities.[90] Academic literature describes a phenomenon of dissociation from important resources for food and nesting. Stated more simply, when foraging and nesting resources are no longer within flying distance (which can be less than a few hundred meters, depending on body size), bees don't have enough to eat or enough places to breed. This problem could become particularly acute in Israel, where open spaces are lost from year to year.

Israel has a large population of domesticated honeybees whose colonies are maintained at high densities. This is another source of vulnerability. The Israel Honey Board reports ninety-five thousand colonies in Israel—with only sixty-one hundred forage points on an area which is less than seven thousand square kilometers.[91] Most of the points are located in the country's open spaces and forests.

There has been a continuous decline in Israeli honey bee numbers over the years. Until recently, this was largely attributed to reduced profitability for beekeepers, many of whom closed up shop. There still appears to be an ample supply of hives for the country's farmers. While there is some concern that continued use of chemicals could reduce population levels, no clear signs of stress, comparable to the international dynamics, have been recorded.[92]

Agricultural officials devote considerable attention to maintaining the overall abundance of bees, but preserving their diversity is no less important. Israel, it turns out, has a remarkably diverse bee population in the wild, with more than one thousand species identified thus far.[93] No real data have ever been generated about the size of wild bee populations or patterns at the community level.[94] Israel's indigenous bees tend to be solitary, ground-nesting species, most active in the spring when flowering peaks throughout the Mediterranean region.[95] For the extraordinary local bee diversity to survive, populations must enjoy access to a rich variety of plants and flowers. Some bees are "generalists," who can enjoy a multitude of environmental conditions and a varied diet. But others are specialists, who have evolved to forage and seek nectar and pollen from a very narrow range of plants.

It is not clear why these bee species are so picky, but it has been hypothesized that some pollens contain compounds that require physiological adaptations by adult or larvae to properly digest them. Alternatively, the very shape of a flower might deter visitation by some bees as opposed to others.[96] Whatever the underlying mechanisms, as a rule, the greater the number of bee species, the greater the variety of flowers required to support them. Researchers in the Carmel Forest found bee species diversity to be linked directly to floral variety (especially of annual plants) and availability of diverse sources of nectar.[97]

The transformation of Israel's open spaces into agricultural holdings cannot but have affected the country's bee population. All told, a quarter of total lands in Israel are farmed,[98] with a far higher percentage of the nondesert, northern regions intensively cultivated.[99] One study evaluating bee populations in Israel's arid regions found that while bees flourished in local gardens, these were largely generalist species and did not represent the more rarefied species that are found in natural habitats. The study concluded that desert gardens and agriculture may increase overall abundance of bee populations, but can also change community composition and negatively affect species richness.[100]

The population dynamics and diversity of Israel's native bees raise an important policy question regarding the siting of hives for honey bees near forests or other Israeli nature reserves. Ever since Darwin speculated about the possible phenomenon in 1876, there has been concern that the massive populations of honey bees used commercially would become "superior competitors" for native species. This would mean that that rarer, specialist species need to change their foraging patterns after being crowded out by the more numerous and robust honey bees, causing their populations to dwindle.[101] During the dry seasons, when nectar and pollen sources are at a minimum in Israel, this dynamic would be particularly acute.

A team from Haifa University conducted research in the Carmel Forest and the Yad Hanadiv nature reserve to assess the "honey bee as bully" hypothesis in Israel's forests and reserves.[102] The results were not entirely conclusive. Amid the many sites evaluated, four cases showed a significant decrease in native bee visitation, while in seven cases there appeared to be no significant effect. The findings were significant enough to lead to unequivocal recommendations: "We therefore, as a safety measure, highly recommend keeping the nature reserves in Israel out of bounds for honeybee hives."[103] If forests are going to support bee diversity, such a policy should be implemented in all sensitive habitat areas.

This presents a challenge to forest managers. Can they design a forest so that it is not only pleasant to its human visitors but also to the bees that call it home? When a forest is young and constitutes a "successional, open habitat," the collection of flowering plants is more diverse. Bees seem to thrive in fledgling forests. As the forest matures and the canopy fills out, many bees are crowded out. It is not just food they miss; bees also need nesting sites. It turns out that Israel's solitary bees mostly nest on the ground, others find a home in dead or empty shells scattered across the land or amid live vegetation. Recent studies reported that nesting resources might be more limiting to bee populations than foraging resources, especially in agricultural regions.[104]

Open spaces are typically rich in soil and stems for nesting bees. Forests can also provide these havens if they contain openings that allow for smaller plants to grow and blossom. For instance, when dominating, tall evergreen trees were removed from Israel's Meron Forest, it took little time for the original rich assemblage of shrubs and plants to return to newly unshaded lands.[105] The existing JNF decision rule to leave from 10 to 15 percent of forests in Israel as patches of open space, in order to enrich biodiversity,[106] suggests that such findings have already translated into management practices.

Another team of Hebrew University researchers conducted a study that looked at five pine forests in central Israel and found that wild bee abundance and species richness was highest in natural forest stands, as opposed to pine plantations. In the wilted days of autumn, when there was little blooming, bee numbers dropped in all habitats. Yael Mandelik, who headed the team, believes that monoculture pine forests in general are not a good substitute for more open and diverse natural habitats because they lack sufficient foraging resources in the driest seasons. At the same time, she reckons that those forests that allow for openings and natural regeneration have more "bloom" and provide better bee habitats than natural dense shrub lands. This, however, is typically not the case in arid southern stands where little flowering takes place in the limited understory.[107] All told, relative to the overgrazed state of many open spaces, the growth of forests constitutes a net gain for wild pollinators in Israel. But woodlands are most welcoming to wild bees when natural successional processes are allowed to unfold.[108]

Aware of the pollination challenge, the JNF funded research by biology professors Tamar Keasar and Avi Shmida to systematically consider forest composition and how it might be upgraded to support bee populations. The operational goal of the study was to identify which indigenous tree and shrub species had the highest foraging value for bees and which of these might be introduced into Israel's forests. A list of 562 potential honey-foraging plants was condensed down to 266 relevant species, which were then ranked based on observations and elicitation of expert judgment. One criterion of particular importance was performance during the dry months in Israel, when little if anything is flowering and bee populations have difficulty surviving. Conversely, during the rainiest months, bees often face a dearth of foraging sites. The scientists prioritized plants that flower during this "dearth period," along with other parameters such as number of flowers per plant, nectar production rates, and so on.[109]

Ultimately, the study recommended expansion of specific trees such as almonds, *Ziziphus spina christi,* and eucalyptus. All are good bloomers and copious producers of nectar and pollen. As foresters seek to expand the ecosystem services contributed by Israeli woodlands, beyond the usual tree-replacement schedules, they need to adopt supplementary planting programs in poor-bloom forests, including promotion of wild weeds and perennial plants. This should enlarge the number of nesting sites for bee populations. As ecologist Mandelik explains, the key lies in increasing season-long floral availability in summer until the rains return.[110]

Ironically, it would seem that one of the happy outcomes of Israel's frequent forest fires is the patchiness and openings in the forest that they create, which facilitate vegetative diversity.[111] A survey of the Carmel Forest summarized: "Immediately following a substantial burn there is a catastrophic loss of both flowers and bees, but of an extremely short duration. The majority of flower species are fire-adapted as post-fire resprouters or seedling recruiters. Bee and flower communities are most diverse during the first two years after a burn and follow the same asymptotic decrease during the next five decades until they reach a minimum in mature sites [which] mirrors that of the flora."[112]

Clearly, it would be best to create a heterogeneous woodlands mosaic systematically, using controlled burnings and harvesting schedules to produce appropriate habitats. In the meantime, ironically, while arsonists seek to destroy local ecology, they may actually contribute to a richer landscape and local bee diversity.[113]

Pollination is among the more valuable services that nature provides humans that is taken for granted. While no precise estimate of the value of pollination by *wild* bees in Israel exists, a 1999 economic analysis did assess *honey* bee pollination's value to agriculture. Based on yields for sixteen key fruits and vegetables, the value of the honey bee pollination was estimated at 811,853,000 shekels—over 200 million dollars at the time.[114] This is clearly an ecosystem service worth paying attention to.

Looking ahead, there is a significant human dimension to this challenge. Although Israel is blessed with a high density of gifted natural scientists and

One of many pollinator habitats in the forest near the Shikma stream. (Iliyah Onn, KKL-JNF Photo Archives)

ecologists, there are many gaps in local expertise. One of these gaps involves bee systematics. Numerous bee species living in Israel have not even been identified yet. Although the country has become far more coordinated in its efforts to monitor nature and biodiversity,[115] there still is no full-time expert who regularly follows the bee community at the national level, checking population trends. No wild bee monitoring program is anticipated any time soon, even as some initial grants provided basic training and a few site-specific studies were funded.[116] In light of the international crisis facing bee populations, it would seem advisable for all countries to invest in such human capacity.

Bees are not the only beneficial insects nurtured by Israel's forests. Zvi Mendel from the department of entomology at the Volcani Institute has assessed whether planted forests serve as refuge for the natural enemies of some of Israel's more deleterious agricultural pests. In particular, he is concerned about scale insects (mealybugs and armored scales) that feed on plant sap. As it turns out, planted pine and cypress stands harbor important parasitoids and predators of this pest group.[117] Forests surrounding fruit orchards are able to provide a virtual arsenal of natural enemies, able to take on the wingless females that can literally suck the life out of fruit trees.

Soil Erosion Prevention

A central motivation behind the British Mandate's commitment to forestry in Palestine was preventing acute soil erosion. Most local soils are susceptible to fertility loss.[118] Lying bare after the desiccating summer season, the land is fully exposed to high-intensity rainstorms that punctuate the winter months. Under such conditions, the long-term *on-site* implications of cultivation practices utilized over many centuries were catastrophic.[119] In the northern Negev, one can see remnants of water holes that clearly indicate historic soil loss of three to four meters. The acute *off-site* impacts of erosion during extreme rain events were no less pernicious, with sediment deposited by heavy storms regularly obstructing roads, causing massive property damage, and even causing dozens of deaths.[120] A description of one 1939 storm by a visiting American soil scientist is instructive: "Here before our eyes the remarkable red earth soil of Palestine was being ripped from the slopes and swept into the blue of the Mediterranean to a dirty brown as far as the eye could see. We could well understand how after many centuries, this type of erosion had wasted the neglected lands. It is estimated that over three feet of soil has been swept from the uplands of Palestine after the breakdown of terrace agriculture."[121]

While better infrastructure prevents much of the worst erosion today, there is growing concern about the off-site impact on water quality. Sediments and contaminants from "nonpoint source" runoff flowing from areas with poor vegetative cover have come to dominate the pollution profiles of Israeli streams.[122] Forests can alleviate this problem.

The ecosystem service of soil conservation involves retention of sediment and nutrients by vegetation. Erosion primarily occurs in Israel during the rainy season, when soil particles become detached from the land by raindrop impact and then are transported off-site by flowing water. (Wind erosion also exists when gales circulate, removing and redistributing soil particles.) It is a physical process that relies on energy. If the energy created by the rains can be dissipated, then erosion can be controlled. Tree canopies break rainfall, reducing its velocity and facilitating the absorption of water. The energy of surface runoff is also attenuated when it flows through woodlands. Understory vegetation can trap sediment. Tree roots hold soils.

Afforestation also leads to recovery of soil fertility, providing an important "supporting ecosystem service" to degraded lands. Over time, forests enhance the underlying structure and aggregate stability, improve water infiltration through the soil, and decrease surface runoff and erosion. Long-term research has confirmed the contribution of trees to soil integrity in a variety of climates, topographies, and geologies around the world.[123] The carbon sequestration in the soils beneath Israeli forests previously described is but one such example.

International experience suggests that there is little need to question the off-site erosion control benefits of local woodlands. When the many ecosystem services provided by afforesting steep lands in New Zealand were compared, the benefits from erosion control provided the most consequential environmental return.[124] Indeed, one study down under suggests that forest cover can reduce landslide probability by 90 percent.[125] In China, loess soils, like those found in Israel's southlands, have also been subject to massive degradation. Chinese soil scientists have calculated that when compared to vegetated lands with relatively few trees, forested areas released only a fifth of the sediment during rainstorms. Afforestation was recommended as the most promising measure for addressing the country's severe erosion problem.[126] Subsequent Chinese studies reached the same conclusions, calling for bans on clear-cutting on erosive terrain.[127]

Similar conclusions emerged from research throughout the Mediterranean Basin. One study in southern Italy concluded: "Where erosion on agricultural land is particularly severe, land use change and afforestation are frequently seen as the most appropriate means of reducing erosion risk."[128]

When the State of Israel was founded, the government turned to the United Nations for help in designing a national policy to address this scourge. Along with malaria, erosion and soil loss were considered to be among the nascent state's most pressing environmental priorities. A UN committee was appointed to review the situation and recommend changes. It was headed by American soil scientist Walter Clay Lowdermilk, who had led the fact-finding delegation from the U.S. Department of Agriculture in 1939 that considered land-management practices in the Near East.[129] Pursuant to the UN committee's recommendations in 1953, the Soil Conservation Department of the Israel Ministry of Agriculture set up an applied research center which is still located alongside the Rupin Agricultural College on the central coastal plain.

Over the years, the Erosion Research Station's professional staff has been reduced. Yet, it continues to conduct surveys and applied research projects to address the most pressing problems associated with Israel's soil erosion.[130] From its inception, the station focused on support for agricultural operations, promoting a long list of conventional soil-conservation practices, such as bench terraces and contour ploughing.[131] Forest management was never part of its operational agenda. The contribution of woodlands to soil conservation was taken for granted.

As a government agency that needed to set priorities, this made perfect sense. When lands were exposed during storm events, several Israeli stream measurements showed that dislodged soil sediment reached 20 percent of total flow volume in water. Forests prevent this by providing the necessary cover. Except for the most torrential downpours, measurable flow can hardly be found at the edge of woodlands. After decades in the field, the Erosion Research Station's director, Shmuel Arbel, is confident that the soil eroded off Israel's forest lands is trivial. This is why the research agenda of his research station remains agricultural. The only currents running off forested areas are on roads which serve as de facto drainage systems that collect runoff.

Both in the denser forests in the north of Israel and in the more dispersed woodlands in the south, erosion control is extraordinary. Arbel recalls being in the field near the Yatir Forest during a particularly violent rainstorm. Practically no runoff was found around the forest, while in the downstream sections outside the forest, for a square kilometer area, flow volume reached ten thousand liters of water per second.[132] The few Israeli surveys that do consider the soil-conservation performance of forest stands are unfailingly consistent with Arbel's sanguine operational assumption.

When the Be'er Sheva suburb of Meitar was built, for instance, it was sited in a valley encircled by rocky knolls. The JNF, brought in as a partner in the early planning stages, planted a terraced forest in the surrounding hillsides with extensive draining infrastructure to harvest rainwater. A central gorge runs through the city, so it was decided *not* to allow houses there due to the flood risk. Rather, the land supported an urban forestry project, creating a pedestrian link to the forests that encircle the town.[133]

The Erosion Research Station, along with the JNF's southern region, decided to evaluate the effect of the forest on runoff there. Although the area does not get much precipitation, when it does rain in Meitar it truly pours. Precipitation generally arrives in droves, causing massive sediment transport and floods. Because of its location, the town is particularly vulnerable to massive property damage and personal injury during storm events. The Meitar Forest successfully prevented any flooding, even as surrounding communities showed high levels of erosion during the periodic desert storms. In one particularly virulent flash flood that occurred during the course of the study,[134] the nearby townships of Omer and Be'er Sheva, situated on much safer plains, were inundated with runoff, causing considerable property damage. In the same deluge Meiter was protected by the trees and the terrace infrastructure.[135]

Further north, in the Givat HaMoreh Forest in Israel's heartlands, near Afula, between 2006 and 2007 measurements were taken during storm events at the edge of the forest. Again, the forests appear to provide underlying soils with almost impermeable protection. No flow or nonpoint source runoff could be measured throughout two rainy seasons. The situation was totally different in the adjacent agricultural land located in the Yechezkel Drainage Basin. Soil loss here during the same rain events was massive (four thousand cubic meters per square kilometer). Forest lands located on far more vulnerable hilly terrain showed no net soil loss.[136]

In another, more comprehensive three-year study in thirteen watersheds, the Erosion Research Station's scientists contrasted three types of land use throughout the south of Israel to see how they responded to rainfall events. The first land use consisted of forests that harvested water using the JNF's signature terrace and swale systems. The second group involved agricultural cultivation. The third group consisted of rangelands where extensive grazing took place. Once again, the forested lands showed practically no erosion, while the other two areas discharged copious quantities of soil sediment into the surrounding seasonal streams.[137]

Even in arid regions, trees can play a constructive role in protecting the soil and stabilizing erosive conditions. The riverbeds in the south of Israel are generally dry, but due to historic overgrazing and other disturbances, they are often barren of vegetation and can easily be washed away with the occasional winter tempests. Drought-resistant trees, when planted alongside the riverbeds, change this. A survey of over twenty streambeds showed that not only did the banks successfully support trees, but they enjoyed greater stability than treeless control groups.[138]

Soil-conservation services are particularly important after forest fires. In Israel, the patchy nature of lands after major conflagrations highlights the relative advantages of canopy cover for soil conservation. After the 1989 forest fires in the Carmel, Haifa University scientists Lea Wittenberg and Moshe Inbar compared runoff from burned plots with those where tree cover remained intact. Runoff and sediment yield on the exposed lands were *two orders of magnitude* higher than those on the "control plots" where tree cover remained intact after the fire. That is one hundred times more erosion! It is worth mentioning that even very young stands can provide excellent protection against sediment loss from rainfall. Typically, within one winter of a forest fire, the infant seedlings and herbaceous plants provide enough vegetative cover to reduce severe erosion.[139]

The magical protective shield that trees offer the soils below is due to their ability to create a canopy in the rain which slows flowing water and disperses it, neutralizing its abrasive impact. This is a service that in the future will be more important than ever, because abrasive impacts are about to get much worse—the greenhouse effect will see to that. Israel's precipitation patterns during the past decades suggest that global warming is not a futuristic scenario but already constitutes a new climatic reality for the country.[140] With the ferocity of rainstorms increasing, there is an expectation that erosion will only grow more pernicious with time. Contoured terraces slow the flow on slopes, allowing the trees planted there to take maximal advantage of rainfall while protecting against future storms. This has convinced soil-conservation officials to make forests a significant part of the country's "climate change adaptation" strategy.[141]

There are innumerable studies about the economic benefits accruing to individual farmers who adopt soil-conservation practices. At a macrolevel, one study by the world's leading soil scientists suggests that the total loss from erosion of cultivated soils globally costs four hundred billion dollars a year.[142] Yet these projections only consider agricultural operations. Surprisingly few international assessments on the subject have been published; certainly no study in Israel has quantified the sediment-retention benefits provided by forests.[143]

Theoretically, the gains from forest cover can be estimated by calculating what the cost of mitigating or repairing erosion damages would have been without the service—as a surrogate value for this protective function.[144] At a minimum, this should include estimates of soil-fertility loss, property damage, and water-quality deterioration. Even if there are no hard "numbers" to put on the table, these manifold benefits need to be accessible and part of the standard presentation when highlighting the forests' many blessings.

Noise Prevention

Anyone living on a densely planted tree-lined boulevard knows that in a cacophonous city street, trees can provide an important service as live acoustic barriers. While "noise prevention" is not widely considered in reviews of ecosystem services,[145] in urban settings or alongside deafening thoroughfares, noise reduction constitutes a highly appreciated contribution to quality of life. This "regulating" function does not happen by itself and requires careful planting of appropriate trees. Studies show that the size of leaf area is directly correlated with the degree of noise absorption. Dense planting prevents openings through which sound waves can slip.

In one early Israeli study, using a generator to produce a constant and replicable source of noise, different forest types were compared in the Carmel, in the Modi'in region, and in the "Ilanot" forestry research center along the Mediterranean coastal plain. Decibel levels were contrasted at varying distances from the noise source in different types of forest, with topographically indistinguishable control sites where there were no trees to intercept sound waves. Not all sounds are the same. Low-frequency noises were hardly reduced at all by the forests. Yet, as sound frequencies rose, they were readily absorbed by the trees.[146]

Trees' performance as acoustic barriers is far from uniform. Pine stands, for example, appear to provide much less acoustic protection than broader-leafed species. Indeed, the thicker the foliage the more efficient the noise-reduction effect. This is especially true for the understory in conifer stands. As is often the case, nonnative species can offer particularly efficient ecosystem services, with ficus trees proving to be noise-reduction champions, and some eucalypt species also showing strong acoustic-absorption qualities. Pine trees, however, have only half the noise-dampening capabilities of the more efficient foreign species, with local oaks doing somewhat better.[147] When noise reduction is an absolute priority, there may be advantages to careful utilization of nonindigenous species.

Serving Up Culture

Notwithstanding the forests' many contributions to human economic ventures and activities, the thing people most value about the woods are often the intangible but very real experiences they provide. Two hundred years ago, inspired by the fictional adventures of *Robinson Crusoe,* the French philosopher Jean Jacque Rousseau proposed that boys not learn to read until age twelve. Rather they should attain wisdom through direct communion with nature and the forests.[148] Rousseau articulated a Romantic intuition that forests offer a hidden wealth of wisdom and ethical insights.

Today this view has considerable empirical support. Children who frequent forests perform better on a range of cognitive tests than comparable youngsters who spend the same time in a nonnatural environment.[149] Visitors leave even brief visits to forests with significantly less stress than they were suffering upon arrival.[150] Many forests are considered sacred, holding great spiritual value even if they are seldom visited. For many people, religious, social, or cultural life is unthinkable without the shade of local woodlands.[151]

In the new nomenclature of *natural capital,* these edifying qualities are called "cultural ecosystem services." The Millennium Ecosystem Assessment defines them as the "nonmaterial benefits people obtain from ecosystems through spiritual enrichment, cognitive development, reflection, recreation, and aesthetic experiences."[152] Anyone who has been stimulated, stirred, energized, or spiritually refreshed by a walk in the woods understands how powerful these services are. Indeed, if calculations are done properly, cultural services are often the most valuable.

Counting "cultural" benefits, however, is not a simple matter. People expend substantial resources to travel to and hang out in the woods. They donate money to plant trees and support preservation efforts. They pay more for real estate that is located in and around the woods than for homes without tree cover. Economists have any number of ways for teasing out these "revealed preferences." They can also use questionnaires to project what the public is willing to pay to preserve forests and the experiences they provide. Price estimates can vary dramatically, depending on the kind of survey instruments used.[153] Frequently many of the intangibles go unrecognized.[154] For instance, the "option value" of leaving forests to future generations or the "existence value" associated with knowing that a particular environmental resource endures, may constitute the greatest single "benefit" of all. But they are rarely calculated. This highlights a fundamental dilemma. It is commonly held that cultural ecosystems are among those things that make life worth living. This makes them essentially priceless. It is argued, therefore, that society should not even try to put a price on the Carmel Forest or the Meron Forest, lest it belittle them. And yet, if they are not enumerated and highlighted, they may be dismissed or even worse, go unnoticed altogether.

In Israel, there are many ways that forests serve local culture and religion. For example, it is estimated that half a million Israelis — almost 10 percent of the country — come out every February to plant and celebrate the traditional birthday of the trees during the annual celebration of Tu Bishvat.[155] The holiday (literally translated: the fifteen day of the Jewish month of Shvat) originally appears in the Talmud, the ancient compilation of religious legal debates. The date is one of the four which compete for the title of "new year" on the Jewish calendar.[156] At the end of summer, Rosh Hashanah celebrates the traditional day marking the world's creation, as the Jewish New Year. The winter date of Shvat commemorates the new year for trees.[157]

Prior to the holiday, JNF nurseries are humming with young and old, coming to pick up seedlings for the ceremonies across the country. JNF trucks are busy delivering saplings to twelve hundred communities.[158] The present JNF chairman, Efi Stenzler, promotes a new slogan: a "tree for every resident." All Israelis are supposed to get in on the act. A wide selection of trees is available free of charge at the nurseries: including oaks, carobs, almonds, and cypresses, but also more obscure varieties. In most planting ceremonies, small pits are dug in advance, before the people arrive to ensure quality control. The proceedings usually include brief speeches, Bible readings, an occasional song, and a special blessing for the trees. Citizens flock to these events and enjoy the edifying experience of getting their hands dirty, taking a seedling with its roots embedded in dirt, laying it in place, and covering it with the topsoil conveniently lying alongside the predug hole.

It is sad to think that as the number of people in cities on the planet comes to outnumber rural residents,[159] more and more women and men go through life without ever planting a tree. This is not the case in Israel. Former JNF chairman Yehiel Leket describes the experience as a sensual one, creating an organic bond between humans and the earth. "When you plant the tree there is a physical connection. If you take the metaphors of land and forests you will find out there are many such metaphors. . . . Indeed, many people say, 'I want to hug the trees' or 'the trees hug me.' You take a tree and put your hand in the soil — it's a physical intimacy."[160] For impressionable children, the cumulative effect of planting a tree year after year is impossible to quantify, but it is surely formidable.

There are dozens of other countries that commemorate a national arbor day, celebrating trees and their many wonderful gifts: from China (where civilians are required by law to participate in planting activities) to Niger (where every citizen plants a tree on Independence Day as part of the national battle to combat desertification). None of these celebrations have their origins in tax collection, as Israel's does.

The fifteenth day of Shvat falls in the dead of winter, which the ancients considered the end of one tree-planting season and the start of another. In other words, it is the end of the arboreal "fiscal year." In days of old, Jews were expected to deliver 10 percent or a "tithe" of their agricultural harvest to God and the needy via their priests. The day of Tu Bishvat originally set the temporal boundaries that separated fruits that needed to be delivered during the previous fiscal year from the new one.[161] When this historic context is highlighted, celebrating a birthday for the trees of the land can also be seen as a commemoration of Jewish social responsibility.

When the Jews were exiled by the Romans, their connection to agriculture and tree planting was curtailed and the holiday hardly commemorated. Yet, like the resurrected forests, the return to Israel reinvigorated the holiday that emerged as a general-purpose Earth and Arbor Day. New rituals sprang up, including the revival of the mystical, nocturnal Tu Bishvat feast, first observed in medieval times in the Galilee city of Safed. Like the Passover seder, the ceremony includes four glasses of wine, but the menu is strictly vegan, comprised of the fruits and nuts that flourish on the trees in the land of Israel. The highpoint of the Tu Bishvat celebrations, however, involves tree-planting ceremonies.

In recent years, another, completely different, non-Jewish festival has emerged in Israel's forests celebrating the olive-picking season in autumn. As olive trees and their produce have always been a central symbol of the Palestinian life cycle, the harvest has traditionally been a family affair for Arab families. Due to the *bustan* program, there are now hundreds of mature olive groves interspersed among Israel's forests. For the first time the vast majority of families who do *not* own ancestral plots with olive trees have a place to pick olives. The event's popularity has been steadily increasing, attracting more Arab and Jewish families.[162] It is a refreshing turnaround. Rather than serving as a battleground for nationalist ideologies, the forests provide a platform to celebrate a time-honored Palestinian tradition and catalyze better understanding with Israel's Arab citizens. Recently Suohil Zedan, the talented JNF fruit-tree maven, launched yet another forest festivity in the Harod stream area based on picking pomegranates—a traditional fruit for the Jewish New Year.[163]

As Israel becomes increasingly urban, it is crucial that shaded, quiet places remain accessible so that inhabitants can get away, refresh their spirits, and enjoy the direct embrace of nature. Trees affect physical and mental health. Nazareth Elite is a small Israeli city, where 70 percent of the lands are forested. Its mayor argues that the astonishing cohort of 103 residents over the age of a hundred is directly attributable to the salubrious effect of living among trees.[164]

Suohil Zedan, the tree-transplant maven who brought fruit orchards to Israel's forests, sitting on the world's oldest known olive tree. (Photo courtesy Suohil Zedan)

Many natural landscapes can nurture special, authentic interactions between people as well as offer a place for the contemplative solitude needed to make sense of the world. The flora, fauna, and panoramas of forests offer such a sanctuary. It is true that recently planted plantations lack the authenticity of wilderness. There are certainly those Israelis who prefer the inspiration of "creation" in the unaltered vistas of the arid deserts in the south. Here the fingerprints of human intervention are less discernible. But as the successional process unwinds in Israeli woodlands and trees and understory return, spontaneous and free, forests increasingly offer the "real deal."

There are several empirical studies which suggest that while there is room for improvement, Israel's forests successfully deliver a range of cultural services. In one survey of 317 visitors at five forests across Israel, people were asked about their motivation for visiting. The two major justifications expressed were a desire to enjoy nature and a need to break routine and see a change in landscape. Only 45 percent mentioned hiking as a chief reason for visiting, but 70 percent did mention "enjoying" nature.

Israeli forests are often stereotyped as a destination for large, tribal-like barbeques on weekends and holidays. Yet, only 48 percent wrote picnicking and barbequing as the primary purpose for visiting.[165] Of particular interest is the high general level of satisfaction among visitors: over 85 percent expressed full satisfaction with the forests. Only 34 percent felt that Israel's forests were monotonous and lacked diversity (as opposed to 41 percent of the respondents in a control group surveyed by telephone who do not visit forests at all). Critically, over 95 percent of visitors and nonvisitors believed it important to preserve the forests in Israel for future generations. Of course, such answers might be expected from a questionnaire that contained no real tradeoffs, broadcasting certain expectations and bias to respondents. But even as lip service, it is an important statement.

Heritage is one of the hypothesized cultural ecosystem services. Embedded in many of the forests are remarkable archeological relics and places from the country's past. In Nir Becker's and Yael Choresh's 2005 study, forest visitors were asked to rank their preferences for the different elements present in the Biriya Forest. The vistas and observation points received highest ranking, with picnic sites, hiking trails, and scenic roads receiving fairly good grades as well (eight of ten). Religious sites (gravesites of *tzadikim,* or saintly rabbis) received lower marks, but that might be because many surveys were completed on Saturdays, when religious Israelis, who refrain from driving on the Sabbath, do not frequent the forest.[166] In their summary, the economists single out the ruins of the ancient fortress in the forest as one of the most captivating attractions (ranked by visitors far ahead of restaurants, guided tours, or activities for children). They calculated over a five million dollar annual value (twenty million net value at present), representing the public's willingness to pay for the inspirational experience of visiting.[167]

Israel's forests may be at a distinct disadvantage when it comes to offering certain cultural ecosystem services. The vast majority of stands are new and there has not been time for traditions and rituals to develop around them. To be sure, the Bible authenticates the long history of trees and woodlands in the land of Israel. But only recently have Israelis had an opportunity to establish a real relationship with them. There are no enduring ancestral connections that can be tapped. Although many contain sites with antiquities, the ecosystems are new. Like the Jews who planted them, the woods are just getting used to the new climate.

For Arab Israelis, taking advantage of the forests' many cultural services may not be so easy. Beyond olive cultivation, there is little indigenous Arab culture associated with woodlands, as by the eighteenth century forests had simply ceased to be an important part of Palestine's landscape. During the

British Mandate, afforestation was largely unpopular among many local Arabs, many of whom perceived it as a threat to their land rights, expressing their displeasure by burning reserves. During the 1950s, trees were enlisted by the new Jewish state in the great ethnic conflict of the past century. Many new forests cover abandoned Palestinian villages and the memories they hold, compounding Arab discomfort. The national forestry agency remains the *Jewish* National Fund—albeit the Hebrew translation, Keren Kayemeth L'Yisrael, uses the word *Israel.* It may be too much to expect Israel's Arab citizens to be enthusiastic about reconnecting to the woods as part of a new cultural identity. And yet, in many parts of the country, visitation censuses suggest that they have become the most regular and enthusiastic callers.

The inchoate nature of Israel's forests offers an opportunity to build a new culture of symbiosis between humans and nature. They are a work in progress and there is room for improvement. There is an openness among many forest managers to try new things. Some of the management measures for upgrading cultural services are seemingly trivial. For instance, in surveys, the most common complaint of visitors is the level of cleanliness in Israel's forests. Although only a small percentage of Israelis engage in serial littering, it is enough create a major mess and nasty nuisance.[168] Other common suggestions include better access to restrooms, garbage cans, and running water. Such adjustments require resources, but there has been slow albeit steady progress. More difficult and profound changes involve steps to enhance the spiritual and educational experience of visitors, facilitating a deeper connection to the forests. The menu of guided tours and ranger presentations, for example, could be expanded.

Recreation for many Israeli families focuses around *chanyonim*—literally "parking areas" that contain basic picnic and playground infrastructures. In order to maximize forests' recreational services, these leisure facilities should be planned meticulously. Researcher Gabriel Schiller argues that most forest recreational sites were selected on an intuitive or even random basis, without clear criteria. When these sites are located inappropriately, soils became compacted from overuse, leading to reduced water percolation and the withering of trees.[169]

The most uncomfortable result involves unpleasant temperatures. In a hot country that is growing hotter, forest facilities should maximize climatic benefits that the forest cover provides. Beyond scenic values and accessibility, a key criterion that should drive site selection is *thermal comfort.* This is a subjective factor driven by humidity, evaporative heat loss, and convection. While this parameter has been widely developed for designing indoor environments, it has not enjoyed as much attention at outdoor recreational locations.[170] In the 1970s, Schiller along with forestry pioneer René Karschon, conducted a

Israel's forests' recreational services: young cyclists in the Carmel. (Paul Orliev, KKL-JNF Photo Archives)

series of studies assessing heat stress and thermal comfort in Israeli woodlands. They showed that proper shading and ventilation reduce the effective temperature on warm days by as much as eight degrees Celsius.[171] These findings have since been confirmed by other researchers throughout the country.[172] Recreational facilities should be planned carefully to optimize thinning and pruning, providing the best possible mix of shading, ventilation, and coverage.[173]

Forest infrastructure at present primarily supports sedentary experience. But a growing number of Israelis seek active recreational outlets in the woods. There has been a steady increase in hiking and bike trails, but maintenance is spotty. Playgrounds in most forests are fairly uninspired. Some argue that playgrounds belong in city parks and not in forests. Activities like horseback riding or swimming are unknown. And there is nothing to encourage people to climb the trees themselves.

An estimated fourteen million people visit Israel's forests each year—about two visits on average for every citizen. Obviously, many citizens are more frequent clients than others. They represent a cross-section of Israel's diverse demographic mosaic. (Russian Israelis are particularly avid consumers of the cultural ecosystem services.)[174] Because entry is free—as it should be—forest popularity does not translate into a monetized ecotourism bonus. But that does not mean that visits should not be tallied in a system that measures the public interest or an index that helps set funding priorities.

To a large extent, with a few impressive exceptions, the popularity of the forests has been a lost educational opportunity. There is little information available at the popular picnic spots about the surrounding natural history, flora, or outstanding qualities of the site. Rather than deepening the connection to the forest, the experience remains superficial, with many picnic sites serving as little more than outdoor dining rooms. Signs highlighting natural amenities, visitor's centers, or eco-art can enhance the meaningfulness of visitation.

The potential of forests to provide edifying educational experiences is most profound among the young. In 2008, San Diego journalist Richard Louv drew attention, in his book *Last Child in the Woods,* to the growing epidemic of "nature-deficit disorder" among young people in the United States.[175] Cable television, the Internet, shopping malls, urbanized lifestyles, and parental safety hysteria inhibit children's "going outside to play." This exacerbates the disconnect between nature and young people.

Israel's objective reality is somewhat different. For many soldiers, military service takes place almost exclusively outdoors. Because apartments are small and extended clans large, many children's most formidable holiday and weekend experiences involve family gatherings in forests. If there is a unique local antidote to the nature-deficit disorder available to Israel's children, it can be found in the country's youth movements and the outdoor, self-reliant ethos that they espouse. Hiking, camping skills, and environmental activities are just part of the package whose conscious and subconscious educational message involves discovery and appreciation of nature. Close to two hundred thousand children, from ages nine to fifteen, belong to youth groups: from the Israeli scouts, to the Labor-affiliated "Working Youth," to informal educational frameworks for Orthodox, Conservative, and Reform Jewish children. There is even a youth movement for children with mental disabilities. It is true that youth movements do not reach every child. Only 20 percent of youngsters joining youth movements come from poor socioeconomic backgrounds.[176] But relative to other societies, young Israelis are no strangers to nature.

One of the distinguishing features of Israeli youth movements is "peer leadership." At age fifteen, members take a leadership training course and then become counselors. In this role, they inculcate a message of independence, love of nature, and patriotism to children only a few years their junior. Israel's forests provide these organizations with a place to conduct their activities. The

JNF Education Department reports over 170,000 youth coming to "informal education" activities in its forests each year. (This is in addition to comparable numbers that attend school sponsored tree-planting programs, educational field studies, or "cleanup" campaigns.)[177] During the summer, forty thousand youngsters spend a week or more camping out in forests. As it almost never rains in July and August, the children sleep under the stars and branches at night, while enjoying the trees' protection from the searing, sunny daylight. Collapsing exhausted into their sleeping bags at day's end, they fall asleep to the raucous chirping of crickets and occasional howl of the jackals proclaiming their territory.[178] The forest offers an authenticity that more than compensates for the lack of creature comforts.

Psychologists researching the restorative powers of forests believe that humans have evolved with an unconscious autonomic response to natural elements. Forests, like many natural places, create a sense of safety and healing power for people under stress.[179] A 2007 study reported that visitors experienced an immediate physiological response to the woods within just fifteen minutes of moving from an urban to a forest setting. This was confirmed by measurable changes in salivary cortisol concentration, diastolic blood pressure, and pulse rates.[180] To maximize the solace they provide, forests should be designed to accentuate those soothing qualities that captivate people when they visit the woods. This includes an ability to feel immersed in nature, a sense of escape from day-to-day reality, and an interesting landscape that effortlessly captures visitors' attention.[181] Forests need to offer more isolated pockets for picnicking, to minimize the disquiet of overcrowding on busy days.

Policy Prescriptions

Foresters need to worry about recreation, carbon sequestration, grazing, soil conservation, not to mention biodiversity. And yes, to the extent possible, why not take advantage of the timber which trees provide. Clearly, it is impossible to do everything. Just as this chapter surveys only part of Israel's forest ecosystem services, trade-offs and hard choices must be made. Priorities must be set. There will be sensitive habitats where biodiversity preservation needs to be boosted. Soil conservation might take precedence on steep, vulnerable lands. In forests surrounding farms, wild bee populations' foraging and nesting sites may need to drive forest design. In this balancing act, Israel's forests offer one key advantage. In many countries, exhausting timber resources undermines the quality of other services like pollination or picnicking. With wood production a low priority in Israel, this dilemma is not as troublesome.

To navigate the compromises and reach optimal strategies, Israel needs to adopt a new model for decision making—one that understands the variety and value of what forests have to offer. Forest ecosystem services need to be monitored periodically so that they can be ranked and optimized. This requires considerable in-house expertise, or at the very least, budgets for outsourcing. The process should not constitute an economic exercise per se. But it does need to be systematic and dispassionate. Economic benefits of different services can and should be on the table.

Once a particular ecosystem service is selected, the management implications should be identified and fully understood. For instance, the move to low-density planting makes sense in terms of species richness, but will typically lead to a flourishing (and flammable) vegetative understory. If a forest is designed to highlight cultural services like recreation and ecotourism, clearing must be more aggressive.[182] As Israel's new forests move from a pioneer, adolescent phase to a more mature one, it is time to raise the bar. Keeping trees alive is important but it is not enough. Preserving and strengthening priority ecosystem services constitutes the next phase in the country's ecological restoration, making forests the best they can be and maximizing their benefits for people.

Epilogue

All the Trees of the Forest

The mountains and hills will burst into song before you, and all the trees of the field will clap their hands.

—Isaiah 55:12

A Mountain and an Ideological Divide

The Gilboa Mountain is a small range by international standards, with its highest peak reaching only 536 meters. Yet its eighteen undulating square kilometers contain profound historic meaning. It was on top of the Gilboa that Saul, Israel's first and tormented king, launched his final, doomed stand against the Philistines. The book of Kings describes it: "And it came to pass on the morrow, when the Philistines came to strip the slain, that they found Saul and his three sons fallen in Mount Gilboa."[1] Even though Saul had betrayed, persecuted, and hounded him, his eventual successor to the throne, King David was despondent upon receiving the news and issued his famous curse on the mountain: "Ye mountains of Gilboa, let there be no dew, neither let there be rain upon you, nor fields of offerings: for there the shield of the mighty is vilely cast away, the shield of Saul, as though he had not been anointed with oil."[2]

The iconic Gilboa Iris, which appears to have reached symbiosis with the surrounding woodlands and a new generation of admirers. (Yossi Zamir, KKL-JNF Photo Archives)

The underlying dolomite and hard limestone geological formations along with overgrazing may have as much to do with the historically scant vegetation on the hillsides as David's damnation. Regardless of the cause, the Gilboa was always considered a treeless, unadorned landmark. And yet the mountain was anything but empty. It was home to fourteen small and scattered villages and any number of wildflowers that created colorful canopies of psychedelic colors, the most notable blossom being the spectacular Gilboa Iris (*Iris haynei*). The flower's droopy, deep-purple petals became a symbol of endemic Israeli botanical splendor when it was adopted as the logo for the Society for Protection of Nature in Israel (SPNI). Every year in March and April, hikers and nature lovers make their way to the Gilboa to marvel at the brilliant colors.

With the coming of impatient settlers at the start of the twentieth century and Jewish national independence fifty years later, the Gilboa would not stay that way for long. The clash over its restoration came to symbolize the larger debate about the form that Israel's open spaces should take. JNF strongman Yosef Weitz saw the area as a personal challenge and meticulously followed the heroic efforts of his foresters to establish trees on the inhospitable hillsides. In 1950 he built a village alongside the twelve remaining Arab towns to provide a home for newly arrived Jewish emigrants from Yemen. Then he sent them forty-seven thousand saplings to plant. Despite the promising start, Weitz would write resignedly in his diary: "Complete failure here in afforestation over the past two years—in 1951 due to drought and 1952 due to excessive rain. Very few plants surviving. Tens of thousands of trees dried up and dead. Is the curse still active? Gave instructions to fill everything in during 1953 and to advance to the east." Weitz was not a quitter. His crusade continued and by 1960 he could report 355,000 trees planted on 170 dunams with only a 2 percent mortality rate: "We have at last freed the Gilboa from the desert's grip," he ecstatically wrote.[3]

More than fifty years later, with great acuity but considerably less elation, Azariah Alon remembers the Gilboa afforestation project quite differently. As Israel's most celebrated naturalist for over sixty years and one of the two founders of Israel's leading conservation organization, the SPNI, he has a personal stake in the project. Alon lives across the street from the Gilboa on Kibbutz Beit HaShita: "The foresters were total barbarians," he reminisces. "First they came in and started burning. They simply liquidated all natural vegetation. It was a total annihilation. Then they brought in pines and eucalyptus."[4] The actual planting was overseen by Weitz's son Sharon, who headed the northern region of the Forestry Department at the time. When Alon could stand it no longer, he crossed the highway and confronted Sharon Weitz in person. The naturalist not only brought his vaunted personal stature to the argument as host of a popular national radio show about nature. He also brought along some "*muscle*" in the form of burly nature lovers from the local community. Sharon Weitz agreed to send the issue back to his father, Yosef, at the JNF central offices in Jerusalem, who ultimately proposed a compromise.[5] In the upper region of Barkan, a biologically rich area marking the transition to the Mediterranean botanical zone, tree planting was stopped. Here, the wild wheat and irises are most plentiful and they were left in peace.

Alon refers to the modus vivendi as the best deal he could strike at the time, even though new forest still obscured many iris beds. He also takes satisfaction in the fact that many of the pine trees suffered an unnaturally early demise: "You can't plant them 'abnormally' at such a high density—where they have no room to develop so they grow as thin as twigs without branches or limbs—and expect them to survive normally." The forest that ultimately endured is in fact a blessing compared to other development projects that recently came to the Gilboa, like the twenty-million-dollar "Ski Gilboa" recreational complex. The 230-meter, artificial white Astroturf surface now carpets a corner of the mountain twelve months a year, letting tourists ski down the side into the Jezreel Valley.[6] Azariah calls the attraction "a supernatural horror."[7]

Most of the Gilboa remains undeveloped. The stark green lines along the summit demarcate the border between lush forest and the comparatively moderate vegetation in nature reserves. It offers a physical microcosm for the competing views about the form that ecological restoration should take.

A view from the Gilboa ridge, where the tree line at right marks the ideological divide between Israel's forestry and nature conservation communities. (Yossi Zamir, KKL-JNF Photo Archives)

Israel's forests have an ancient past that offers a vision as to how their future might look. Lessons from the forest's more recent history explain how to better reach that future. Very different ideas exist about what these lessons are.

A New Forestry Bible

Few people have succeeded in viewing the conflict over open-space management from both sides of the Gilboa. But Avi Perevolotsky's professional career has seen him travel the distance across the ideological divide within Israel's environmental movement. When Israel still held the Sinai Peninsula during the 1970s, he headed the SPNI field school that was built near the foot of Mount Sinai. Its stunning panorama and Perevolotsky's charismatic wisdom made the remote school an elite training ground for many gifted young conservation leaders. Later he would complete a doctorate and receive a position at the Ministry of Agriculture's academic center, the Volcani Institute, as an ecologist. His Hebrew ecology textbook has been widely praised by a typically nitpicky local academic community.[8] With a full gray beard, twinkling eyes, and consistently insightful presentations, he can speak to Israel's many ecological circles.

In 1998 Perevolotsky was asked to serve as the chief scientist at Israel's Nature and Parks Authority, an organization that is responsible for over four hundred reserves covering some 25 percent of the country's lands. It enjoys a mandate to put nature's interests over those of humans. There he staked out positions that were often at odds with the traditional ecological orthodoxy of the agency and pushed it to negotiate a more conciliatory relationship with the Jewish National Fund. For instance, he lobbied hard to temporarily classify sensitive lands as forests to expedite their immediate protection, rather than wait for the protracted process associated with declaring lands as protected nature reserves.

Perevolotsky began his tenure during the awkward days when the National Parks and the Nature Reserve Authorities were merging. As the forests inside National Parks at the time were managed by the JNF, he got to know the foresters and noted their growing commitment to ecologically sustainable management programs. A decade later, when the JNF foresters sought an academic to help them articulate these practices in a new *Forestry Bible,* they turned for advice to Perevolotsky and forestry researcher Yagil Osem. The new manual seeks to apply the vast accumulated knowledge in the field of silviculture and conservation biology to the unique conditions in Israel. Perevolotsky sees the project as part of an important juncture in the country's natural history:

> Israel's forests today reflect the final outcome of a battle between two ideologies, each with its own political and scientific perspective. There were those who said, "Let nature do what it can," and there was the other view that said, "Let us improve the landscape." It was not a professional, but an ideological fight between two schools, and Israel's open spaces were divided up according to the outcome. Today, this battle is behind us. We're not going to see a doubling of the forests or of Israel's nature reserves. If you figure in the country's agricultural lands, the entire cake has been divided and, at best, we are now arguing over the crumbs. The problem is that people were so busy fighting over the borders of the lands they sought to control they had no time to think about how to manage the lands that had been set aside. There was no clear strategy for managing forests or nature reserves.[9]

Perevolotsky believes that a new *ecosystem-based approach* to forestry management is needed. He sees this as the antithesis of the original "commercial" approach that informed Israel's first foresters and forests. To some extent he caricaturizes the country's original foresters as driven by timber-maximization impulses, prioritizing a few fast-growing erect trees with minimal branches. By contrast, the ultimate objective of his alternative approach is to allow the ecosystem structure of woodlands to return to a state "as similar as possible" to its natural one, while rehabilitating the system's natural interactions and processes.[10]

An underlying rule for making forest-policy decisions, from this perspective, is whether a given management practice serves the overarching aim of ecosystem rehabilitation. Accordingly, planting nonnative species runs counter to the natural assemblage and would be unacceptable. Ideal forest management is minimalistic or "extensive," allowing nature to do what it knows best as opposed to the costly, "intensive" management of traditional forestry. Yet, in some areas, for example where visitation is concentrated and relentless, maintenance is required, and oversight may be more intensive than ever.

Given the emphasis on returning the forests of old, a prerequisite to evaluating strategies involves defining what the natural forest and climax ecosystems that preceded human activities were.[11] The problem is that humans chopped the first trees down in Israel some ten thousand years ago, and local vegetation has changed beyond our ability to precisely reconstruct it. Perevolotsky espouses a hands-off approach to management that allows natural ecosystems to slowly reestablish themselves. He anticipates the results as a healthy mix of oaks, pines, and other native species: "With time, we will come to know this natural state and its unique characteristics."[12]

Other forestry experts find his "ecological anarchy" a sure formula for disaster.[13] In a spirited exchange in the Israeli professional journal *Forest*, Professor Zvika Mendel counters that Perevolotsky has created a "straw man" by criticizing a timber obsessed forestry program that never existed. For example, clear-cutting was unknown in Israel, except in extreme cases of pest infestation or following fires. He argues that the Israeli method for forestry coalesced over decades: "Based on the most successful forests' ostensibly impossible locations . . . it maintains that planting should be dense and that the lands should be prepared before planting. . . . Israel's planted forests largely imitate the natural regeneration of pine forests after a fire."[14] Mendel cites international acclaim for Israel's first generation of forests, but acknowledges that, in retrospect, two main problems plagued performance: the lack of effective scientific tools to address the Matsucoccus pest outbreak in pine forests, and chronic failure to conduct the selective thinning so important for young pine forests. These were fundamentally management *not* conceptual failings.

Mendel is correct in asserting that Israel's foresters gave up timber production as a primary objective earlier than is commonly recognized.[15] By the 1980s, Israel's foresters welcomed the steady return of other native species, especially regeneration of oak,[16] recognizing it as the "climax species" that would take its natural place with the help of shade from "pioneer" pine trees.

He concludes that the debate ultimately is aesthetic or even ethical. The question is: "How should planted forests in Israel look?" Mendel explains: "There are those who dream of the chaparral scrublands. These plant groups are indeed an expression of the natural landscape today. But this landscape is a manifestation of retreat from the original towering vegetation that was once Israel's forest; a result of hundreds and thousands of years of destruction and soil erosion. I believe that most of us want a different vegetative assemblage, one whose dominant component is a tall and shady forest that only experienced foresters can grow in our not so simple habitats."[17]

It is certainly true that, given the degraded baseline and marginal lands on which Israel's forest are located, a laissez-faire "ecological" approach to forest management will not lead to optimal results and realization of ecosystem service potential. Letting nature take its course will short-change the carbon sequestration, grazing, habitats, pollination, erosion control, recreation, and even cultural enrichment that forests can provide. Nurturing different services requires different interventions. This is particularly true in Israel's semi-arid southern forests, where most planting now takes place. The forests there are wonderful but decidedly unnatural. In order to produce meaningful services, cautious and selective utilization of nonnative species may be in order. In a country where a quarter of the land is already designated as nature reserves and human interference an anathema, the 10 percent of lands zoned as forests need not conform to the same creed and constraints of Orthodox preservation.

While these philosophical arguments rage on, the trees keep on growing, making reality even more complicated. Take the matter of assemblage. Neat divisions according to forest type make sense on maps and the safe polygons of GIS software, reflecting an agreeable, theoretical mixed network of patches. Such a rational societal compromise was codified as Israel's forestry master plan. But no one bothered to consult with the trees. One of the surprises that foresters increasingly encounter is the obstinacy and persistence of Aleppo pines.[18]

Long ago, this particular conifer fell out of favor among nature lovers who had grown alienated from the sterile, crowded "pine deserts" favored by the JNF foresters. The better-informed grudgingly accepted them as indigenous trees. But the assumption was that—at best—Aleppo pines were a pioneer species that would establish itself, create a shaded understory, and eventually give way to slower-growing, worthier oak trees which would claim the privileges coming to a climax species. While foresters were never as intolerant as the nature preservationists, by the 1980s they too had come to dismiss Aleppo pines, in favor of the more pest-resistant *brutia,* the neighboring, but ultimately nonnative conifer species. The Aleppo pines, however, were never aware of the new discriminatory policies and happily spread their seeds on every conceivable soil type. Research corroborates that by the time pine trees are twenty to twenty-five years old and reach full sexual maturity, they start a quiet invasion process in nearby areas, with seeds sprouting as much as a hundred meters from the mother tree.[19] Across the center and the north of the country where rainfall exceeds three hundred millimeters per year, beginning in the 1970s new Aleppo stands began to germinate without fanfare in nature reserves and broadleaf JNF forests, often to the consternation of local rangers and nature lovers.[20]

This may not be such a bad thing. There are of course many reasons why pines were discredited. Much has to do with the pitiful "scarecrow" trees that emerged from unnaturally dense plantings. When not stifled in a crowded monoculture, pine trees can live long and prosper. They happily coexist with other broad-leaf trees species. No other indigenous species has yet been found that can produce lumber meaningfully in Israel's dryland climate. A pragmatic approach can capitalize on the many ecosystem services that these robust pines provide through smart management and selective breeding. Just as Israel's delicious and bountiful vegetables represent the successful adaptation of foreign species to harsh local conditions, Aleppo pine evolution can be coaxed along.

Perevolotsky watches the Aleppo pine's exceeding its anticipated role as pioneer species with amusement. It is only one example of expert tendencies to remain locked in default paradigms of forestry theory, even when they may prove quite incongruous in Israel. For example, with commercial timber production a low management objective, it is not clear why foresters still feel compelled to ensure that trees grow so straight. As Perevolotsky says: "I ask: why? It's not going to a saw mill. This approach is anachronistic, a remnant of a different world and an older age. We shouldn't expect Israel's forests to look like classical ones. It should be a much more diverse and interesting landscape."[21]

There are areas where the transition to the new ideology is complete. The question of the indigenousness of the local forest—in the past a source of heated controversy—has long since been resolved. The Ilanot Forest Research Station highlights the conversion. During the Mandate years, scientists began testing the adaptation of hundreds of trees from around the world to Israel's somewhat inhospitable physical conditions at Ilanot. Today, the lovely thirty-hectare sanctuary, located east of Netanya on Israel's central coast, is a lush and bucolic arboretum, home to splendid, frequently idiosyncratic trees. It also houses Israel's national seed bank that preserves the genetic makeup of hundreds of indigenous plant species. But there are no longer any researchers on staff in the arboretum; Ilanot's exotic species simply are no longer candidates for planting. While eucalypts have a senior enough status in Israel to still be widely utilized in the south, there is legitimate criticism that recently they have become too prevalent. There are amorphous plans to turn the Ilanot center into a tourist site with a strong pedagogical program that tells the unique story of each foreign tree. But the *exotic* specimens in such a tree museum will primarily be a testimonial to Israel's abiding commitment to *local* vegetation.

The optimal assemblage of Israel's tree species in the forest is less clear. Biodiversity is usually the best management strategy to ensure provision of ecosystem services. But Perevolotsky is quick to emphasize that an ecosystem based forestry management definitely will *not* preserve *maximal* biodiversity. But that is not its aim. Rather, it is designed to preserve the "oak-pine" balance characteristic to Israeli forests.[22] And while everyone agrees that Israel's new forests will be far more diverse than the preceding generation, diversity can take many forms. There is an ongoing internal debate among JNF foresters as to whether there needs to be any homogeneity at all in forest stands. In the past, forests painstakingly planted as diverse stands with an assortment of species had difficulty getting establishing. Some had survival rates as low as 5 percent.[23] The broadleaf species simply could not compete with the faster-growing Aleppo pine trees. Oaks, for example often require irrigation during their early years, something which is more feasible to carry out when they are not intermingled with conifers.[24]

There is a certain logic, therefore, behind patchwork forests, comprised of continuous stands of uniform tree assemblages, selected especially for soil conditions and topography. These forests offer relative convenience when it comes to thinning in the early years or even harvesting when they reach the end of their cycle. That is why there remains a school of Israeli forestry experts who still argue forcefully for mixed forests, based on homogeneous patches. This position is informed by conventional forestry theories that may not be designed for timber production, but still contain a presumption of prolonged intensive management. A patchwork of homogenous stands presumably avoid interspecies competition and is more easily maintained.

Omri Bonneh, the JNF northern region director, has challenged stand diversity pessimists ever since he received a commendation in 1983 for successfully planting his first groves on the edge of the Carmel with a hodgepodge of local species. Thirty years later a diverse forest thrives. Now as the organization's chief scientist, Bonneh continues this pathway, introducing mature, twenty to fifty-year-old oak trees that were salvaged from imminent urban and infrastructure development and transplanted onto empty hills or existing monoculture conifer patches. The technique for the logistically and botanically complicated transplants was perfected by fruit tree maven Suohil Zedan. It comes at an average cost of 250 to 400 dollars per tree, requiring three to four years of irrigation after the transplant.[25]

There are less expensive ways to upgrade and diversify pioneer monocultures. A recent study reports surprisingly successful acorn planting in a *thinned* pine forest. After three years without any additional treatments, the 4 percent survival rate under the gentle protective shade of the thinned pine forests was as high as comparable open "batha" scrublands, and considerably better than in natural oak forests.[26]

Even Azariah Alon can't help but offer a word of praise for Bonneh's efforts to diversify the forest—a rare moment of generosity from an old soldier who spent much of his many years battling JNF foresters. It is important to recall that diversification of monoculture forests is limited to young forests in the center and north of Israel, where there is sufficient rainfall to support diverse understory. Most of these pioneer woodlands are already well established. To a large extent, nature has taken over and mooted the debate. The second-generation forests produced by natural successional process tend to be crowded with a diverse understory that is difficult to penetrate. This makes systematic management difficult. When compared to conditions in a well-managed, low-density pine plantation, the biomass along the forest floor presents a fire risk.[27] But "ready or not," the next stage in the great experiment is off and running.

While Perevolotsky offers important theoretical underpinnings, the shift to an ecologically aware forestry program was a pragmatic "bottom-up" affair, led by foresters and managers in the field. Once the entire orientation of Israel's forestry program had already shifted, and the academic community added its blessing,[28] the political leadership eventually moved to ratify the change. In 2005, JNF chairman Yehiel Leket appointed a special "sustainable subcommittee" comprised of JNF board members and professional staffers to officially formulate new sustainable policies for forestation, stream restoration, reservoir construction, and public participation. It was a three-year process that involved consultation with outside experts and government officials, and a great deal of drafting. When the debates and the compromises were finally complete, the position papers were unanimously approved by the board of directors in 2007. The forestry policy opens with a general set of principles that are a total departure from those that guided the workers who set fire to the natural Gilboa vegetation in the 1950s to make way for pine trees:

> The JNF's afforestation policy is designed to serve all of Israel's citizens today and in the future, and shall be based on the principles of sustainable development. Consistent with these principles, JNF forest management will entail:
>
> 1. An ecological approach that will be implemented consistent with the natural ecosystem, and not counter to it;
> 2. Societal considerations according to which the forest serves the public and the community;

3. Promotion of economic uses and initiatives that include tourism development, grazing, timber industries, etc.;
4. Opening the forests to the public for free, with unrestricted access, in keeping with the principle of intra-generational equity; and
5. Preserving the extent of Israel's woodlands and their quality out of concern for future generations.[29]

The policy painted a vision of the country's forests in very broad strokes. Four years later, a team of senior foresters led by Yisrael Tauber and Hanoch Zoref along with Perevolotsky and forestry researcher Yagil Osem submitted the first draft of a much more detailed management guide called *The Forestry Bible in Israel.*[30] The document offers greater resolution, translating the high-minded political objectives into the nitty-gritty decisions which foresters in the field need to make: from baseline tree densities and grazing schedules to irrigation and pesticide practices. The *Forestry Bible* even blesses prescribed fires as a management tool. It strives to create a forest that involves *minimum* human intervention in management and *maximum* vegetative diversity—ensuring uneven age and multispecies forests. Forests are to be broken into patches, with thinning and harvesting used to ensure contrasts in the field. The new gospel of forestry also emphasizes the preservation of natural amenities, scenic landscapes, and historic sites. For example, planting must not damage existing geophytes and seed banks.[31]

Azariah Alon is also pleased with the 180-degree change in postfire, restoration-management practices. The JNF's new written policies contain many manifestations of an "ecosystem based approach," but fall short of a full embrace. Should trees in a forest die naturally or from human disturbance, it is fine to remove burnt residues out of concern for safety, but there is no immediate imperative to plant.[32] The oaks, pines, and other local species usually are capable of natural rejuvenation. Natural succession is deemed ecologically healthier and more cost-effective than rehabilitative plantation plantings. Foresters need to wait two full years after a fire before they even think about thinning the seedlings that naturally reemerge, and only after they reach a height of 0.2 to 1.5 meters.[33]

Yet, the *Forestry Bible* rejects a laissez-faire, hands-off forest-management strategy. It is not only possible but important to upgrade the quality of woodlands. Perevolotsky claims that some JNF foresters may have even gone overboard in their ecological conversion. Many can be sentimental about individual trees, resisting any harvesting not deemed absolutely essential. "Today we need to cull 50 percent of the country's pine trees because of the prolonged drought," he explains. "If we want to save some of the trees, we have to make sure that they have enough space to extract the water they need. But JNF foresters can't accept this decision. In some places it will lead to certain death for all."[34]

Forests and the Future

The future of Israel's forests is far from certain. It is clear that the woodlands are leaving an initial phase that began eighty years ago where the focus was on planting. The new phase emphasizes management. Already, some 25 percent or more of Israel's tree planting involves maintaining previously planted forests.[35] This percentage will only increase with time.

Not only are the forests in flux, but Israel's climate appears to be changing.[36] Referring to "drought" conditions may constitute a misnomer. These were not anomalous dry years.[37] Two decades of data suggest that the country needs to get used to new, lower levels of precipitation. Global warming threatens to be one of the forests' greatest future challenges, as the *Forestry Bible* describes: "Following successive drought years that marred the first decade of the twenty-first century we are witnesses a growing mortality rate among trees in the forest. This situation requires a new synchronization between the structure of the forest and the way it is managed for water scarcity." The fire in the Carmel painfully drove this point home.

Because Israel is located on an extremely steep rain gradient, it is experienced managing forests under a range of climatic conditions. Many of the adjustments that will be needed for adaptation are self-evident. For instance, irrigation at present is typically utilized once or twice during a seedling's first summer to ensure it survives the dry season. Fruit trees and more sensitive species such as cedars require more prolonged support.[38] As the country grows warmer and drier, watering frequency may have to be increased and extended. Thinning young woodlands will be more important than ever and tree density may drop further to reduce fire risks. Greater genetic selection for drought resistance will also be needed for trees to survive in the arid southlands. Optimizing the timing and dosages for inoculating seedling roots with mycorrhiza fungi to facilitate better mineral supply to trees might become commonplace.[39]

Most measures are less technical, involving common sense or the application of adaptive management principles. Pragmatism and intelligence continue to drive Israeli forestry policy. For example, a recent study showed that despite the ancient tradition that favors planting saplings during the peak of the winter rains, the success rate among autumn seedlings is far higher.[40] It turns out that warm root-zone temperatures are more important than soil moisture.[41] As a result, planting schedules have been modified. "Assuming that supplementary irrigation will be impossible, planting should take place as early as possible in the wet season, when the soil reaches dampness levels of a receiving field."[42]

Such ongoing adjustments can be seen in tree density. Planting in the rainier northern parts of Israel, which used to be extremely dense (four thousand to five thousand stems per hectare) now begin at about a thousand stems per hectare (roughly three meters distance between seedlings). These are subsequently thinned, with very clear schedules prescribed. In the semi-arid regions, tree density today starts at a modest hundred to two hundred stems per hectare. It is interesting to note that these were the precise standards adopted by the British and later Israeli government forestry service under Amihud Goor,[43] and are within the recommended range set by veteran researcher Gabriel Schiller.[44] Because of the extensive site preparation required to harvest runoff to irrigate saplings and retain rainwater for maximum utilization, the cost of lower-density southern stands is higher (ten thousand dollars per hectare) than conventional planting (seventy-five hundred dollars per hectare).

The importance of reducing tree density was highlighted in an experiment run by forestry researcher Yagil Osem, whose team looked at the effect of different thinning regimes on the Kedoshim Forest, a relatively dense woods located in the Judean Hills. Stands that had reached the age of forty were culled down to a hundred trees per hectare and three hundred trees per hectare, and then compared to a control group with five-hundred-trees-per-hectare densities. Within two years, statistically significant improvements in the water availability and growth of trees in the low density stands were measured. Even at age forty, trunk radius immediately began to expand. The lower the density the better the performance and the lower the mortality rate.[45]

Biodiversity recovery was no less remarkable. All the forested plots shared the same, relatively sparse level of biodiversity prior to the experiment. A pine monoculture, the forest had only half the species richness of the adjacent *unforested* open spaces. Within two years of thinning, in response to the drop in tree density, the number of species increased dramatically: parcels with a hundred trees per hectare boasted 119 plant species, closing the gap with the 141 species in open spaces. Oak trees rebounded in the moderately shaded openings — reaching four hundred sprouts per hectare — even as they remained completely absent in the adjacent open spaces. One plot in the forest was completely clear-cut. Within two years its species richness was indistinguishable from that of open spaces. The ability of forty-year-old trees to respond to the opportunities provided by newly available sunlight, water, and space and the rapid reappearance of native plants in understory confirms the astonishing resilience of nature when given a little help.

The dramatic conversion in Israeli forestry's ecological dogma should also be reflected in a parallel institutional transition. A change in the overseeing ministry needs to be considered. As long as the primary activity of foresters was planting, the Ministry of Agriculture was a natural ministry for oversight of JNF forestry. Today, when foresters need to maximize ecosystem services, protect wildlife, and augment biodiversity, the Ministry of Environment would seem to be a better fit. For almost a decade, new versions of Israel's forestry law have been considered which would create an independent council to oversee forest policy. Ultimately, the last word in forest policy should be held by professional appointees of democratically elected officials who represent all Israeli citizens.

The forests in the land of Israel constitute a grand experiment. Few if any countries have systematically tried to plant as many tree types on such diverse topographies along such a steep rain gradient. The experiment continues. Attempts to reintroduce cedar trees proved to be a mighty failure. Forest researchers identified a correlation between soil type and tree mortality, with light rendzina soils proving to be fairly toxic to cedars.[46] Based on the findings, forest lands will need to be better characterized before introducing cedars in the future.

Long before this grand experiment reaches conclusions, animals are already finding Israel's "ecologically imperfect" forests a safe haven. As urban centers sprawl outward, they simply have nowhere else to hide. A survey of the largely conifer forests around the area of Modi'in reveals a rich variety of wildlife thriving in the woodlands. The census included gazelles, jackals, foxes, mongoose, hyena, wild cats, badgers, porcupines, hedgehogs, and hares.[47] Hawks (*Accipiter nisus*), which had disappeared entirely from the Israeli landscape, suddenly began to appear in Israel's planted forests, with dozens now nesting in the pines.[48] Other raptor species, apparently refugees from lost habitat in the Judean Hills, appear happy to share the branches of young woodlands planted in the Judean plains.[49] Wild boars can damage young saplings, but generally have integrated into the growing zoological community that live in Israel's forests. Just as it is possible to create improved conditions for wild bee foraging, forests can offer hospitable homes for these animals. Creating the open spaces in forests to meet different species' particular proclivities is important. As more people fill the land, ecological corridors narrow as do habitat alternatives.

Adaptive management makes sense with all natural resources. It requires that lessons learned be applied in real time. This requires that high-quality information be available and that an ongoing interchange exists between forestry experts and field workers. Until 2012 no formal scientific department existed in the JNF bureaucracy to oversee this experiment. With the right organizational culture, the absence of a single, in-house "scientific authority" might be considered a benefit. An entrenched scientific bureaucracy can lead to conceptual ossification, and the present system allows for outsourcing of research to a broad range of scientists with disparate disciplinary perspectives. But this requires a highly *professional* and fully-staffed forestry department, with the ability to integrate findings into updated practices. The present managerial framework of JNF, with its semi-autonomous regions, creates room for diverse approaches to management. Yet, it also undermines the authority of the JNF Forestry Branch and the ability of its professional staff to ensure the application of new information across the country. The 2013 appointment of veteran forester Omri Bonneh as the organization's first "chief scientist" should help ensure the scientific and ecological integrity of forestry practices for the foreseeable future.

Foresters must never lose sight of the fact that science is critical for long-term innovation and improved management. Budgets must include ample support for forestry research, far beyond the annual allocations of recent years, which never exceeded one million dollars. Without meaningful funding, there will be no cadre of independent Israeli academicians to continue the country's splendid heritage of forestry research. There also needs to be an openness among managers to new ideas and applying empirical results on the ground. Even if their formal job descriptions do not include research, foresters need to be increasingly well trained, possessing the interdisciplinary literacy required to oversee a complex forest.

The most immediate step in implementing an effective adaptive management strategy for forestry is deciding what the dependent variables in the grand experiment should be so that the lessons learned are absolutely clear. Here Israel needs to sharpen its objectives, which remain too generic and unfocused. The ecosystem services in forests are not well characterized and are poorly integrated into the adaptive management that the *Forestry Bible* advocates. While the centrality of ecosystem services is widely acknowledged among forestry management, they have only a vague sense of what these are, much less whether a given forest is providing them and what should be done to increase them. Quantifiable, site-specific performance indicators need to be established based on the sundry services different stands should provide. It may well be that after twenty years, the rough designations of the national master plan need to be formally or informally upgraded to include transparent ecosystem service objectives. This would be a tedious exercise, but there may be no real shortcuts.

Like it or not, the trees keep growing. As the forests mature, so do the services they provide, which can bring happy surprises. The most recent chapter in the Gilboa saga offers one such discovery: When the JNF first planted pine trees and eucalyptus over significant chunks of the Gilboa iris habitat, it was Azariah Alon's worst nightmare. Fifty years later, something strange is happening in the nearby Gilboa nature reserve. The sanctuary was zoned to capture the richest concentration of flowers and ensure their preservation. Afforestation was banned there. But recent years have not been good for irises in the almost treeless Gilboa reserve, and the flowers have almost completely disappeared. Azariah Alon acknowledges the disturbing trend but can offer no explanation, responding sheepishly: "No one knows why."

Ironically, irises flourish in the contiguous JNF pine forest—so much so that bulbs have been transferred to be reestablished in the depleted nature reserve. In ecological terms, the "sink" has become the "source." It is a phenomenon that has yet to be fully explained. David Brand, head of the JNF Forestry Department, has intuited that the continuation of grazing in the forest has something to do with iris prosperity. Perhaps the age-old equilibrium that evolved between humans, their livestock, and wildflowers was destroyed when the park rangers limited grazing in the reserves.[50] Professor Gidi Ne'eman recalls that a park ranger used to quietly do his own flower pollination by hand without asking for permission. The ranger presumably has moved on and the protected flowers no longer benefit from the intervention. An alternative explanation is the throngs of people who came to admire the flowers and who inadvertently trampled them. Ne'eman speaks well of the relatively open forest in the Gilboa, where light can penetrate and support flourishing understories and annuals.[51] Whatever the reason, this much-maligned forest provides a critical service, serving as a partner in the overall quest for botanical diversity.

Israel's forests are designed to serve people. The fourteen million visits to the country's disparate woodlands each year offer incontrovertible evidence of meaningful, nonmaterial benefits derived by the public. Entry to the forests remains free, one of the few public resources in Israel which has not yet been privatized to one degree or another.[52] This sends a message that human beings are more than welcome to get to know their wooded heritage and that they have a God-given right to enjoy the natural world. In the shade of forests people find the quiet space to think, or celebrate in the picnic grounds. Here people can recreate: pick fruit, ride a bike, plant a tree, hike, hear a concert, barbecue, climb a jungle gym, wander, or just breathe fresh air.

Many people in the planet can no longer enjoy such experiences, but this doesn't mean such opportunities are lost forever. The world needs to be far more resolute in its efforts to repair the planet's ecosystems and replace the natural forests that have paid the price for human progress. Deforestation is one environmental problem that is largely reversible. Ecological conditions differ around the planet and there is no single magic formula for rehabilitation. There should also be no illusions that the new stands that people plant

UN Secretary General Ban Ki-moon's 2007 visit to Israel begins by planting a tree with JNF's international chairman, Efi Stenzler. (Zvi Hayuy, KKL-JNF Photo Archives)

will be replicas of the original woodlands they replace. Especially at first, they may appear sparse and dull. But if thoughtfully planned, these novel ecosystems can work their own magic. Most important, afforestation can happen anywhere.

The Israeli forestry project is encouraging precisely because it is taking place in relatively arid conditions that are so inhospitable to trees. Some 45 percent of the planet earth faces water scarcity, relentless heat, and degraded landscapes. Hundreds of millions of humans live on the these drylands. Progress made in less than a century can offer them encouragement. In 2010, for the first time, the JNF board of directors approved a pilot project in Rwanda. Hopefully, this will lead to a new organizational direction, where successful forestry techniques will be shared with other African countries as they set out on their own road to reforestation. Israel's experience suggests that even countries facing existential challenges can find the resources and political will to reverse two-thousand-year trends and take the path of renewal.

Ecological indicators worldwide are largely discouraging. For people who consider "the big picture" and think beyond isolated "micro"-successes, it is often difficult to find reasons for optimism. So when there is good news, it is important to hail environmental achievements that show that, in fact, there are alternatives. A more sustainable course really does exist. Unlike other countries in the eastern Mediterranean that accepted the loss of their forestry heritage as preordained, Israel chose to change natural history and replant. Initial efforts may have been clumsy and overzealous, but there was sufficient humility to learn from mistakes and seek a more sustainable model.

Israel's climax species: typical oak stand in the Tivon Forest. (Michael Huri, KKL-JNF Photo Archives)

The Psalmist wrote: "Then, all the trees of the forest shall rejoice."[53] After a long absence, and a slow but stubborn start a century ago, many trees in the land of Israel again rejoice as they deliver a host of wonderful services to the inhabitants. The new generation of forest remains a work in progress. There is still much work left to be done. The woodlands have only begun their successional passage to greater complexity, stability, and beauty. Israel's forestry policies and management strategies should be scrutinized, criticized, and improved. But this dispassionate discourse should never forget that the overarching context is one of proactive engagement. Loss of woodlands need not be an ineluctable part of modern civilization, even in drylands. Human beings can choreograph a better reality. If all the trees of the forest once again are rejoicing, surely humans can also celebrate with them in this dance of restoration.

Notes

Chapter 1. Degradation and Restoration

1. Clare Billington et al., *Estimated Original Forest Cover Map: A First Attempt* (Cambridge: WCMC, 1996).

2. The National Geographic website cites the World Resources Institute's claim that more than 80 percent of the earth's natural forests already have been destroyed, see: "Deforestation and Desertification," www.nationalgeographic.com/eye/deforestation/effect.html (last accessed January 11, 2013). Another comprehensive analysis sets this figure at 20 percent: Michael Williams, *Deforesting Earth: From Prehistory to Global Crisis* (Chicago: University of Chicago Press, 2003), 689; also IUCN, "Remaining Original Forest Cover by Basin," in *Earth Trends: Environmental Information* (Washington, DC: World Resources Institute, 2003), 9.

3. Millennium Ecosystem Assessment, "Forests and Woodland Systems," in *Ecosystems and Human Well-Being,* vol. 1, *Current State and Trends* (Washington, DC: Island Press, 2005), 588.

4. Lester Brown, the noted analyst of global environmental trends, writes: "At the beginning of the twentieth century, the earth's forested area was estimated at 5 billion hectares. Since then it has shrunk to just under 4 billion hectares." Lester Brown, *Plan B, 3.0* (Washington, DC: World Resources Institute, 2008), 86.

5. Food and Agricultural Association, *Global Forest Resources Assessment 2010: Main Report* (Rome: United Nations, 2010).

6. Ibid., xiii.

7. Sadanandan Nambiar and Ian Ferguson, eds., *New Forests: Wood Production and Environmental Services* (Melbourne: CSIRO, 2005).

8. Ibid.

9. Ecosystems are typically defined as "the organisms living in a place and their interactions with the physical environment and with one another." Gretchen Daily, personal communication, March 20, 2012.

10. Colin Tudge, *The Tree: A Natural History of What Trees Are, How They Live, and Why They Matter* (New York: Three Rivers Press, 2005), 9–10.

11. Ibid., xvi–xvii.

12. Brian Groombridge and Martin Jenkins, *World Atlas of Biodiversity: Earth's Living Resources in the 21st Century* (Berkeley: University of California Press, 2002).

13. U.S. Forest Service, *State of Knowledge: Ecosystem Services from Forests* (Washington DC: USFS, 2006). For a particularly poetic description of the ecosystem services provided by forests see: Gretchen Daily and Charles Katz, *The Power of Trees* (San Antonio: Trinity Press, 2012).

14. B. Britton et al., "Weak Northern and Strong Tropical Land Carbon Uptake from Vertical Profiles of Atmospheric CO_2," *Science* 316, no. 5832 (2007): 732.

15. Joseph Weitz, *Forests and Afforestation in Israel* (Jerusalem: Masada Press, 1974), 196.

16. Forestry Department, Food and Agriculture Organization of the United Nations, *Global Forest Resources Assessment, 2010, Country Report, Israel* (Rome: FAO, 2010), 6–7, www.fao.org/docrep/013/al536E/al536E.pdf (last accessed January 30, 2013).

17. Earthtrends, "Country Profiles," *Forests, Grasslands, and Drylands: Jordan* (Washington, DC: World Resources Institute, 2003), www.fao.org/forestry/country/57478/en/jor (last accessed February 15, 2013).

18. Food and Agriculture Organization of the United Nations, *Country Profiles* (2010), FAO, Rome, www.fao.org/forestry/country/57478/en/jor (last accessed January 11, 2013).

19. Nili Lipschitz and Gideon Biger, *Green Dress for a Country: Afforestation in Eretz Israel, the First Hundred Years, 1850–1950* (Jerusalem: Ariel, 2004), 23; also Ingrid Schneider and G. Wesley Burnette, "Protected Area Management in Jordan," *Environmental Management* 25, no. 3 (2000): 241–46.

20. Kathrin Wiss, *A Thousand Hills for 9 Million People: Land Reform in Rwanda; Restoration of Feudal Order or Genuine Transformation* (Bern, Switzerland: Swiss Peace Foundation, 2006).

Chapter 2. Israel's Forests

1. Ecological restoration is defined as an "intentional activity that initiates or accelerates the recovery of an ecosystem with respect to its health, integrity and sustainability." Society for Ecological Restoration, SER International Primer on Ecological Restoration, International Science and Policy Working Group (Tucson: SER, 2004), www.ser.org/resources/resources-detail-view/ser-international-primer-on-ecological-restoration (last accessed February 15, 2013). See, generally, Alon Tal, "Restoration of Desertified Ecosystems," in *Encyclopedia of Soil Science,* ed. Rattan Lal (Oxford: Taylor and Francis, 2010).

2. R. Shimelmitz, R. Barkai, and A. Gopher, "Systematic Blade Production at Late Lower Paleolithic (400–200 kyr) Qesem Cave, Israel," *Journal of Human Evolution* 61, no. 4 (2011): 458–79. Also, William Lee Adams, "World's Oldest Human Remains Found in Israel," *Time Magazine,* December 27, 2010.

3. Guy Rotem, Amos Bouskila, and Alon Rotshchild, *Afforestation in the Northern Negev and the South Hevron Hills: Ecological Implications* (Tel Aviv: SPNI, 2012).

4. Jewish National Fund, "Functional Restoration of Desert Ecological Systems in the Northern Negev" (Eshtaol: JNF, 2013) (draft copy with author).

5. Ze'ev Naveh and J. Dan, "The Human Degradation of Mediterranean Landscapes in Israel," in *Mediterranean Type Ecosystems: Origin and Structure,* ed. Francesco di Castri and Hal Mooney (Berlin: Springer, 1973), 373–90.

6. No'am Seligman, "The Environmental Legacy of the Fellaheen and the Bedouin in Palestine," in *Between Ruin to Restoration: An Environmental History of Israel,* ed. Char Miller, Daniel Orenstein, and Alon Tal (Pittsburgh: University of Pittsburgh Press, 2013), 34–35.

7. Avi Perevolotsky and No'am G. Seligman, "Role of Grazing in Mediterranean Rangeland Ecosystems: Inversions of a Paradigm," *BioScience* 48, no. 12 (1998): 1007–17.

8. It is particularly ironic that the *sabra* cactus, which became synonymous with native-born, Jewish Israelis who were "prickly on the outside and sweet on the inside" was later adopted as a symbol of indigenousness by Palestinian nationalism. In fact, the Sabra is a relative newcomer to Israel—with very little historical seniority. A North American subspecies of the prickly pear cactus, Sabras were brought out of Mexico by the Spanish conquistadors to Spain, where they eventually arrived at sundry lands, including Ottoman Palestine. The precise date that Sabras arrived in Israel is unknown, although there are many who claim it was as late as the nineteenth century!

9. Nasser Abufarha, "Land of Symbols, Cactus, Poppies, Orange, and Olive Trees in Palestine," *Identities* 15, no. 3 (2008): 343–68.

10. Irus Braverman, "Planting the Promised Landscape: Zionism, Nature, and Resistance in Israel/Palestine," *Natural Resources Journal* 49 (2009): 341.

11. In Palestine Remembered, a partisan website, there is an entire section, "Quoting Mark Twain out of Context," which attempts to discredit the accuracy of the famous travel log of the American author: www.palestineremembered.com/Acre/Articles/Story845.html (last accessed January 11, 2013).

12. Yoav Sagi, interview by author, Tel Aviv, August 11, 2011.

13. Anthony C. Woodbury, "Counting Eskimo Words for Snow: A Citizen's Guide" (July 1991), www.princeton.edu/~browning/snow.html (last accessed February 15, 2013).

14. *The Jewish Encyclopedia,* s.v. "Forests" (1901), www.jewishencyclopedia.com/articles/6230).

15. In the Septuagint, the ancient Greek translation of the Old Testament, Sharon is translated as "forest."

16. Saraleh Oren, *The Terms for Forest in the Scriptures* (Neot Kedumim, Israel, 2013). Copy with author.

17. These include pine, sandalwood, terebinth, oak, cedar, tamarisk, mulberry, cypress, gopher, styrax, oil tree, willow, plane, juniper, lotus, poplar, acacia, box tree, elm, and ilex.

18. Noga HaReuveni, *Shrub and Plant in the Israel's Heritage* (Jerusalem: Neot Kedumim, 2002), 25 (in Hebrew). All translations are mine unless stated otherwise.

19. Genesis 35:8 describes the final resting place for Deborah, the matriarch Rebecca's nurse, under an oak tree.

20. Midrash HaGadol, "Breishit—Noah," 28.

21. Irving Koller, "Each Tree As It Is: Either Bad or Good" (1952) (in Hebrew), paper copy with author.

22. Jerusalem Talmud, Peah 7:3.

23. Talmud, Baba Bathra 91B.

24. Talmud, Baba Kamma 91B.

25. Pirkei d'Rabbi Eliezer, chap. 34, 121 English translation is available at: www.usc.edu/projects/pre-project (last accessed January 30, 2013).

26. Dr. Elaine Solowey, personal communication, September 8, 2011.

27. These include *teashur, gopher, etz shemen, tirza, tidhar, bacha, sneh, armon, algumim, hovnim, zeelim, aviyonah.* Nili Lipschitz, *Timber in Ancient Israel: Dendroarchaeology and Dendrochronology* (Tel Aviv: Tel Aviv University Press, 2007), 5.

28. HaReuveni, *Shrub and Plant,* 25.

29. Michael Zohary, *Geobotanica* (Tel Aviv, 1959), 342–45, as quoted in note 2 in Nili Lipschitz and Gideon Biger, *The Rise and Fall of the Jerusalem Pine as the Main Tree in the Land of Israel* (Jerusalem: JNF, 1994). Also Alon Tal, *Pollution in a Promised Land* (Berkeley: University of California Press, 2002), 83.

30. Joseph Weitz, *Forests and Afforestation in Israel* (Jerusalem: Masada Press, 1974), 213.

31. Ibid., 156.

32. Ofri Ilani, "2,000-Year-Old Date Seed Grows in the Arava," *Haaretz,* February 15, 2007, www.haaretz.com/print-edition/news/2-000-year-old-date-seed-grows-in-the-arava-1.213054 (last accessed January 30, 2013).

33. H. G. Lund, "Definitions of Forest, Deforestation, Afforestation, and Reforestation," Forest Information Services (2007), available at http://home.comcast.net/~gyde/DEFpaper.htm (last accessed January 30, 2012).

34. Food and Agriculture Organization of the United Nations, "Annex 2: Terms and Definitions Used in FRA 2010," *Global Forest Resources Assessment 2010,* FAO Forestry Paper 163 (Rome: FAO, 2010), 209.

35. "Definition from Report of the Conference of the Parties on Its Seventh Session, Held at Marrakesh from 29 October to 10 November 2001: Addendum. Part two: Action taken by the Conference of the Parties," vol. 1, Annex: Definitions, modalities, rules, and guidelines relating to land use, land-use change, and forestry activities under the Kyoto Protocol. FCCC/CP/2001/13/Add.1, http://unfccc.int/resource/docs/cop7/13a01.pdf (last accessed February 15, 2013).

36. Quoted in Weitz, *Forests and Afforestation,* 23–24.

37. Hanah Margalit, "The Cultural Landscape of the Land of Israel" (Ph.D. diss., Hebrew University of Jerusalem, 1955), as reported in Ruth Kark and Noam Levin, "The Environment in Palestine in the Late Ottoman Period, 1798–1918," in *Between Ruin to Restoration: An Environmental History of Israel,* ed. Char Miller, Daniel Orenstein, and Alon Tal (Pittsburgh: University of Pittsburgh Press, 2013), 1–28.

38. Weitz divides the forests described in the Bible into four geographical categories:

the forests of the Negev, the forests of Ephraim, the Carmel, and the stands of Lebanon. Weitz, *Forests and Afforestation,* 12.

39. Among examples of books dedicated to the trees that appear in the ancient texts of Israel: Lipschitz, *Timber,* lists the four volumes by Low, *Die Flora der Juden* (Vienna, 1926–34); H. N. Moldenke and A. L. Moldenke, *Plants of the Bible* (Waltham, MA, 1952); U. Feldman, *The Plants of the Mishna* (Tel Aviv, Dvir, 1956); D. A. Anderson, *All the Trees and Woody Plants of the Bible* (Waco, Texas, World Books, 1979); J. Felix, *Fruit Trees in the Bible and Talmudic Literature* (Jerusalem: Rubin Mass, 1994); and J. Felix, *Trees: Aromatic, Ornamental and of the Forest in the Bible and Rabbinic Literature,* (Jerusalem: Rubin Mass, 1997).

40. Michael Zohary, *Plant Life of Palestine* (New York: Ronald Press, 1962), 68.

41. Aharon Horowitz, "Human Settlement Pattern in Israel: A Discussion of the Impact of Environment," *Expedition* (Summer 1978): 55–57.

42. Nili Lipschitz and Gideon Biger, "Ancient Dominance of the *Quercus calliprinos–Pistacia palaestina* Association in Mediterranean Israel," *Journal of Vegetation Science* 1, no. 1 (1990): 67–70.

43. Lipschitz, *Timber,* 167.

44. Avi Perevolotsky, "Forestry in Israel: The Ecological Alternative," *Forest* 12 (2011): 69 (in Hebrew).

45. The Talmud for example expands the prohibition to breaking vessels, tearing garments, destroying buildings, clogging wells, and wasting food. (Kiddushin 32a). See generally, Eilon Schwartz, "Bal Taschit: A Jewish Environmental Precept," *Environmental Ethics* 19, no. 4 (1997): 355–74.

46. Flavius Josephus, *The War of the Jews,* bk. 6, chap. 1, in *The Complete Works of Josephus,* trans. William Whiston (Grand Rapids, MI: Kregel Publications, 1981), 570–71.

47. Ibid., 557.

48. Thomas Wright, ed., *Early Travels in Palestine, Comprising the Narratives of Arculf, Willibald, Bernard, Sigurd, Bejamin of Tudella, Sir John Maudeviolle, De La rocquiere and Manundrell* (London: Henry G. Bond, 1878), 7.

49. Ibid., 9.

50. Hans Prutz, *Kulturgeschichte der Kreuzzuge* (Berlin, 1883), as quoted in Weitz, *Forests and Afforestation,* 21.

51. Weitz, *Forests and Afforestation,* 320.

52. *Jewish Encyclopedia,* s.v. "Moshe ben Nachman" (1901).

53. Bernhard von Breydenbach, *Pilgrims and Pilgrimages,* ed. Hugh W. Davies (London: J. and J. Leighton, 1911), available at Internet Archive, http://archive.org/details/bernhardvonbreydoodavi (last accessed January 30, 2013).

54. Helen Gertrude Preston, "Rural Conditions in the Kingdom of Jerusalem during the Twelfth and Thirteenth Centuries" (Ph.D. diss., University of Pennsylvania, 1903), http://archive.org/details/ruralconditionsioopresrich (last accessed January 30, 2013).

55. C. F. Volney, *The Ruins of Empire; or, Meditation on the Revolutions of Empires* (L'Academie Française, Et De Plusieurs Autres Sociétés Savantas, 1793). The English version (New York: Truth Seeker, 1896) is reprinted at Project Gutenberg, http://www.gutenberg.org/files/1397/1397-h/1397-h.htm (accessed February 14, 2013).

56. Even this forest would soon be harvested by the Egyptian local ruler Ibrahim Pasha who supplied his father, Mohammed Ali the timber he needed for boats and heating. Lipschitz and Biger, *Green Dress.*

57. Tal, *Pollution,* 36–39.

58. Roza I. M. El-Eini, "British Forestry Policy in Mandate Palestine, 1929–1948," *Middle Eastern Studies* 35, no. 3 (1999): 72–155; 77.

59. Ibid., 77.

60. J. V. Thirgood, *Man and the Mediterranean Forest: A History of Resource Depletion* (London: Academic Press, 1981), 115.

61. Seligman, "Environmental Legacy," 22.

62. David Schorr, "Forest Law in the Palestine Mandate: Colonial Conservation in a Unique Context," in *Managing the Unknown,* ed. Uwe Luebken and Frank Uekoetter (Brooklyn, NY: Berghahn Books, forthcoming).

63. Mark Twain, *The Innocents Abroad* (New York: Oxford University Press, 1996), 555.

64. George Perkins. Marsh, *Man and Nature; or, Physical Geography as Modified by Human Action* (Seattle: University of Washington Press, 1965), 369.

65. Seligman, "Environmental Legacy," 30–31.

66. Nili Lipschitz and Gideon Biger, *Green Dress for a Country: Afforestation in Eretz Israel, the First Hundred Years, 1850–1950* (Jerusalem: Ariel, 2004), 42–43.

67. Weitz, *Forests and Afforestation,* 36–37.

68. Lipschitz and Biger, *Green Dress,* 44.

69. Ibid., 47–48.

70. Aaron Aaronsohn, *On the Deforestation of Eretz Israel and a Suggestion for Its Renovation,* submitted according to invitation by His Majesty Backri Bay, the General Governor of Beirtu, on occasion of his visit to the agricultural research station in October 1913 (Zichron Yaakov, Beit Aaronson Public, 1913; in Hebrew), as quoted in Lipschitz and Biger, *Green Dress,* 54–55.

71. Claude Reignier Conder and Horatio Kitchener, *The Survey of Western Palestine* (London: Committee of the Palestine Exploration Fund, 1883).

72. Anderlind, *Einfluss der Gebirgswaldungen im Nördlichen Palästina* (Z.D.P.V., 1885), as quoted in the *Jewish Encyclopedia.*

73. Claude Reignier Conder, *Tent Work in Palestine,* vol. 2 (1879; repr., Charleston: Bibliolife South Carolina, 2004), 320–23.

74. E. R. Sawer, *A Review of the Agricultural Situation in Palestine* (Jerusalem: Department of Agriculture and Fisheries, 1923), 19, as reported in El-Eini, "British Forestry Policy," 138.

75. Ilan Pappe, *A History of Modern Palestine,* 2nd ed. (Cambridge: Cambridge University Press, 2006), 64.

76. Yosef Weitz, *HaYa'ar V'haYiur B'Yisrael* (Ramat Gan: Masada, 970), 95–96 (in Hebrew).

77. Ruth Kark and Noam Levin, "The Environment in Palestine in the Late Ottoman Period, 1798–1918," in *Between Ruin and Restoration: An Environmental History of Israel,* ed. Char Miller, Daniel Orenstein, and Alon Tal (Pittsburgh: University of Pittsburgh Press, 2013), 1–28.

78. Walter Clay Lowdermilk, *Palestine: Land of Promise* (New York: Harper and Brothers, 1944). See also: Adolph Reifenberg, *The Struggle between the Desert and the Sewn* (Jerusalem: Hebrew University Press, 1955).

79. Jared Diamond, *Collapse* (New York: Penguin, 2006) 118.

80. Weitz, *Forests and Afforestation,* 196.

81. The U.S. Library of Congress, for example, has a marvelous series of some three hundred photographs taken between 1917 and 1926, by photographers John David Whiting, Lewis Larsson, and G. Eric Matson, available at World War I and the British Mandate in Palestine, www.loc.gov/pictures/item/2007675259 (last accessed January 30, 2013).

Chapter 3. A Mandate for Trees

1. Herbert Samuel, "An Interim Report on the Civil Administration of Palestine, during the period 1st July, 1920–30th June, 1921," as printed in *The United Nations Information System on the Question of Palestine* (UNISPAL), http://unispal.un.org (last accessed February 3, 2013).

2. Covenant of the League of Nations, Article 22, Creation of Mandates, June 28, 1919.

3. General Allenby, for example, warned that the appointment would lead to violence and British ally Arabian King Faisal considered it a provocation. See Howard Sachar, *History of Israel: From the Rise of Zionism to Our Time* (New York: Knopf, 2007), 124.

4. Samuel, "Interim Report."

5. Ordinance to Provide for the Regulation of Forest Lands and the Protection of Trees (1920), reprinted in *Land Legislation in Mandate Palestine* (Cambridge: Cambridge University Press, 2009), 4: 71–80. Sec. 1 offers a shorter title: "Woods and Forest Ordinance."

6. Ibid., 72.

7. Samuel, "Interim Report."

8. Irus Braverman, *Planted Flags: Trees, Land, and Law in Israel/Palestine* (Cambridge: Cambridge University Press, 2009), 32–33.

9. "Even these modest numbers dwarf . . .": Joseph Weitz, *Forests and Afforestation in Israel* (Jerusalem: Masada Press, 1974), 194. Lipschitz and Biger list annual plantings indicating that prestate JNF afforestation that cumulatively sum up to 23,390 dunams. Nili Lipschitz and Gideon Biger, *Green Dress for a Country: Afforestation in Eretz Israel, the First Hundred Years, 1850–1950* (Jerusalem: Ariel, 2004). The disparity between their figure and the official JNF number may be due to Weitz's not counting trees planted in the West Bank which was not in pre-1967 Israel or stands that were subsequently destroyed, before and during the War of Independence. It is highly unlikely that Weitz, a consummate stickler for details, was imprecise! "Some 6,000 hectares . . .": Lipschitz and Biger, *Green Dress,* 162.

10. Yosef Weitz, *HaYa'ar V'haYiur B'Yisrael* (Ramat Gan: Masada, 1970), 302 (in Hebrew).

11. See the Palestine Government, *Annual Report of the Forestry Service* (Jerusalem: Palestine Government Printing Press, 1931–36), and the *Annual Report of the Depart-*

ment of Forests (Jerusalem: Palestine Government Printing Press, 1936–39, 1939–45, 1946–47).

12. Nili Lipshitz and Gideon Biger, "The Forestry Policy of the British Government in the Land of Israel," *Horizons in Geography* 40–41 (1994): 5–16 (in Hebrew); Roza I. M. El-Eini's 1999 work is particularly comprehensive and scholarly: see: Roza I. M. El-Eini's "British Forestry Policy in Mandate Palestine, 1929–1948: Aims and Reality," *Middle Eastern Studies* 35, no. 3 (1999): 72–155, as well as: her book on the same topic: Roza I. M. El-Eini *Mandated Landscape: British Imperial Rule in Palestine, 1929–1948* (New York: Routledge, 2006). Also David Schorr, "Forest Law in the Palestine Mandate: Colonial Conservation in a Unique Context," in *Managing the Unknown,* ed. Uwe Luebken and Frank Uekoetter (Brooklyn, NY: Berghahn Books, forthcoming, 2013).

13. This is one of the central themes in David Schorr's piece "Forest Law in the Palestine Mandate": "From the point of view of both the British and the Zionists, the effects of deforestation were pernicious, not only because of the alien landscape, the paucity of forest products, the unproductive soil, and the sand dunes and swamps which threatened to overwhelm arable lands. The harsh climate made settlement of Europeans a dubious proposition. . . . A newly green and lush Palestine would even, many believed, lead to a change for the better in the climate, making at a cooler, wetter, more temperate land—something like Northern Europe." Schorr, "Forest Law in the Palestine Mandate."

14. Indeed in the nineteenth century, the British priest-naturalist Tristram wrote in his *Natural History of the Bible* that if the land became more wooded: "many light clouds which now pass over from the west would then be attracted and precipitated in rains over the highlands." H. B. Tristram, "Meteorology of Palestine," *Natural History of the Bible: A Review of the Physical Geography, Geology, and Meteorology of the Holy Land* (London: Society for Promoting Christian Knowledge, 1883), 32.

15. In 1946 this was reframed as: "a) To re-afforest the hilly and waste non-agricultural land in order to conserve water supplies, to prevent soil erosion and denudation and to afford shelter and protection to adjoining crops and orchards; b) to curtail the encroachment of sand dunes; and c) to bring into economic use land not suited to agriculture and horticulture by the production of timber, fuel and other forest produce."

16. Palestine Government, *Annual Report* (1947), 1.

17. Gilbert Sale, "Forestry and Soil Conservation in Palestine," summary, November 28, 1944, as quoted by El-Eini, "British Forestry Policy," note 97.

18. Walter Clay Lowdermilk, *Palestine: Land of Promise* (New York: Harper and Brothers, 1944), 5.

19. Harry Charles Luke and Edward' Keith-Roaca, *Handbook of Palestine* (London: MacMillan, 1922), 186.

20. Ken Stein, "Palestine's Rural Economy, 1917–1939," *Studies in Zionism* 8, no. 1 (1987): 25–49. Stein reports that in 1920 a hectare of wheat in Palestine produced an average of 593 kg while the same area in Egypt yielded 1,793 kg. See generally, Baruch Kimmerling and Joel Migdal, *Palestinians: The Making of a People* (New York: Free Press, 1993), 23.

21. Noam Levin, "The Palestine Exploration Fund Map (1871–1877) of the Holy Land as a Tool for Analyzing Landscape Changes: The Coastal Dunes of Israel as a Case Study," *Cartographic Journal* 43 (2006): 45–67.

22. N. Tiberski, *Rishon L'Zion Is Seventy,* pamphlet (1952), 97–98, as quoted in Weitz, *HaYa'ar V'haYiur,* 94.

23. Palestine Government, *A Survey of Palestine: Prepared in December 1945 and January 1946 for the Information of the Anglo-American Committee of Inquiry* (Washington, DC, Institute of Palestine Studies, 1946), 426–27.

24. El-Eini, "British Forestry Policy," 115.

25. H. Taylor, *Save Our Soil: A Booklet Explaining the Dangers of Soil Erosion Which Threaten the Prosperity of Palestine and the Remedies Which Can Cure It* (Government of Palestine, Soil Conservation Board, 1940), as quoted in Schorr, "Forest Law in the Palestine Mandate."

26. Communication of District Engineer of the Lydia District to Palestine Attorney General, February 20, 1935, as quoted in El-Eini, "British Forestry Policy," note 494.

27. Pua Bar (Kutiel), Oded Cohen, and Maxim Shoshany, "Invasion Rate of the Alien Species *Acacia saligna* within Coastal Sand Dune Habitats in Israel," *Israel Journal of Plant Sciences* 52 (2004): 115–24.

28. See table in R. S. Troup, *Colonial Forest Administration* (Oxford: Oxford University Press, 1940), 120.

29. Palestine Government, *Annual Report* (1946), 1.

30. Lipschitz and Biger, *Green Dress,* 172.

31. Gershon Avni, interview with author, Ben Shemen Forest, August 7, 2011.

32. Ibid.

33. Amihud Grasovksy, *Some Aspects of Light in the Forest* (New Haven, CT: Yale University Press, 1929).

34. Yerahmiel Kaplan, interview with author, Rehovoth, May 5, 1999.

35. El-Eini, "British Forestry Policy," 106.

36. Dan Goor, personal communication, September 16, 2011.

37. Avni, interview.

38. Amihud Goor, *Tree Planting Practices for Arid Areas* (Rome: FAO, 1955), 78.

39. Ibid., 1–2.

40. These include: *Quercus coccifera, Q. aegilops var. ithaburensis, Pistachia terebinthus,* and *P. Palestina, Ceratonia siliqua, Phyllera media, Laurus noblis, Arbutus Andrachne.* Palestine Government, *Annual Report of the Department of Forests,* 9.

41. Palestine Government, Department of Forests, "Lists of Trees and Shrubs Native to Palestine, Appendix II," *Annual Report* (1947), 7–8.

42. Ibid.

43. Gideon Biger and Nili Lipshitz, "Attempts to Acclimate Exotic Trees in the Land of Israel during the Mandate Period," *Ketadra* 85 (1997): 123–64 (in Hebrew).

44. Palestine Government, *Annual Report* (1935), 10.

45. Ibid. (1936), 16–23.

46. Haim Blass, personal interview with author, Herzliyah Pituach, September 3, 1997.

47. J. F. Tear, 1931, "Note on the Need for an Expanded Program of Afforestation in Palestine," as reported in Lipschitz and Biger, *Green Dress,* 177.

48. Tom Segev, *One Palestine, Complete: Jews and Arabs under the British Mandate* (New York: Metropolitan, 1999) 171.

49. Palestine Government, *Annual Report* (1936–39), 16.

50. Palestine, *Report on Immigration, Land Settlement and Development, Sir John Hope Simpson, Command Paper No. 3686, London, His Majesty's Stationery Office* (AKA: Hope-Simpson Report, 1930), http://unispal.un.org/UNISPAL.NSF/0/E3ED8720F8707C9385256D19004F057C (last accessed February 3, 2013), chap. 7, "Agricultural Development."

51. A table comparing the key resources and areas of the two forest services can be found in the detailed 1933 memo by Dawe, to the chief secretary, as quoted in El-Eini, "British Forestry Policy," note 83.

52. Palestine Government, *Annual Report* (1934), 91.

53. "Explanatory Notes to the Woods and Forest Ordinance (1920), sec. 7 as quoted in Schorr, "Forest in the Palestine Mandate."

54. Sec. 4, Ordinance to Provide for the Regulation of Forest Lands and the Protection of Trees, 4: 72.

55. Sec. 3, Definition of a Forest Reserve, ibid.

56. "Forestry and Conservation," in Palestine Government, *A Survey of Palestine*, 425–27.

57. Sec. 15, Ordinance to Provide for the Regulation of Forest Lands and the Protection of Trees, 4: Laws of Palestine, vol. 1, cap. 61, p. 710.74.

58. Ibid., sec. 14, 74. A similar version of this provision was later copied in the 1926 version of the law in section 5.

59. Ibid., sec. 16, 75.

60. Palestine Government, *Annual Report* (1931 and 1932), 7.

61. Ibid. (1931 and 1936–39), 14.

62. The 1934 report complains, "During the year, the Forest Surveyors have carried out a considerable amount of work for other branches of the Department, involving 72 days of their time and dislocation of their forestry survey programme. This is regrettable in view of the fact that it will take some years to complete the programme of forest survey and demarcation required in connection with reservation, apart from current requirements for the survey of plantations and coupes." Palestine Government, *Annual Report* (1934), 91.

63. Palestine Government, *Annual Report* (1931, 1932), 6.

64. Palestine Government, "Forestry and Conservation," in *A Survey of Palestine*, 426.

65. Ibid.

66. Schorr, "Forest Law in the Palestine."

67. Ibid.

68. R. S. Troup, *Colonial Forest Administration* (Oxford: Oxford University Press, 1940), 224, 229, as quoted by Schorr, "Forest Law in the Palestine."

69. Palestine Government, *Forest Ordinance* (1926). An ordinance to repeal the Woods and Forests Ordinance (1920), and to Make Other Provision in Lieu hereof, reprinted in: *Land Legislation in Mandate Palestine*, 216–25.

70. Ibid., sec. 13, 221.

71. Palestine Government, *A Survey of Palestine*, 426.

72. League of Nations, Permanent Mandates Commission, 1929, "Minutes of the Fifteenth Session including the Report of the Commission to the Council and Comments

by Various Accredited Representatives of the Mandatory Powers," *Palestine* (Geneva, July 1–19, 1929).

73. Palestine Government, *Annual Report* (1947), 4.

74. Ibid. (1931, 1932, 1934, 1935).

75. Zvi Mendel, interview with author, September 27, 2011.

76. To manage closed reserves effectively, "one woodman" was needed for each fifty hectares of closed forest, a level of staffing that proved unrealistic given budgetary constraints. Palestine Government, *A Survey of Palestine,* 427.

77. United Kingdom Government, 1937, *Report of the Palestine Royal Commission* (Peel Commission), presented by Earl Peel, the Secretary of State for the Colonies to the United Kingdom Parliament by Command of His Britannic Majesty (July 1937), available at Jewish Virtual Library, www.jewishvirtuallibrary.org/jsource/History/peel1.html (last accessed February 3, 2013).

78. Kimmerling and Migdal, *Palestinians,* 33.

79. Census of Palestine (Jerusalem: Government Printing Office, 1931), 1: 398.

80. Palestine Government, *Annual Report* (1936–39), 11.

81. *The Palestine Gazette, 1918–1919,* as quoted in Lipschitz and Biger, *Green Dress.*

82. Sec. 17, Ordinance to Provide for the Regulation of Forest Lands and the Protection of Trees, *Laws of Palestine,* vol. 1, cap. 61, 710.

83. F. J. Tear, 1931, "Note on the Need for an Expanded Program of Afforestation in Palestine," as reported in Lipschitz and Biger, *Green Dress,* 177.

84. Palestine Government, *Annual Report* (1934), 91.

85. John Hough, "Obstacles to Effective Management of Conflicts between National Parks and Surrounding Human Communities in Developing Countries," in *Environmental Conservation,* ed. Karl Zimmerer and Kenneth Young (Madison: University of Wisconsin Press, 1988), 15: 129–36. For an interesting case study in the Philippines, see Howie G. Severino, *Opposition and Resistance to Forest Protection Initiatives in the Philippines,* United Nations Research Institute for Social Development, Discussion Paper 92 (Geneva: UNRISD, 1998).

86. Palestine Government, *Annual Report* (1936–39), 5.

87. "Opposition on the part of the villagers has, however hardened and claims have multiplied to such an extent that into appears that the Forest Department should suspend all further attempts to constitute new reserves until state ownership has been determined." Report by His Majesty's Government in the United Kingdom of Great Britain and Northern Ireland to the Council of the League of Nations on the Administration of Palestine and Trans-Jordan for the year 1936. Ibid., 5.

88. Palestine Government, *Annual Report* (1936–39), 98–99.

89. Lipschitz and Biger, "Forestry Policy," 5–16.

90. Lipschitz and Biger, *Green Dress,* 175.

91. Yerahmiel Kaplan, interview with author, October 19, 2011.

92. Palestine Government, "Report On Palestine Administration" (1922), http://unispal.un.org/UNISPAL.NSF/0/A682CABF739FEBAA052565E8006D907C (last visited February 3, 2013).

93. League of Nations, *Report of His Majesty's Government in the United Kingdom of Great Britain and Northern Ireland to the Council of the League of Nations on the*

Administration of Palestine and Trans Jordan, December 31, 1936, http://unispal.un.org/UNISPAL.NSF/0/FD4D250AF882632B052565D2005012C3 (last accessed February 3, 2013).

94. Yerahmiel Kaplan, interview with author, October 1, 2011.

95. "It (pine processionary moth) was first discovered in the winter of 1936–37 in a pine wood at Um-Safa (west of Ramallah) which was then six years old. Mr. Sale, the chief forestry inspector, was called out to the site and was later followed by a party of entomologists headed by the late Professor Bodenheimer. The entomologists found numerous nests of the insect in the area and recommended that they be removed and burned. These instructions were apparently not carried out very strictly." Report of Joseph Halperin, quoted by Weitz, *Forests,* 402.

96. United Kingdom Government (1937), *Report of the Palestine Royal Commission* (Peel Commission) presented by the Secretary of State for the Colonies to the United Kingdom Parliament by Command of His Britannic Majesty (July 1937).

97. Howard Sachar, *History of Israel: From the Rise of Zionism to Our Time* (New York: Knopf, 2007), 199.

98. The letters home of Palestine policeman Sydney Burr provide an explicit personal account of police brutality: "It is the only way with these people." Extrajudicial executions, torture, beatings, and general violence were commonplace among the British Palestine police officers with whom Burr worked during the Arab revolt. Burr discusses the "third degree" dished out to Arab suspects along with general beatings and trashing of Arab shops and houses in almost every letter home. Much of the brutality was casual and wantonly destructive, described by the police and soldiers in terms akin to a good, fair fight—rebel "hunting is still the great sport"—enjoyed by all concerned. See Mathew Hughs, "The Banality of Brutality: British Armed Forces and the Repression of the Arab Revolt in Palestine, 1936–39," *English Historical Review* 74 (2009): 313–54.

99. Palestine Government, *Annual Report* (1936–39), 11.

100. Ibid., 15.

101. Ibid., 10.

102. Ibid., 24.

103. League of Nations (1936), sec. 30.

104. El-Eini, "British Forestry Policy," 91.

105. Ibid.

106. United Kingdom Government (Peel Commission).

107. Palestine Government, *Survey of Palestine,* 433.

108. El-Eini, "British Forestry Policy," 103–4.

109. Palestine Government, *Annual Report* (1939–45), 14.

110. Ibid., 5.

111. Ibid., 7.

112. Palestine Government, *Survey of Palestine,* 423–34.

113. Ibid., 426–27.

114. Palestine Government, *Survey of Palestine.*

115. Ibid., 433.

116. Yoav Sagi, interview with author, Tel Aviv, August 11, 2011.

117. Kaplan, interview.

118. Lipschitz and Biger, *Green Dress,* 225.

Chapter 4. Enthusiastic Saplings

1. Joseph Halperin, "Part 2: The Truth Is Not Immediately Revealed, the Matsucoccus Case," in *People, Insects, and Forests: Memoirs of an Entomologist, Parts One and Two* (Nes Tsionah, Israel: Aviv Press, 2008), 9 (in Hebrew).

2.Haim Blass, personal interview with author, Herzliyah Pituach, September 3, 1997.

3. Hugh Wilcox, *Report on the Die-off of Pinus halepensis in the Shaar Hagai Forest* (Syracuse: State University of New York, College of Environmental Science and Forestry, 1975).

4. "An Insect Killed the Pine Trees in the Shar Hagai Forests: The Conclusion of the American Scientist Professor Wilcox," *Maariv,* April 17, 1975 (in Hebrew).

5. Ze'ev Naveh et al., "Photochemical Air-Pollutants: A Threat to Mediterranean Coniferous Forest and Upland Ecosystems," *Environmental Conservation* 7, no. 4 (1980): 301–9.

6. Zvi Mendel, interview with author, September 27, 2011.

7. Joseph Halperin worked as a government, research entomologist with a focus in forestry, for more than forty years, beginning in the late 1950s. In his memoirs, he remains quite bitter about what he believes was a fallacious assessment of the Shaar Hagai tree collapse pathology. Halperin argued that in this particular case, it was the crowding of the trees that was responsible more than any infestation per se. The pest, in his view, was not a deadly problem when its dimensions were controlled by natural predators leaving overall density low, as he felt the case to be in Shar Hagai. However, spraying pesticides in cotton or even in the forest upset the natural balance and often created conditions where the scales could flourish without natural enemies. See Halperin, *People, Insects, and Forests,* 12–17.

8. Yossi Riov, personal communication, December 27, 2011; also Mendel, interview.

9. Raymond Young and Ronald Giese, *Introduction to Forest Ecosystem Science and Management,* 3rd ed. (Hoboken, NJ: Wiley and Sons, 2003), 170.

10. Grant Sharpe, John Hendee, and Wenonah Sharpe, *Introduction to Forests and Renewable Resources,* 7th ed. (New York: McGraw-Hill, 2003), 128.

11. H. Jactel et al., "Habitat Diversity in Forest Plantations Reduces Infestations of the Pine Stem Borer *Dioryctria sylvestrell,*" *Journal of Applied Ecology* 39, no. 4 (2002): 618–28.

12. I. A. S. Gibson and T. M. Jones, "Monoculture as the Origin of Major Pests and Diseases," *Origins of Pest, Parasite, Disease, and Weed Problems: 8th Symposium of the British Ecological Society, Bangor, April 12–14, 1976,* ed. J. M. Cherrett and G. R. Sagar (Oxford: Blackwell Scientific, 1977), 139–61.

13. Nili Lipschitz and Gideon Biger, *The Rise and Fall of the Jerusalem Pine as the Main Tree in the Land of Israel* (Jerusalem: JNF, 1994), 17 (in Hebrew).

14. Joseph Halperin, *Forest-Entomological Research in Israel* (Bet Dagan, Israel The Volcani Institute of Agricultural Research, 1969), 329–30 (in Hebrew).

15. Zvi Mendel, "Biogeography of *Matsucoccus josephi* (*Homoptera: Matsucoccidae*) as Related to Host Resistance in *Pinus brutia* and *Pinus halepensis*," *Canadian Journal of Forest Research* 28 (1998): 323–30.

16. The male's forewings are also twice as long as its body as are the fine waxen filaments that protrude off the body. Fritz Simon Bodenheimer and Soloman Neumark, *The Israel Pine Matsucoccus ("Matsucoccus josephi nov. spec.")* (Jerusalem: Kiryat Sefer, 1955).

17. Zvi Mendel and Gabriel Schiller, "Biogeography of *Matsucoccus josephi Bodenheimer et Harpaz* in Crete and Mainland Greece," *Annals of Science Forestry* 50 (1993): 383–88.

18. Ibid.

19. Z. Madar, Z. Solel, and M. Kimchi, "Enhancement of Sphaeropsis Canker of Aleppo Pine by the Israeli Pine Bast Scale," *Phytoparasitica* 3, no. 1 (2005): 28–32.

20. Oppenheimer report, October 30, 1939, as quoted in Lipschitz and Biger, *Rise and Fall,* 19.

21. Israel Gindel, letter to Yosef Weitz (1938), Central Zionist Archives document JNF 10/10524, as quoted in Nili Lipschitz and Gideon Biger, *Green Dress for a Country: Afforestation in Eretz Israel, the First Hundred Years, 1850–1950* (Jerusalem: Ariel, 2004), 80–82.

22. Paul Ginsburg, personal communication, October 9, 2011.

23. Zvi Mendel, "Major Pests of Man-Made Forests in Israel: Origin, Biology, Damage and Control," *Phytoparasiticia* 15, no. 2 (1987): 683.

24. Halperin, *People, Insects, and Forests,* 16.

25. Zvi Mendel, personal communication, October 9, 2011.

26. Joseph Weitz, *Forests and Afforestation in Israel* (Jerusalem: Masada Press, 1974), 402–3 (in Hebrew).

27. Halperin, *People, Insects, and Forests,* 17.

28. Mendel, "Major Pests," 685.

29. Halperin, *People, Insects, and Forests,* 24.

30. Ibid., 18.

31. Howard Sachar, *History of Israel: From the Rise of Zionism to Our Time* (New York: Knopf, 2007), 353.

32. Quotations in the preceding section are from Yosef Weitz, *My Diary and Letters to the Children* (Ramat Gan: Masada, 1965), 4: 47–48 (in Hebrew).

33. David Ben-Gurion, as cited in *Divrei HaKnesset* (Proceedings of the Knesset), November 7, 1949 (Jerusalem: Proceedings of the Knesset, 1950), 8 (in Hebrew).

34. Alon Tal, *Pollution in a Promised Land* (Berkeley: University of California Press, 2002), chap. 4, 69–112.

35. Lipschitz and Biger, *Green Dress,* 57.

36. Weitz, *Forests and Afforestation,* 48.

37. Lipschitz and Biger, *Green Dress,* 304.

38. Walter Clay Lowdermilk, *Palestine: Land of Promise* (New York: Harper and Brothers, 1944), 101–2.

39. Weitz, *Forests and Afforestation,* 4.

40. Yerahmiel Kaplan, interview with author, October 19, 2011.

41. Dan Goor, personal communication, September 16, 2011. Dr. Goor still has the canceled check.

42. Weitz, *Forests and Afforestation,* 3.

43. Halperin, *People, Insects, and Forests,* 14.

44. Blass, interview.

45. Zvika Avni, "Government Afforestation in the State of Israel: An Epilogue," in *Forest,* 5–6 (2004): 3-11 (in Hebrew). Avni points out an important caveat: several areas that are only counted once in terms of territory may have been replanted two and three times.

46. Israel Central Bureau of Statistics, "Planted Forest Area, *Statistical Abstract of Israel,* (2011), Jerusalem, www.cbs.gov.il/shnaton62/st19_07.pdf (last accessed on February 3, 2013).

47. Mordechai Ruach, "The Forest and Afforestation in Israel Since Independence," *Horizons in Geography* 35–36 (1992): 7 (in Hebrew).

48. Weitz, *Forests and Afforestation,* 203.

49. Mordechai Ruach, interview with author, Beit Zayit, Israel, September 9, 1997.

50. Yosef Weitz, "Afforestation in Israel," *La-Yaaran* 9 (1959): 9–12 (in Hebrew).

51. Avni, "Government Afforestation," 5.

52. Kaplan, interview.

53. Halperin, *People, Insects, and Forests,* 14.

54. Avni, "Government Afforestation," 6–7.

55. Ibid., 7.

56. Ibid., 9.

57. Joseph Weitz, diary entry from May 18, 1958, as quoted in Avni, "Government Afforestation," 7.

58. Moshe Amon, "How I Became a University Professor," *Israel, Globalization, Autobiography, Biographical Stories, and Photographs* (personal blog, 2008), www.mosheamon.com/index.php?option=com_content&task=view&id=36&Itemid=36 (last accessed on December 29, 2012).

59. Jewish Telegraphic Agency, "Israel Ministry Charged with Mismanagement of State Forests," *Daily News Bulletin,* August 17, 1959.

60. Avni, "Government Afforestation," 10.

61. Weitz, diary entry (volume 5) as quoted in ibid., 11.

62. Letter originally printed in Argentinian Jewish press and quoted in ibid., 9.

63. Kaplan, interview.

64. Halperin, *People, Insects and Forests,* 13.

65. Ibid., 14.

66. Kaplan, interview.

67. David Ben-Gurion, as cited in *Divrei HaKnesset,* December 11, 1962 (Jerusalem: Proceedings of the Knesset, 1963), 473–74 (in Hebrew).

68. Lipschitz and Biger, *Green Dress,* 67.

69. Leah Goldberg, "Pines," *Collected Poems,* trans. Rachel Tzvia Back (Tel Aviv:

Iachdav/Writers Association, 1970), reprinted in Poetry International Web, www.poetryinternationalweb.net/pi/site/poem/item/3405 (last accessed on February 3, 2013).

70. Claude Reignier Conder and Horatio Kitchener, *The Survey of Western Palestine* (London: Committee of the Palestine Exploration Fund, 1881).

71. As cited in Zvi Mendel, "Survey of the Development of Forest Research in the Land of Israel during the Twentieth Century," unpublished manuscript, available with author.

72. Weitz, *Forests and Afforestation,* 256–57. Much like the parallel invention of calculus, two minds worked in parallel at almost exactly the same time. The first was one H. G. Spafford, an attorney who lived in Ottoman Jerusalem, whose diary from January 1883 documents the receipt of a package from an Australian friend "Cotton," which contained eucalyptus seeds and a prayer that "a great blessing would accrue to the Holy Land from these seeds." Spafford at the time worked as a teacher at the Alliance school, and he passed the seeds on through the school principle, Nissim Behar to the Mikveh Israel, agricultural school. There they were planted in spring. See Bertha Spafford Vester, *Our Jerusalem: An American Family in the Holy City, 1881–1949* (Garden City, NY: Doubleday, 1950), 100. This was confirmed in a 1952 letter sent from the Tasmanian Commissioner of Forests that confirmed that Cotton had visited Palestine in 1882, was affected by the absence of trees, and sent the package of eucalyptus seeds.

The other claim is that Karl Netter, the founder of the Mikveh Yisrael agricultural school, himself brought the seeds to Israel in the hope that they would produce the wood for orange crates. Subsequently Dr. Karschon resolved the apparent discrepancy by explaining that the Spafford and Cotton had sent blue gum (*Eucalyptus globulus labill*) which are typically unsuccessful in Israel. Indeed the resulting seedlings failed to survive. The "red gum" eucalyptus—which ultimately thrived at Mikveh Yisrael does not even grow in Tasmania and probably came from Algeria, given Netter's close association with French colleagues.

73. Weitz, *Forests and Afforestation,* 286 n. 5.

74. Ibid., 269.

75. Nurit Kliot, "Afforestation for Security Objectives: Spatial Geographical Aspects," *Horizons in Geography* 35–36 (1992): 23–34 (in Hebrew).

76. Israel Central Bureau of Statistics, "Planted Forest Area," *Statistical Abstract of Israel, 2011* (Jerusalem: ICBS, 2011), www.cbs.gov.il/shnaton63/st19_07.pdf (last accessed February 3, 2013).

77. Gwen Kinkead, "At Age 4,600-Plus, Methuselah Pine Tree Begets New Offspring," *New York Times,* June 17, 2003.

78. Weitz, *Forests and Afforestation,* 274–79.

79. Fernando Maestre and Jordi Cortina, "Are *Pinus halepensis* Plantations Useful as a Restoration Tool in Semi-Arid Mediterranean Areas?" *Forest Ecology and Management* 198 (2004): 303–17.

80. Michael Zohary, *Vegetation of Israel and Adjacent Areas* (Weisbaden: Reichert, 1982), 43–44.

81. Nili Lipschitz and Gideon Biger, "Past Distribution of Aleppo Pine (*Pinus halepensis*) in the Mountains of Israel," *Holocene* 11, no. 4 (2001): 427–36.

82. Joseph Weitz, "Forest Trees in the Land of Israel: The Jerusalem Pine" (1936), as quoted in Lipshitz and Biger, *Rise and Fall,* 11.

83. Weitz, *Forests and Afforestation,* 43–45.

84. Kaplan, interview.

85. Israel Gindel, *Forests and Afforestation in Israel* (Tel Aviv: Shoham, 1952), 22 (in Hebrew).

86. Omri Bonneh, "Management of Planted Pine Forests in Israel, Past, Present, and Future," in *Ecology, Biogeography, and Management of "Pinus halepensis" and "R brutia" Forest Ecosystems in the Mediterranean Basin,* ed. Ne'eman Trabaud (Leiden: Backhuys, 2000), 377–90.

87. Halperin, *People, Insects, and Forests,* 13, 18.

88. Bonneh, "Management of Planted Pine."

89. Aviva Rabinovitch, "Affidavit to Israel's Supreme Court," January 4, 2000 (copy with author).

90. The main species formations planted in the hill country were *Pinus halepensis* and oaks; *Acacia cyanophylla, Pinus halepensis,* Stone pine (*Pinus pinea*), *Acacia saligna* (golden wreath wattle), and Zizyphus in Tiberias; as well as *Acacia saligna, Pinus halepensis,* and Stone pine on the coast in the area of the sand dunes.

91. State of Israel, *Annual Report, Department of Forests, 1948–49* (Netanya, Israel: Ministry of Agriculture, 1950) (in Hebrew).

92. Gershon Avni, interview with author, Ben Shemen Forest, August 7, 2011.

93. Gabriel Schiller, personal communication, October 23, 2011.

94. Grant Sharpe, John Hendee, Wenonah Sharpe, *Introduction to Forests and Renewable Resources*, 7th ed. (New York: McGraw Hill, 2003), 128.

95. David Brand, personal communication, October 23, 2011.

96. Mendel, interview.

97. Gabriel Schiller, interview with author, Beit Dagan, Israel, August 7, 2011.

98. Gabriel Schiller, "Inter- and Intra-Specific Genetic Diversity of Pinus Halepensis Mill, and P. Brutia," in *Ecology, Biogeography and Management of Pinus halepensis and R brutia Forest Ecosystems in the Mediterranean Basin,* ed. Ne'eman, Trabaud, (Leiden: Backhuys Publisher, 2000), 13–37. Also: Gabriel Schiller Leonid Korol, Galina Shklar, "Genetic Variation within *Pinus halepensis* Mill: Provenances Growing in Different Microenvironments in Israel," *Israel Journal of Plant Sciences* 50, no. 2 (2002): 135–43.

99. Bonneh, "Management of Planted Pine," 386.

100. Ofer Regev, *Forty Years of Blossoming* (Tel Aviv: Society for the Protection of Nature in Israel, 1993).

101. Gidi Ne'eman, interview with author, Moshav Amirim, August 10, 2011.

102. Chaim Weizmann, *Trial and Error: The Autobiography of Chai Weizmann* (New York: Harper, 1949), 371.

103. Shaul Amir and Orly Rechtman, "The Development of Forest Policy in Israel in the 20th Century: Implications for the Future," *Forest Policy and Economics* 8 (2006): 43, 47.

104. For example, in 1920 we have testimony to the destruction of the Beer Tuviah forest by armed Arab thugs. See Weitz, *Forests and Afforestation,* 132.

105. Shaul Ephraim Cohen, *The Politics of Planting* (Chicago: University of Chicago Press, 1995).

106. Irus Braverman, *Planted Flags: Trees, Land and Law in Israel/Palestine* (Cambridge: Cambridge University Press, 2009).

107. Martin Bunton, "Inventing the Status Quo: Ottoman Land-Law during the Palestine Mandate, 1917–1936," *International History Review* 21, no. 1 (1999): 28–56.

108. A translation of the Ottoman Land Code from 1858 is available: trans. F. Ongley (London: William Clowes and Sons), and can be found online at the Internet £8Archive, www.archive.org/stream/ottomanlandcodeooturkuoft/ottomanlandcodeooturkuoft_djvu.txt (accessed February 14, 2013).

109. Raja Shehadeh, "The Land Law of Palestine: An Analysis of the Definition of State Lands," *Journal of Palestine Studies* 11, no. 2 (1982): 87–88.

110. Lipschitz and Biger, *Green Dress,* 77.

111. Blass, interview.

112. Nur Masalha, ed. *Catastrophe Remembered Palestine, Israel, and the Internal Refugees: Essays in Memory of Edward W. Said* (New York: Zed, 2005).

113. Joan Peters, *From Time Immemorial* (New York: Harper Row, 1984).

114. The official UN projection of refugee numbers from the period is 711,000. United Nations, *General Progress Report and Supplementary Report of the United Nations Conciliation Commission for Palestine, Covering the Period from 11 December 1949 to 23 October 1950, 1950,* UN Conciliation Council, October 23, 1950 (UN General Assembly Official Records, 5th session, supplement n. 18, document A/1367/Rev. 1).

115. Noga Kadman, *The Depopulated Palestinian Villages of 1948 in the Israeli Discourse* (Jerusalem: November Books, 2008).

116. Ariel Hirschfeld, "Erasing Words, Names, and a Way of Life," *Haaretz,* December 19, 2008.

117. Braverman, *Planted Flags,* 99.

118. Tom Segev, *1949: The First Israelis* (New York: Free Press, 1986), 29–30.

119. Weitz, *My Diary,* 358.

120. Benny Morris, "Yosef Weitz and the Transfer Committees, 1948–1949," *1948 and After, Israel and the Palestinians* (Oxford: Clarendon, 1990), 123.

121. To understand Weitz's perspective, it is also crucial to understand the context for Weitz's position on the Transfer Committee. The War of Independence had claimed the lives of 1 percent of Israel's population, including his youngest son. Palestinian Arab leaders had supported the Nazis, rejected the United Nations proposed Partition plan, and indiscriminately attacked Jewish settlements across the land. Not just Weitz, but Jewish leaders from both left- and right-wing parties considered this "fifth column" the paramount security threat to the nascent and vulnerable nation. As the military situation changed, so did Weitz's position. Ra'anan Weitz recalled that after the 1967 war, his father took an unpopular dovish stance against keeping the newly occupied territories. The Gush Etzion region had been conquered from Jordan and with the reestablishment of Jewish settlements there, a big celebration was held. As Yosef Weitz had initiated the original Etzion settlements at the JNF, it was natural to make him an honored guest. He refused to come to the dedication, however, explaining, that he wasn't interested in coming to settle in a place we have no business being. Ra'anan Weitz, interview with author, Jerusalem: January 12, 1998.

122. Avni, interview.

123. Blass, interview.

124. Quoted in Weitz, *Forests and Afforestation*, 249.

125. Blass, interview.

126. Weitz, *Forests and Afforestation*, 371.

127. Weitz lecture entitled "Twenty-five Years of Forming Forestry," given on January 11, 1945, as quoted in Lipschitz and Biger, *Green Dress*, 75–76.

128. Mendel, interview, quoting his teacher René Karschon.

129. David Brand, interview with author, July 27, 2011.

Chapter 5. Sustainable Forestry

1. Haim Zaban, interview with author, Tel Aviv, August 14, 2012.

2. Grant Sharpe, John Hendee, and Wenoah Sharpe, *Introduction to Forests and Renewable Resources*, 7th ed. (New York: McGraw-Hill, 2003), 473–74.

3. Omri Bonneh, interview with author, Qiriyat Haim, Israel, August 10, 2011.

4. Avi Perevolotsky, interview with author, Latrun, Israel, August 3, 2011.

5. Jewish National Fund, Land Development Authority, *Updated Program for Managing and Treating the Planted Forests, January, 1990* (Jerusalem: JNF, 1990), 1 (in Hebrew).

6. Mordechai Ruach, "The Forest and Afforestation in Israel since Independence," *Horizons in Geography* 35–36 (1992): 7–12 (in Hebrew).

7. Food and Agriculture Organization, *Global Forest Resource Assessment 2005, Progress towards Sustainable Forest Management*, FAO Forestry Paper 147 (Rome: FAO, 2005).

8. Jewish National Fund, *Updated Program*, 2–3.

9. Moti Kaplan, *National Master Plan for Forests and Afforestation, NOP 22, Policy Document* (Jerusalem: JNF, 2011), 51.

10. Fabrice DeClerck, Michael Barbour, and John Sawyer, "Species Richness and Stand Stability in Conifer Forests of the Sierra Nevada," *Ecology* 87 (2006): 2787–99.

11. Zaban, interview.

12. Ibid.

13. Ibid.

14. Israel Tauber, interview with author, Tel Aviv, August 7, 2011.

15. Yitzschak Moshe, interview with author, Gilat JNF Offices, August 8, 2011.

16. Menahem Sachs, interview with author, Eshtaol, Israel, October 28, 1998.

17. Imanuel Noy-Meir, "An Ecological View point on Afforestation in Israel: Past and Future," *Algemeine Forst Zeitschrift* 24–26 (1989): 614.

18. Ibid., 616.

19. Ibid., 616–17.

20. Ibid., 617.

21. Ibid., 618.

22. Avshalom Rokach and Haim Zaban, "Forty Years of Land Development and Afforestation in Israel," *Ariel: A Review of Arts and Letters in Israel* 75 (1989): 23.

23. Ibid., 23.

24. Omri Bonneh, "Management of Planted Pine Forests in Israel, Past, Present, and Future," in *Ecology, Biogeography, and Management of "Pinus halepensis" and "R brutia" Forest Ecosystems in the Mediterranean Basin,* ed. Ne'eman Trabaud (Leiden: Backhuys, 2000), 377–90.

25. Jewish National Fund, *Updated Program,* 3–4.

26. Yoav Sagi, interview with author, Tel Aviv, August 11, 2011.

27. Aviva Rabinovitch, *Parent Rock, Soil, and Vegetation in the Galilee* (B'nei Brak: Kibbutz Hameuchad, 1986), 41–42.

28. Uriel Safriel, interview with author, Jerusalem, August 9, 2011.

29. Michael Zohary, *Vegetation of Israel and Adjacent Areas* (Weisbaden: Reichert, 1982), 44.

30. Jean-Michel Carnus et al., "Planted Forests and Biodiversity," *Journal of Forestry* 104, no. 2 (2006): 65–77.

31. Sachs, interview.

32. Yoram Goldring, "Regeneration of the Natural Flora in a Planted Pine Forests" (M.Sc. thesis, Hebrew University of Jerusalem, 1977), as cited in Bonneh, "Management of Planted Pine."

33. Nir Har et al., "Factors Influencing Survival and Development of Transplanted Tabor Oak Trees," *Forest* 11, (2009): 32–34.

34. Milena Holmgren, Marten Scheffer, and Michael A. Huston, "The Interplay of Facilitation and Competition in Plant Communities," *Ecology* 78 (1997): 1966–75.

35. Thomas Smith and Michael Hustun, "A Theory of the Spatial and Temporal Dynamics of Plant Communities," *Vegetatio* 83, nos. 1–2 (1989): 49–69.

36. Shani Ben-Yair, "Spatial and Temporal Effects on Recruitment of the Palestine Oak *Quercus calliprinos* within Pine Plantations in Israel" (Master's thesis, Ben-Gurion University, 2010).

37. Y. Harif, "Development of Elements of the Main Woodlands in Their First Years and the Significance of Successional Lines in the Hills of Judea" (Ph.D. diss., Hebrew University of Jerusalem, 1974), as cited in ibid.

38. Ornea Reisman-Berman et al., "Forest or Gap: The Establishment of Broadleaved Woody Native Chaparral Species within Conifer Plantations of Israel: The Case Study of *Quercus calliprinos,*" *Ecology and the Environment* 1: (2010): 38–46 (in Hebrew).

39. Zohary, *Vegetation of Israel,* 57.

40. Zohar Litmanovitch et al., "Direct Seeding of *Quercus ithaburensis*–Tabor Oak for Reforestation," *Forest* 7 (2005): 35–38 (in Hebrew).

41. Bonneh, "Management of Planted Pine," 378–79.

42. J. W. Hauswirth and L. T. Wetzel, "Toxicity Characteristics of the 2-Chlorotriazines Atrazine and Simazine," chap. 29 in *Triazine Herbicides: Risk Assessment,* ed. L. Ballantine et al., ACS Symposium Series (Washington, DC: American Chemical Society, 1998), 370–83.

43. Bonneh, "Management of Planted Pine," 385–87.

44. Although more expensive, there are many advantages cited in the silviculture literature for container-grown nursery stock as opposed to bare-root planting. These include: ability to plant over an extended season with less need to strictly control dormancy; protection from mechanical injury and desiccation; provision of a rooting me-

dium that remains around roots at planting providing an enhanced microsite; easy adaptation to mechanized handling and to certain kinds of machine planting, without bunching and twisting roots. See Ralph Nyland, *Silviculture: Concepts and Applications* (New York: McGraw-Hill, 2002), 136.

45. Joseph Weitz, *Forests and Afforestation in Israel* (Jerusalem: Masada Press, 1974), 355–57.

46. Bonneh, "Management of Planted Pine," 383–84.

47. Jewish National Fund, *Updated Program,* 7.

48. Fernando T. Maestre and Jordi Corina, "Are *Pinus halepensis* Plantations Useful as a Restoration Tool in Semi-Arid Mediterranean Areas?" *Forest Ecology and Management* 198 (2004): 304.

49. Bonneh, "Management of Planted Pine," 384.

50. Jewish National Fund, *The Bible of Afforestation in Israel: Policy and Directives for Forest Planning and Management,* draft version (May 5, 2011), 19.

51. Yossi Riov, personal communication, December 27, 2011.

52. Paul Ginsburg, interview with author, Moshav Amirim, August 10, 2011.

53. Ibid.

54. Ibid.

55. Ibid.

56. Moti Kaplan, interview with author, Jerusalem, August 9, 2011.

57. Ibid.

58. State of Israel, *Sefer HaChokim* (Book of Laws) (Jerusalem: Government Printing Office, 1965), 307 (in Hebrew).

59. Nurit Alfasi, "Is Public Participation Making Urban Planning More Democratic? The Israeli Experience," *Planning Theory and Practice* 4, no. 2 (2003): 185–202.

60. For a review of the Israeli approach see: Arie Shachar, "Reshaping the Map of Israel: A New National Planning Doctrine," *Annals of the American Academy of Political and Social Science* 555 (1998): 209–18.

61. Eran Feitelson, "Muddling toward Sustainability: The Transformation of Environmental Planning in Israel," *Progress in Planning* 49 (1998): 1–53.

62. Kaplan, *National Master Plan for Forests,* 11.

63. Zaban, interview; Dina Rachevsky, informal written recollections of Master Plan 22, undated, copy with author.

64. Rachelle Alterman, *Planning in the Face of Crisis: Land Use, Housing, and Mass Immigration in Israel* (London: Routledge, 2002).

65. Dina Rachevsky, interview with author, January 29, 2013.

66. Aqiva Etinger, memo, from February 2, 1921, file JNF3/136 in Zionist Archives, Jerusalem, quoted in Nili Lipschitz and Gideon Biger, *Green Dress for a Country: Afforestation in Eretz Israel, the First Hundred Years, 1850–1950* (Jerusalem: Ariel, 2004), 68–69.

67. Gilbert Sale, memorandum from March 31, 1935, quoted in Lipschitz and Biger, *Green Dress,* 179.

68. After conferring with Prime Minister Ben-Gurion, Weitz deducted the 425,000 dunams of existing stands and estimated that there were probably a bit more than three million dunams of suitable land "in need of afforestation." For this he estimated that

roughly eight hundred million trees would be required. Weitz, *Forests and Afforestation,* 195–98.

69. Kaplan, *National Master Plan for Forests,* 89–90.

70. Kaplan, interview.

71. Ibid.

72. Ibid.

73. Ibid.

74. Rachevsky, interview.

75. There were also a few "misunderstandings" where JNF always saw itself as the legitimate overseer—such as the Beit Keshet forests. Perevolotsky, interview.

76. Israel Ministry of Interior, "Planning and Building Law, 1965: National Masterplan for Forests and Forestry," submitted to government, August 1995, published in *Yalkut Pirsumim,* no. 4363 (Jerusalem: Government Printing Office, December 19, 1995).

77. Kaplan, interview.

78. Zaban, interview.

79. Paul Ginsberg, "Afforestation in Israel: A Source of Social Goods and Services," *Journal of Forestry* 98 (2000): 32–36.

80. Mordechai Shechter, Benjamin Reiser, and Natalia Zaitsev, "Measuring Passive Use Value: Pledges, Donations, and CV Responses in Connection with an Important Natural Resource," *Environmental and Resource Economics* 12, no. 4 (1998): 457–78.

Chapter 6. Dryland Forests and Their Natural Enemies

1. Yitzschak Moshe, interview with author, Gilat JNF Offices, August 8, 2011.

2. Ibid.

3. Dirk W. Van Der Zel, "Sustainable Industrial Afforestation in South Africa under Water and Other Environmental Pressures," in *Sustainability of Water Resources under Increasing Uncertainly* (proceedings of the Rabat Symposium S1, April 1997), International Association of Hydrological Sciences Publication no. 240 (1997), 217–25, http://iahs.info/redbooks/a240/iahs_240_0217.pdf (last accessed February 15, 2013). One site in Barberton reported successful planting at 697 millimeters.

4. Yongqiang Liu, John Stanturf, and Houquan Lu, "Modeling the Potential of the Northern China Forest Shelterbelt in Improving Hydroclimate Conditions," *Journal of the American Water Resources Association* 44, no. 5 (2008): 1176–92.

5. Uriel N. Safriel et al., "Soil Erosion-Desertification and the Middle Eastern Anthroscapes," in *Sustainable Land Management: Learning from the Past for the Future,* ed. Selim Kapur, Winfried Blum, and Hari Eswaran (Heidelberg: Springer, 2010), 60.

6. Yitzschak Moshe, personal communication, May 1, 2012.

7. Abed Abulkian, interview with author, Hura, Israel, August 7, 2011.

8. Moshe, interview.

9. Michael Evenari, Leslie Shanan, and Naphtali Tadmor, "Ancient Runoff Agriculture in the Negev," *The Negev: The Challenge of a Desert* (Cambridge, MA: Harvard University Press, 1982), 95–119.

10. Ibid., 97.

11. Yitzchak Moshe, "Plantings by Run-off Harvesting in the Negev," *Algemeine Forst Zeitschrift* 24–26 (1989): 640–41.

12. Safriel, "Soil Erosion-Desertification," 72.

13. Interview with Danny Ben David, Gilat, Israel, August 7, 2011.

14. Abulkian, interview.

15. Safriel, "Soil Erosion-Desertification," 72.

16. Yossi Riov, personal communication, December 27, 2011.

17. David Lamb, Peter D. Erskine, and John A. Parrotta, "Restoration of Degraded Tropical Forest Landscapes," *Science* 310, no. 5754 (2005): 1628–32, www.sciencemag.org/content/310/5754/1628 (last accessed February 3, 2013).

18. Safriel, "Soil Erosion-Deserti?cation," 99.

19. Ibid., 92.

20. Ibid., 95.

21. Tarin Paz-Kagen et al., "Evaluating Ecological Function and Services: Assessing Signs of Soil Quality and Primary Production in Desertified Regions in Israel," in *Ecological Services in Israel: Present Picture, Proceedings* (Jerusalem: National Academic Academy, April 16–17, 2012), 46–47.

22. Jewish National Fund, *Savanization: An Ecological Answer to Desertification* (Jerusalem: JNF, 1994).

23. Menachem Sachs and Itshack Moshe, "Savannazation: An Ecologically Viable Approach to Desertified Regions," *Arid Lands Management: Towards Ecological Sustainability,* ed. Thomas Hoekstra and Moshe Shachak (Champaign: University of Illinois Press, 1999), 248–53.

24. Jewish National Fund, "Functional Restoration of Desert Ecological Systems in the Northern Negev" (Eshtaol: JNF, 2013) (draft copy with author.)

25. Anat Gold, introduction to Iris Bernstein, *The Limans of the Negev: A Policy Paper* (Jerusalem: JNF, 2010).

26. Moshe Shachak, Menachem Sachs, and Itshack Moshe, "Ecosystem Management of Desertified Shrublands in Israel," *Ecosystems* 1 (1998): 475–83.

27. Nikolai Orlovsky, "Preventing Desertification in Israel," in *The Socio-Economic Causes and Consequences of Desertification in Central Asia,* ed. Roy Behnke (Dordrecht, Netherlands: Springer, 2006), 221.

28. Ely Kligler, "Planting Techniques in the Semi-Arid and Arid Negev Regions," *Algemeine Forst Zeitschrift* 24–26 (1989): 636–37.

29. Danny Ben David, interview with author, Gilat, Israel, August 8, 2011.

30. Moshe, interview.

31. Uriel Safriel, personal communication, April 16, 2012.

32. Iris Bernstein, *The Limans of the Negev: A Policy Paper* (Jerusalem: JNF, 2010). Patricia Golan, "Redeeming the Desert: Successful Experiment in Reversing the Process of Desertification in Israel's Negev," *Israel Environmental Bulletin* 13, no. 3 (1990): 21–22.

33. Shmuel Arbel, "The Influence of Water Harvesting by the JNF on the Hydrological System of the Negev," *Research Abstracts, JNF 2011 Research Symposium* (May 9, 2012, Beit Dagan, Israel), 14.

34. Moti Kaplan, *National Master Plan for Forests and Afforestation, NOP 22, Policy Document* (Jerusalem: JNF, 2011), 61.

35. Zafar Adeel et al., Millennium Ecosystem Assessment, *Ecosystems and Human Well-Being: Desertification Synthesis* (Washington, DC: World Resources Institute, 2005), 1–8, www.unep.org/maweb/documents/document.355.aspx.pdf (last accessed February 3, 2013).

36. Alon Tal, "Degraded Commitments: Reviving International Efforts to Combat Desertification," *Brown Journal of International Affairs* 13, no. 2 (2007): 187–97.

37. Ahuva Bar-Lev and Gabi Bron, "The International Arena, Innovations, and Achievements," *Jerusalem Post,* January 19, 2010.

38. Orr Karassin, "KKL-JNF Sharing Knowledge about Adaptation to Global Warming," *Jerusalem Post,* January 18, 2010.

39. Alon Rothschild, interview with author, Tel Aviv, August 3, 2011.

40. Ofer Regev, *Forty Years of Blossoming* (Tel Aviv: Society for the Protection of Nature in Israel, 1993), 18–23 (in Hebrew).

41. Yosef Weitz, "The Redemption and Settlement of the Huleh Valley," *The Huleh: An Anthology* (Jerusalem: World Zionist Organization, 1954), 114, 119.

42. Azariah Alon, "This Is Not the Forest We Sought," *Yarok Kachol Lavan* 8 (May–June 1996): 338–40 (in Hebrew).

43. Alon Tal, *Pollution in a Promised Land* (Berkeley: University of California Press, 2002).

44. Norman Myers et al., "Biodiversity Hotspots for Conservation Priorities," *Nature* 403 (2000): 853–58.

45. Uri Roll, Lewi Stone, and Shai Meiri, "Hot-Spot Facts and Artifacts: Questioning Israel's Great Biodiversity," *Israel Journal of Ecology and Evolution* 55 (2009): 263–79.

46. Yoram Yom-Tov, "Changes in the Distribution and Abundance of Vertebrates in Israel during the Twentieth Century," in *Between Ruin and Restoration: An Environmental History of Israel,* ed. Char Miller, Daniel Orenstein, and Alon Tal (Pittsburgh: University of Pittsburgh Press, 2013), 53–81.

47. Society for Protection of Nature in Israel, ed., *The Red Book: Vertebrates in Israel* (Tel Aviv: SPNI, 2002).

48. Simon N. Stuart, Craig Hilton-Taylor, and Jonathan E. M. Baillie, *A Global Species Assessment* (Gland, Switzerland: IUCN, 2004).

49. Benny Shalmon, "Mammals in Israel," in *Red Book,* www.teva.org.il/?CategoryID=949 (last accessed February 3, 2013).

50. Asaf Mayrose and Dan Alon, "Birds in Israel," in *Red Book,* Tel Aviv, SPNI, www.teva.org.il/?CategoryID=946 (last accessed February 3, 2013).

51. Malcolm McCallum, "Amphibian Decline or Extinction? Current Declines Dwarf Background Extinction Rate," *Journal of Herpetology* 41, no. 3 (2007): 483–91.

52. Zafrir Rinat, "Long Thought Extinct: Hula Painted Frog Found Once Again in Israeli Nature Reserve," *Haaretz,* November 17, 2011.

53. Sarig Gafny, "Amphibians in Israel," in *Red Book,* www.teva.org.il/?CategoryID=944 (last accessed February 3, 2013).

54. Amos Bouskila, "Reptiles in Israel," in *Red Book,* www.teva.org.il/?CategoryID=945 (last accessed February 3, 2013).

55. Menachem Goren, "Freshwater Fishes in Israel," in *Red Book,* www.teva.org .il/?CategoryID=943 (last accessed February 3, 2013).

56. Rothschild, interview.

57. Ibid.

58. Ibid.

59. Nir Hason, "The New Afforestation Systems Prefer Ecology over Conquering the Desert; The Greens: The Present Planting Systems by the JNF Still Damage the Landscape and the Vegetation," *Haaretz,* December 30, 2005.

60. Alon Rothschild, "The *Bible of Forestry,* JNF Draft Document, Response of the Society of the Protection of Nature in Israel," May 12, 2011 (in Hebrew), available with author.

61. Guy Rotem, Amos Bouskila, and Alon Rotshchild, *Afforestation in the Northern Negev and the South Hevron Hills: Ecological Implications* (Tel Aviv: SPNI, 2012).

62. Dror Hawlena et al., "Ecological Trap for Desert Lizards Caused by Anthropogenic Changes in Habitat Structure That Favor Predator Activity," *Conservation Biology* 24, no. 3 (2010): 803–9.

63. In fairness, however, when the seed banks of the two areas are compared, the gap drops to a trivial difference of 109 versus 100. Safriel, "Soil Erosion-Deserti?cation," 118.

64. Ibid., 106.

65. Ibid.

66. Ibid., 107.

67. Zuzeh Greenzweig and Tamar Amit, "The Biotic and Abiotic Influences on Biomass and Species Richness of Shrubs and Grasses in the Yatir Forest," in *Research Abstracts, JNF 2011 Research Symposium* (May 9, 2012, Beit Dagan, Israel), 13 (in Hebrew).

68. J. Buse et al., "Assmann, Saproxylic Beetle Assemblages in the Mediterranean Region: Impact of Forest Management on Richness and Structure," *Forest Ecology and Management* 115 (2010): 1376–84.

69. Uriel Safriel, personal communication, April 16, 2012.

70. Riov, communication.

71. Safriel, interview.

72. Moshe, interview.

73. Safriel, communication.

74. Abulkian, interview.

75. Ben David, interview.

76. Joseph Weitz, *Forests and Afforestation in Israel* (Jerusalem: Masada Press, 1974), 262–63 (in Hebrew).

77. Charles Warren, "Perspectives on the 'Alien' versus 'Native' Species Debate: A Critique of Concepts, Language, and Practice," *Progress in Human Geography* 31, no. 4 (2007): 427–46.

78. Naomi Shemer, "Churshat HaEekaliptus" (The Eucalyptus Grove, 1961), originally written for the musical *How to Break a Heatwave,* first printed in Naomi Shemer, *All the Songs* (Tel Aviv: Lulev, 1967).

79. Gail Rubin, *Psalmist with a Camera* (New York: Abbeville, 1979).

80. Avital Gasith and Vincent Resh, "Streams in Mediterranean Climate Regions: Abiotic Influences and Biotic Responses to Predictable Seasonal Events," *Annual Review Ecological Systems* 30 (1999): 51–81.

81. David Pergament, director of the Yarkon River Authority, personal communication, November 4, 2011.

82. Gidi Ne'eman, interview with author, Moshav Amirim, August 10, 2011.

83. Sharon Udasin, "Environmental Protection Ministry Climate Change Report Predicts Country Will See Decrease in Rainfall, Increases in Temperature," *Jerusalem Post,* January 1, 2012.

84. Riov, communication.

85. Weitz, *Forests and Afforestation,* 52.

86. Nili Lipschitz, *Timber in Ancient Israel: Dendroarchaeology and Dendrochronology* (Tel Aviv: Tel Aviv University Press, 2007), 139–40; Nili Lipschitz, "Recent Distribution of *Pinus brutia* in View of Dendrological Evidence," *Forest* 5–6 (2004): 19–25. A boat in the Mediterranean, that sunk near Kibbutz Maagan Michael, was primarily comprised of *brutia* pine.

87. Nili Lipschitz and Gideon Biger, *Green Dress for a Country: Afforestation in Eretz Israel, the First Hundred Years, 1850–1950* (Jerusalem: Ariel, 2004), 272–73.

88. Mordechai Ruach, "Organization and Activities of the Forest Department," *Algemeine Forst Zeitschrift* 24–26 (1989): 604.

89. Riov, communication.

90. Naftali Karni Gabriel Schiller, "The Occurrence of Mature *Pinus brutia* in the Herzl Forest," *Forest* 8 (2006): 39–40 (in Hebrew).

91. Mark Sogge, Susan Sferra, and Eben Paxton, "Tamarix as Habitat for Birds: Implications for Riparian Restoration in the Southwestern United States," *Restoration Ecology* 16, no. 1 (2008): 146–54.

92. Mark Davis et al., "Don't Judge Species on Their Origins," *Nature* 474 (2011): 153–54.

93. Stephen Jay Gould, "An Evolutionary Perspective on the Concept of Native Plants," *I Have Landed: The End of a Beginning in Natural History* (New York: Harmony, 2002), 343.

94. David Quammen, *The Song of the Dodo: Island Biogeography in the Age of Extinction* (New York: Touchstone, 1997), 321–42.

95. University of Georgia, "Kudzu," *Southeast Exotic Pest Plant Council Invasive Plant Manual,* www.se-eppc.org/manual/kudzu.html (last visited February 3, 2013).

96. Nativ Dudai and Zohar Amar, "Tree Wormwood: A Medicinal Plant Arrived during the Crusades," *Forest* 7 (2005): 35–40 (in Hebrew).

97. "The Appearance of Invasive Trees in Israel," *Forest* 10 (2008): 37–39 (in Hebrew).

98. Mark Williamson, *Biological Invasions* (London: Chapman and Hall, 1997).

99. Amihud Goor, *Tree Planting Practices for Arid Areas* (Rome: FAO, 1955), 94.

100. Amos Sabach, "The Blue Acacia: A Natural Disaster; From a Professional Appendix" (Jerusalem: Nature Reserve and National Parks Authority, 1999), 34–36.

101. Pua Bar (Kutiel), Oded Cohen, and Maxim Shoshany, "Invasion Rate of the Alien

Species *Acacia saligna* within Coastal Sand Dune Habitats in Israel," *Israel Journal of Plant Sciences* 52 (2004): 115–24.

102. IUCN, *IUCN Red List of Threatened Species: Version 2011*, www.iucnredlist.org (last accessed February 3, 2013).

103. O. Cohen et al., "Reducing Persistent Seed Banks of Invasive Plants by Soil Solarization: The Case of *Acacia saligna*," *World Science* 56 (2008): 860–65.

104. Neta Dorchin, "The Possibilities for Biological Control of the Blue Acacia in Israel," *Research Abstracts, JNF 2011 Research Symposium* (May 9, 2012, Beit Dagan, Israel), 11.

105. David Lehrer, Nir Becker, and Pua Bar (Kutiel), "The Economic Impact of the Invasion of *Acacia saligna* in Israel," *International Journal of Sustainable Development and World Ecology* 18, no. 2 (2011): 118–27.

106. Pua Bar Kutiel, personal communication, November 3, 2011.

107. Anat Madmon, "Development of *Tetraclinis articulata* in Israel," *Forest* 12 (2011): 37–45 (in Hebrew).

108. JNF, "KKL Afforestation Objectives: The Preservation of Diverse and Sustainable Forests That Offer Services to the Public," *Sustainability Policies of the JNF*, approved by the JNF International Board (2006), and available at http://192.114.182 .161/kkl/english/main_subject/about_kkl/organization/afforestation%20.x (last accessed on October 3, 2013).

109. Zvika Avni, personal communication, November 9, 2011.

110. H. Charles J. Godfray, "Food and Biodiversity," *Science* 333, no. 6047 (2011): 1231.

111. Ben Phalan et al., "Reconciling Food Production and Biodiversity Conservation: Land Sharing and Land Sparing Compared," *Science* 333, no. 1289 (2011): 1289–91.

112. Joern Fischer et al., "Conservation: Limits of Land Sparing," *Science* 334, no. 6056 (2011): 593.

113. Phalan et al., "Reconciling Food Production," 1289–91.

114. Millennium Ecosystem Assessment Board, *Ecosystems and Human Well-Being*, ed. Rashid Hassan, Robert Scholes, Neville Ash (Washington, DC, Island Press, 2005), www.millenniumassessment.org/documents/document.354.aspx.pdf (last accessed February 14, 2013).

115. Rothschild, interview.

116. Millennium Ecosystem Assessment, *Ecosystems*.

117. Erik Nelson et al., "Terrestrial Biodiversity," in *Natural Capital: Theory and Practice of Mapping Ecosystem Services*, ed. Peter Kareiva et al. (Oxford: Oxford University Press, 2011).

118. Raf Aert and Oliver Honnay, "Seeds of Change for Restoration Ecology," *Science* 333, no. 6039 (2011): 156.

119. David Lamb, Peter D. Erskine, and John A. Parrotta, "Restoration of Degraded Tropical Forest Landscapes," *Science* 310, no. 5754 (2005): 1628–32, www.sciencemag .org/content/310/5754/1628 (last accessed February 3, 2013).

120. Riov, communication.

121. Harold Mooney, interview with author, Stanford, CA, November 7, 2011.

122. Paul Ehrlich and Anne Ehrlich, *Extinction: The Causes and Consequences of the Disappearance of Species* (New York: Random House, 1981).

123. Millennium Ecosystem Assessment, *Ecosystems,* 4.

124. Munyaradzi Chenje and Jennifer Mohamed-Katerere, "Invasive Alien Species," *Emerging Challenges* (Nairobi: United Nations Environmental Program, 2008), 331. See also D. C. Le Maitre et al., "The Impact of Invading Alien Plants on Surface Water Resources in South Africa: A Preliminary Assessment," *Water South Africa* 26 (2000): 397–408.

125. B. I. Nyoka, "Status of Invasive Tree Species in Southern Africa," in *Biosecurity in Forestry: A Case Study on the Status of Invasive Forest Tree Species in Southern Africa* (Rome: FAO, 2003).

126. Safriel, "Soil Erosion-Desertification," 120.

127. Hawlena et al., "Ecological Trap for Desert Lizards," 808.

128. Itzik Moshe, personal communication, February 6, 2013.

129. Richard J Hobbs et al., "Novel Ecosystems: Theoretical and Management Aspects of the New Ecological World Order," *Global Ecology and Biogeography* 15, no. 1 (2006): 1–7.

130. Erle C. Ellis et al., "Anthropogenic Transformation of the Biomes, 1700 to 2000," *Global Ecology and Biogeography* 19 (2010): 589–606.

131. Marilyn Jordan, "You Can't Evolve If You're Extinct: Novel Ecosystems and the Forgotten Food Web," *Science Chronicles* (September 2011): 15–18.

132. René Dubos, *The Wooing of Earth* (New York: Scribner's, 1980).

133. Moshe Shachak, interview with author, Jerusalem, August 9, 2011.

134. American Society of Landscape Architects, "The Rise of Novel Ecosystems," *The Dirt,* posted June 4, 2011, http://dirt.asla.org/2011/04/06/the-rise-of-novel-ecosystems (last accessed February 4, 2013).

135. Jaret C. Daniels et al., *Butterflies and Native Plants: Diversity, Connections, and Opportunities* (Gainesville: University of Florida, Institute of Food and Agricultural Sciences, 2011).

136. Alon Tal, "Restoration of Desertified Ecosystems," *Encyclopedia of Soil Science,* ed. Rattan Lal (Oxford: Taylor and Francis, 2010).

137. Shixiong Cao et al., "Excessive Reliance on Afforestation in China's Arid and Semi-Arid Regions: Lessons in Ecological Restoration," *Earth-Science Reviews* 104, no. 4 (2011): 240–45, www.sciencedirect.com/science/article/pii/S0012825210001571 (last visited February 4, 2013).

138. Xiaohua Wei et al., "The Forest-Streamflow Relationship in China: A 40-Year Retrospect," *Journal of the American Water Association* 44, no. 5 (2008): 1077.

139. Jonathan Watts, "China's Loggers Down Chainsaws in Attempt to Regrow Forests," *The Guardian,* March 11, 2009.

140. Kathleen Farley, Esteban G. Jobba, and Robert Jackson, "Effects of Afforestation on Water Yield: A Global Synthesis with Implications for Policy," *Global Change Biology* 11 (2005): 1565–76.

141. X. Wang et al., "Has the Three Norths Forest Shelterbelt Program Solved the Desertification and Dust Storm Problems in Arid and Semiarid China?" *Journal of Arid Environments* 74 (2010): 13–22.

142. Food and Agriculture Organization, *Global Forest Resource Assessment 2005, Progress Towards Sustainable Forest Management,* FAO Forestry Paper 147 (Rome: FAO, 2006), 27.

143. Ibid., 118.

144. Markku Kanninen, "Plantation Forests: Global Perspectives," in *Ecosystem Goods and Services from Plantation Forests,* ed. Jurgen Bauhus, Peter J. van der Meer, and Markku Kanninen, (London: Earthscan, 2010), 1.

145. Lamb, "Restoration of Degraded Tropical Forest," 1628–32, www.sciencemag.org/content/310/5754/1628.abstract- aff-1 (last visited February 15, 2013).

Chapter 7. Of Fires and Foraging

1. Melanie Lidman, "Teenager Died Trying to Help Put Out Blaze," *Jerusalem Post,* December 5, 2010.

2. H. P. Schwarcz et al., "ESR Dating of the Neanderthal Site, Kebara Cave, Israel," *Journal of Archaeological Science* 16, no. 6 (1989): 653–59.

3. 1 Kings 18:20–22.

4. Ahiya Raved, "Northern Fire: More Than 15,000 Residents Evacuated, Flames near Haifa," *Ynet,* December, 3, 2010.

5. Fadi Eyadat, "Israeli Firefighter Dies of Injuries Sustained in Carmel Blaze," *Haaretz,* December 19, 2010.

6. Naama Tessler et al., "Spatial and Temporal Distribution of Mt. Carmel Forest Fires, 1978–2006," *Forest* 9 (2007): 9–14.

7. Fadi Eyadat, "Israeli Firefighter Dies of Injuries Sustained in Carmel Blaze," *Haaretz,* December 19, 2010.

8. Ibid.

9. Ehud Zion Waldoks, "2010 Was Hottest Year in Israel's Recorded History," *Jerusalem Post,* March 1, 2011.

10. Hanan Greenberg, "IDF Joins Firefighting Efforts in North," *Ynet,* December 3, 2010.

11. Barak Ravid, Anshel Pfeffer, and Natasha Mozgovaya, "European Nations Answer Netanyahu's Call for Help with Carmel Fire," *Haaretz,* December 12, 2010.

12. Calev Ben-David, "Israel Gets Firefighting Aid from Abroad to Help Quell Deadly Haifa Blaze," *Bloomberg News,* December 4, 2010.

13. "Carmel Fire Fully Extinguished," *Ynet,* December 6, 2010.

14. Tomer Zarchin, "The State Comptroller Submits the Report Following the Carmel Fire, Minister of Interior Elli Yishai, Turns to an Attorney," *Haaretz,* August 14, 2011.

15. Uriel Safriel, "The Carmel Fire and Its Conservation Repercussions," *International Journal of Wildland Fire* 7 no. 4 (1997): 277–84.

16. Catherine Belton and Isabel Gorst, "In Moscow, Death Rate Nearly Doubles as Forest Fires Rage On," *Washington Post,* August 9, 2010.

17. David Adam, "Australian Firestorms Prompt Call to Return to Aboriginal Bush Control," *The Guardian,* February 9, 2009.

18. Omri Bonneh, interview with author, August 10, Kiryat Haim, Israel.

19. Tamar Achiron-Frumkin and Ron Frumkin, *Preliminary Assessment of Environmental Damages Induced by the Fighting in Northern Israel-Lebanon,* report prepared for Israel's National Committee of the United Nations Environmental Program (Tel Aviv: Deshe, November 2006), 4–5, www.deshe.org.il/_Uploads/dbsAttachedFiles/Leb2_war.pdf (last accessed February 15, 2013).

20. Dina Kraft, "Dry Forests in Northern Israel Are Damaged as Hezbollah's Rocket Attacks Ignite Fires," *New York Times,* August 6, 2006.

21. Bonneh, interview.

22. Paul Ginsberg, personal communication, November 29, 2011.

23. Bonneh, interview.

24. Paul Ginsburg, interview with author, Moshav Amirim, August 10, 2011.

25. Ibid.

26. Omri Bonneh, personal communication, July 5, 2012.

27. Lea Wittenberg and Moshe Inbar, "The Role of Fire Disturbance on Runoff and Erosion Processes: A Long-Term Approach, Mt. Carmel Case Study, Israel," *Geographical Research* 47, no. 1 (2009): 46–56.

28. Omri Bonneh, Paul Ginsberg, and John Woodcock, *International Forest Fire News,* no. 29 (New York: United Nations, July–December 2003), 74.

29. Pua Kutiel and Ze'ev Naveh, "Soil Properties beneath *Pinus halepensis* and *Quercus calliprinos* Trees on Burned and Unburned Mixed Forest on Mt. Carmel, Israel," *Forest Ecology and Management* 20 (1987): 11–24.

30. Safriel, "The Carmel Fire," 278–79.

31. Gidi Ne'eman, Avi Pervelotsky, and Gabriel Schiller, "The Management Implications of the Mt. Carmel Research Project," *International Journal of Wildland Fire* 7, no. 4 (1997): 348.

32. M. Shechter et al., "On Valuing Natural Resources Damages," in *Preservation of World in the Wake of Challenge,* ed. Y. Steinberger (Jerusalem: ISSES, 1996), 343–49.

33. Gidi Ne'eman, "Regeneration of Natural Pine Forest: Review of Work Done after the 1989 Fire in Mount Carmel, Israel," *International Journal of Wildland Fire* 7, no. 4 (1997): 295.

34. Ne'eman, Pervelotsky, and Schiller, "Management Implications," 346–44.

35. Yohay Carmel, interview with author, Hoshayah, Israel, August 10, 2011.

36. Gidi Ne'eman, "Regeneration of Natural Pine Forest," 295.

37. Shirrinka Goubitz, "Reproduction Ecology of *Pinus halepensis*" (Ph.D. diss., Utrecht University of the Netherlands, 2001).

38. Ran Nathan et al., "Seed Release without Fire in *Pinus halepensis:* A Mediterranean Serotinous Wind-Dispersed Tree," *Journal of Ecology* 87 (1999): 659–69.

39. Ibid., 666–67.

40. Colin Tudge, *The Tree: A Natural History of What Trees Are, How They Live, and Why They Matter* (New York: Three Rivers Press, 2005), 312–13.

41. Aleppo pine seeds are largely absent from the subsurface seed bank. After fires, they germinate quickly, with pine seedlings rarely making additional appearances during subsequent years, in response to favorable environmental conditions. Ido Izhaki, Nava Henig-Sever, and Gidi Ne'eman, "Soil Seed Banks in Mediterranean Aleppo Pine Forests:

The Effect of Heat, Cover, and Ash on Seedling Emergence," *Journal of Ecology* 88, no. 4 (2000): 667–75.

42. Carmel, interview.

43. In the remote White Mountains of Eastern California, bristle cone pine trees have been estimated to reach the unimaginable age of 4,770 years old. C. W. Ferguson, "Bristlecone Pine: Science and Esthetics," *Science* 159, no. 3817 (1968): 839–46.

44. Margarita Arianoutsou and Gidi Ne'eman, "Post-Fire Regeneration of Natural *Pinus halepensis* Forests in the Easter Mediterranean Basin," in *Ecology, Biogeography, and Management of "Pinus halepensis" and "R brutia" Forest Ecosystems in the Mediterranean Basin,* ed. Gidi Ne'eman and Louis Trabaud (Leiden: Backhuys, 2000), 274–75.

45. Avi Perevolotsky, "Forestry in Israel: The Ecological Alternative," *Forest* 12 (2011): 70–71.

46. Paul Johnson, Stephen Shifley, and Robert Rogers, "Regeneration Ecology II: Population Dynamics," *The Ecology and Silviculture of Oaks,* 2nd ed. (Wallingford Oxon, England: CABI, 2009), 151–52.

47. Ze'ev Naveh and Yohay Carmel, "The Evolution of the Cultural Mediterranean Landscape in Israel as Affected by Fire, Grazing, and Human Activities," in *Evolutionary Theory and Processes: Modern Horizons,* ed. S. Wasser (Dordrecht, Netherlands: Kluwer, 2004), 388–89.

48. Gil Sapir and Yohay Carmel, "Predicting Post-Fire Vegetation Regeneration in Planted Pine Forests," *Ecology and Environment* 3 (2010): 14–23 (in Hebrew).

49. Pua Kutiel, "Spatial and Temporal Heterogeneity of Species Diversity in a Mediterranean Ecosystem Following Fire," *International Journal of Wildland Fire* 7, no. 4 (1997): 307–16.

50. R. Velez, "Preventing Forest Fires Through Silviculture," *Unasylva* 162, no. 41 (1990): 10–12.

51. Yossi Riov, personal communication, December 27, 2011.

52. Shaded fuel breaks reduce the risk of fires by selectively thinning and removing the understory vegetation, while leaving large, largely fire-tolerant trees in place. See Craig Chandler et al., *Fire in Forestry,* vol. 2, *Forest Fire Management and Organization* (New York: John Wiley and Sons, 1983), 298.

53. Ibid., 298.

54. Bonneh, Ginsberg, and Woodcock, *International Forest Fire News,* 79.

55. Uriel Safriel, interview with author, Jerusalem, August 9, 2011.

56. Mishnah, Baba Bathra 2:7. A cubit is the distance from the elbow to the index finger or roughly twenty-one inches, so the proscribed distance is a thousand inches or roughly twenty-eight yards.

57. James Ageea et al., "The Use of Shaded Fuel Breaks in Landscape Fire Management," *Forest Ecology and Management* 127 (2000): 55–66.

58. Joseph Weitz, *Forests and Afforestation in Israel* (Jerusalem: Masada Press, 1974), 40.

59. Alon Tal, *The Environment in Israel: Natural Resources, Crises, Campaigns, and Policy from the Advent of Zionism until Twenty-first Century* (B'nei Brak: HaKibbutz HaMeuhad Press, 2006), 172–74 (in Hebrew).

60. Zvika Avni, personal communication, November 30, 2011.

61. Bonneh, Ginsberg, and Woodcock, *International Forest Fire News,* 74.

62. Ralph Nyland, *Silviculture: Concepts and Applications* (Boston: McGraw-Hill, 2007), 109–14.

63. Zvika Mendel, presentation at JNF research symposium, Beit Dagan, Israel, May 9, 2012.

64. Y. Zohar et al., *Ecological Effects of Prescribed Burning in Planted Pine Forests in Israel: Final Report; Prescribed Burning as a Tool in Forest Management* (Beit Dagan, Israel: Ministry of Agriculture, 1996).

65. Bonneh, Ginsberg, and Woodcock, *International Forest Fire News,* 80.

66. Bonneh, personal communication, July 5, 2012.

67. Bonneh, interview.

68. David Brand, presentation at JNF research symposium, Beit Dagan, Israel, May 9, 2012.

69. Weitz, *Forests and Afforestation,* 331.

70. Omri Bonneh, "Management of Planted Pine Forests in Israel, Past, Present, and Future," in *Ecology, Biogeography, and Management of "Pinus halepensis" and "R. brutia" Forest Ecosystems in the Mediterranean Basin,* ed. Gidi Ne'eman and Louis Trabaud (Leiden: Backhuys, 2000), 377–90.

71. Ne'eman, Pervelotsky, and Schiller, "Management Implications," 348.

72. Ido Izhaki and M. Adar, "The Effects of Post-Fire Management on Bird Community Succession," *International Journal of Wildland Fire* 7, no. 4 (1997): 335–42.

73. Yitzschak Moshe and Zvika Avni, "Permanent Infrastructures and Ways to Assist Fire Fighting in the Forest," chap. 7 in *Report on Lessons from the Lebanese War Forest Fires,* ed. Zvika Avni (Eshtaol, Israel: JNF, 2007) (in Hebrew, copy with author).

74. Ibid.

75. Bonneh, Ginsberg, and Woodcock, *International Forest Fire News,* 81–82.

76. Avni, communication.

77. Y. A. Zohar et al., "Fire Behaviour in Conifer Plantations in Israel," *Forêt Méditerranéenne* 10, no. 2 (1988): 423–26.

78. David Brand, Yoram Goldring, and Ami Zahavi, "Utilization of Appropriate Trees to Reduce Fire Risks," chap. 5 in *Report on Lessons.*

79. Avni, communication.

80. Brand, presentation.

81. Colonel Rami Lieberman, Head of Aerial Firefighting, IDF, presentation at JNF research symposium, Beit Dagan, Israel, May 9, 2012.

82. Bonneh, Ginsberg, and Woodcock, *International Forest Fire News,* 83.

83. Bonneh, communication, July 5, 2012.

84. Riov, communication. This view conveniently ignores the dramatic long-term advantages of natural forest stands over first-generation plantations in withstanding fire events.

85. E. R. Sawer, "The Restoration of Palestine's Hill Country: An Address to the Palestine Economic Society on 25th October, 1928" (Jerusalem: Palestine Government, 1928); E. R. Sawer, *The Restoration of Palestine's Hill Country* (Govt. of Palestine Dept. of Agriculture, Forests, and Fisheries, Agricultural Leaflets Series 8, Afforestation, [1929]), CZA box 94, 3630.

86. J. V. W. Shaw, *A Survey of Palestine: Prepared in December 1945 and January 1946 for the Information of the Anglo-American Committee of Inquiry* (Jerusalem: Government Printers, 1946), 428–29.

87. Plant Protection (Damage by Goats) Law 5710-1950, *Laws of the State of Israel* 4 (1949–50): 181.

88. Alon Tal and Jessica Cohen, "Adding 'Top Down' to 'Bottom Up': A New Role for Environmental Legislation in Combating Desertification," *Harvard Journal of Environmental Law* 31, no. 1 (2007): 163–219.

89. J. P. Grime, "Competitive Exclusion in Herbaceous Vegetation," letter, in *Nature* 242, no. 5396 (1973): 344–47. Also Joseph Connell, "Diversity in Tropical Rain Forests and Coral Reefs," *Science* 199, no. 4335 (1978): 1302–10.

90. Richard Hobbs and Laura Huenneke, "Disturbance, Diversity, and Invasion: Implications for Conservation," *Conservation Biology* 6, no. 3 (1992): 324–37.

91. Ze'ev Naveh, "The Evolutionary Significance of Fire in the Mediterranean Region," *Vegetatio* 29 (1992): 199–208; also Ze'ev Naveh and R. H. Whittaker, "Structural and Floristic Diversity of Shrublands and Woodlands in Northern Israel and Other Mediterranean Areas," *Vegetatio* 41 (1979): 171–90.

92. Mario Gutman, Noam Seligman, and Imanuel Noy-Meir, "Herbage Production of Mediterranean Grassland under Seasonal and Year-Long Grazing Systems," *Journal of Range Management* 4 (1990): 64–68.

93. Avi Perevolotsky and No'am Seligman, "Role of Grazing in Mediterranean Rangeland Ecosystems, Inversions of a Paradigm," *BioScience* 49, no. 12 (1998): 1007–17.

94. Ibid.

95. Naveh and Whittaker, "Structural and Floristic Diversity," 171–90.

96. Imanuel Noy-Meir and Talya Oron, "Effects of Grazing on Geophytes in Mediterranean Vegetation," *Journal of Vegetation Science* 12 (2001): 749–60.

97. Perevolotsky and Seligman, "Role of Grazing," 1007–17.

98. J. V. Thirgood, *Man and the Mediterranean Forest* (New York: Academic Press, 1981).

99. Avi Shmida, "Mediterranean Vegetation in California and Israel: Similarities and Differences," *Israel Journal of Botany* 30 (1981): 105–23.

100. Zalmen Henkin, Liat Hadar, and Imanuel Noy-Meir, "Human-Scale Structural Heterogeneity Induced by Grazing in a Mediterranean Woodland Landscape," *Landscape Ecology* 22 (2007): 577–87.

101. Henry Le Houerou, "Land Degradation in Mediterranean Europe: Can Agroforestry Be Part of the Solution? A Prospective Review," *Agroforestry Systems* 21 (1993): 43–61.

102. Suohil Zedan, Michael Weinberger, and Yoram Goldring, "Grazing to Reduce Risk of Forest Fires," chap. 5 in *Report on Lessons*.

103. Avi Perevolotsky, "Livestock Grazing and Biodiversity Conservation in Mediterranean Environments: The Israeli Experience," *Options Méditerranéennes*, series A, no. 67 (2006): 51–56.

104. Safriel, "Carmel Fire," 286.

105. Gideon Kressel, *Let Shepherding Endure: Applied Anthropology and the Preser-*

vation of a Cultural Tradition in Israel and the Middle East (Albany: State University of New York Press, 2003).

106. Liz Wachs and Alon Tal, "Herd No More: Livestock Husbandry Policies and the Environment in Israel," *Journal of Agricultural and Environmental Ethics* 22, no. 5 (2009): 401–22.

107. Bonneh, interview.

108. D. Kaplan et al., *The Influence of the Grazing Season on the Development of "Quercus itaburensis" and "Pinus brutia" Forest,* final report for 1998–99 (Jerusalem: JNF, 1999) (in Hebrew; copy with author).

109. Wachs and Tal, "Herd No More,"401–22.

110. Gidi Ne'eman, interview with author, Moshav Amirim, August 10, 2011.

111. Zedan, "Grazing to Reduce Risk."

112. Bonneh, interview.

113. Gidi, "Management Implications," 348.

114. Paul Ginsberg, Aviram Tzuk, and Elili Benishu, "Preventing Fires in Young Forests," chap. 5 in *Report on Lessons.*

115. Johnson, "Regeneration Ecology II," 150.

116. Ginsberg, Tzuk, and Benishu, "Preventing Fires."

117. Bonneh, Ginsberg, and Woodcock, *International Forest Fire News,* 74.

118. Ibid., 76–77.

119. Alex Moad, U.S. Forest Service, interview with author, Washington, DC, December 9, 2011.

Chapter 8. People and Trees

1. Evegeny Podolsky, interview with author, August 22, 2011.

2. Clinton Bailey, "Dating the Arrival of the Bedouin Tribes in Sinai and the Negev," *Journal of the Economic and Social History of the Orient* 28, no. 1 (1985): 20–49.

3. Emanuel Marx, *Bedouin of the Negev* (Ann Arbor: University of Michigan Press, 1967). Hebrew University geographer Ruth Kark sets the figure somewhat lower. The British had counted some fifty-three thousand, but that was probably an underestimation. Eliyahu Babai, "The Situation of the Bedouins in Israel," *Karka* 42 (October 1996): 79 (in Hebrew).

4. Negev Institute for Strategies of Peace and Development, "The Bedouin-Arab Society in the Negev" (2001), article on website, www.nisped.org.il/info/hebrew/new/hebrew-ajeec-bedouin.htm (last accessed February 4, 2013).

5. Michele Nori, Jason Switzer, and Alec Crawford, *On the Brink: Towards a Global Survey of Pastoral Communities and Conflict* (Geneva: International Institute for Sustainable Development, 2005). Also *African Pastoralism: Conflict, Institutions, and Government,* ed. M. A. Mohamed et al. (London: Pluto Press–OSSREA, 2001).

6. Aref Abu-Rabia, *The Negev Bedouin and Livestock Rearing: Social, Economic, and Political Aspects* (Oxford: Short Run Press, 1994).

7. Babai, "Situation of the Bedouins," 75–80.

8. Yosef Ben-David, "Adaptation through Crisis: Social Aspects of Bedouin Urban-

ization in the Negev," in *The Bedouin Settlement in Israel,* ed. David Grossman (Ramat Gan: Bar Ilan University Press, 1995), 48–49.

9. Gideon M. Kressel, *Let Shepherding Endure: Applied Anthropology and the Preservation of a Cultural Tradition in Israel and the Middle East* (Albany: State University of New York Press, 2003).

10. Some 93 percent of Israel's lands were consolidated under a single publicly managed system during the 1960s. Eliahu Borukhov, *Land Policy in Israel* (Tel Aviv: Tel Aviv University, Center for Urban and Regional Studies, 1975).

11. Podolsky, interview.

12. Ibid.

13. Ibid.

14. Yossi Riov, personal communication, December 27, 2011.

15. Ibid.

16. Cheryl Simrell King et al., "The Question of Participation: Toward Authentic Public Participation Public Administration," *Public Administration Review* 58 no. 4 (1997): 317–26.

17. Nurit Alfasi, "Is Public Participation Making Urban Planning More Democratic? The Israeli Experience," *Planning Theory and Practice* 4, no. 2 (2003): 188–202.

18. Aviva Rabinovitch, interview with author, Kibbutz Kabri, January 11, 1998.

19. Gidi Ne'eman, interview with author, Moshav Amirim, August 10, 2011.

20. Didi Kaplan, "Outlining the Image of Aviva Rabinovitch," *Forest* 10 (2008): 44 (in Hebrew).

21. Rabinovitch, interview.

22. Irus Braverman, interview with author, January 27, 2012.

23. Aviva Rabinovitch, expert opinion for Supreme Court action, January 4, 2000 (copy with author).

24. *Adam Teva V'din and Others v. the Ministry of Interior and Others,* Bagatz 4438/98, submitted on July 13, 1998, petition available with author (in Hebrew).

25. Zafrir Rinat, "Grounds for Suspicion," *Haaretz,* August 14, 1998.

26. *Adam Teva V'din and Others v. the Ministry of Interior and Others,* Bagatz 288/00, August 29, 2001 (in Hebrew). Ruling published in Israel Supreme Court Decisions, (Jerusalem, Government Printers, 2001) (PADI) 55 (5) 673.

27. Ibid.

28. Pinchas Kahana, interview with author, Jerusalem, August 9, 2011.

29. Ibid.

30. Yossi Riov, personal communication, December 27, 2011.

31. Larry Wiseman, presentation, JNF–U.S. Forest Service workshop, Kibbutz Ramat Rachel, June 16, 2011.

32. Pinchas Kahana, personal communication, January 19, 2012.

33. Yehiel Leket, interview with author, Mevaseret Tsiyon, August 9, 2011.

34. Yaacov Shkolnick, *101 Special and Amazing Trees in Israel,* professional staff, Amikam Riklin, Suohil Zedan, and Yoram Goldrin, Professional Editing Staff, (Holon: Am Oved, 2009) (in Hebrew).

35. Amikam Riklin, interview with author, Eshtaol Forestry Center, August 4, 2011.

36. *Adam Teva V'din,* 2001.

37. Riklin, interview. Other quotations from Riklin in this section are from the same interview.

38. Jewish National Fund Inspection Department, *Work Procedures of the Inspection Department of JNF,* September 2003 (copy with author, in Hebrew).

39. For a description of the formation of the Green Patrol, see Alon Tal, *Pollution in a Promised Land: An Environmental History of Israel* (Berkeley: University of California Press, 2002), 347–48.

40. JNF Inspection Department, *Work Procedures,* 12.

41. The employment of private attorneys is not unique to forestry. Israel's Ministry of Environmental Protection has had such an arrangement in place for two decades and is generally satisfied with the efficacy of the system. Alon Tal, Yaara Aharon, and Hadas Yuhas-Peled, "The Relative Advantages of Criminal versus Administrative Environmental Enforcement Actions in Israel," *Journal of Environmental Monitoring* 12 (2010): 813–21.

42. Jewish National Fund, Inspection Unit, *Summary Report of Activities of the Inspection Branch for 2009* (Jerusalem: JNF, 2010) (in Hebrew).

43. A virtual tour of the site can be enjoyed on YouTube at www.youtube.com/watch?v=xnKGiRK9REo as well as teaching by the site's rabbi, Lior Enriques (in Hebrew) at www.youtube.com/watch?v=MvJ6hufNaqo (last accessed on February 4, 2013).

44. Jewish National Fund Inspection Unit, *Summary Report of Activities of the Inspection Branch for 2010* (Jerusalem: JNF, 2011) (copy with author).

45. Riklin, interview.

46. Ibid.

47. Israel, Central Bureau of Statistics, *Statistical Abstract of Israel* (Jerusalem: ICBS, 2011), www.cbi.gov.il.

48. Shaul Ephraim Cohen, *The Politics of Planting* (Chicago: University of Chicago Press, 1995); see also Walter Lehn, *The Jewish National Fund* (London: Kegan Paul International, 1988). See, generally, Oren Yiftachel, *Land, Planning, and Inequality: The Division of Space between Jews and Arabs in Israel* (Tel Aviv: Adva Center, 2000) (in Hebrew).

49. Samer Nachleh, interview with author, Shfaram, Israel, May 14, 2012.

50. Irus Braverman, *Planted Flags: Trees, Land, and Law in Israel/Palestine* (Cambridge: Cambridge University Press, 2009), 3.

51. A. B. Yehoshua, "Facing the Forests," in *Three Days and a Child* (Garden City, NY: Doubleday, 1970), 131–74.

52. Ibid., 169.

53. Abed Abulkian, interview with author, Hura, Israel, August 7, 2011.

54. Jack Khoury and Maya Sela, "Amos Oz: Situation of Bedouin in Negev Is 'Ticking Time Bomb,'" *Haaretz,* August 18, 2010.

55. Among the legal actions filed against the "trespassers" in the Israel's courts are: *Court Decision, Southern Region,* 3569/04; also, *Median v. Israel Lands Authority,* 5898/09, 3705/06.

56. Jack Khoury and Yanir Yagna, "Police Destroy Dozens of Buildings in Unrecognized Bedouin Village in the Negev," *Haaretz,* July 28, 2010; or Ilana Couriel, "El Alarakib Is Destroyed Again: A Fascist Nazi Policy," *Ynet,* January 16, 2011.

57. Ouad Abu Fariach, interview with author, January 23, 2012.

58. Jack Khoury, "Activists Plaster JNF Building with Photos of Razed Bedouin Homes," *Haaretz,* August 29, 2010.

59. Nathan Jeffay, "A JNF Drive to Make the Desert Bloom Means Destruction for a Bedouin Village," *The Forward,* February 9, 2011.

60. Bernhard von Breydenbach, *The Holy Land* (1911). This travel log from 1483–84 was translated and edited by Hugh W. Davies and can be found at http://ia700303.us.archive.org/13/items/bernhardvonbreydoodavi/bernhardvonbreydoodavi.pdf (last accessed February 4, 2013).

61. Daburiah Junior High School, "The Connection Project, 2006–2007," Power Point presentation appearing on the website of Israel's Ministry of Environmental Protection (2008), http://old.sviva.gov.il/Enviroment/Static /Binaries/ModulKvatzim/tavor_daburiya_1.pdf (last accessed February 4, 2013).

62. Suohil Zedan, "Planting and Nurturing Bustanim in the JNF Forests" (Eshtaol, Israel: JNF Forestry Division, 2000).

63. Eli Ashkenazi, "A New Palestinian City Takes Root—with JNF Trees," *Haaretz,* November 29, 2009.

64. Hussein Tarabeih, interview with author, Tel Aviv, July 17, 2012; Abed Namarne, personal communication, July 29, 2012.

65. Iris Hann, "Open Spaces in an Urban Society," in *Between Ruin and Restoration: An Environmental History of Israel,* ed. Char Miller, Daniel Orenstein, and Alon Tal (Pittsburgh: University of Pittsburgh Press, 2013), 146–67.

66. Israel Central Bureau of Statistics, *Statistical Abstract, 2011* (Jerusalem: ICBS, 2011), www.cbs.gov.il/reader/cw_usr_view_SHTML?ID=629 (last accessed February 4, 2013). See also Evgenia Bystrov and Arnon Sofer, *Israel: Demography and Density, 2007-2020* (Haifa: University of Haifa, 2008).

67. Zafrir Rinat, "There Once Was a Forest," *Haaretz,* December 2, 1997.

68. Israel Central Bureau of Statistics, "Population by Population Group," table 2.1 (Jerusalem: ICBS, 2011), www.cbs.gov.il.

69. This common zoological metaphor was most popularized by former Vice President Al Gore in his 1992 book, *Earth in the Balance* (New York: Houghton Mifflin, 1992), and later as a light-hearted, but ultimately disturbing cartoon interlude in his 2006 Academy Award–winning documentary, *An Inconvenient Truth.*

70. Joseph Weitz, *Forests and Afforestation in Israel* (Jerusalem: Masada Press, 1974), 158.

71. JNF, "Community Forests: The Jerusalem Forest" (2012), at www.kkl.org.il/KKL/advanced_template.aspx?id=17257 (last accessed February 4, 2013).

72. The full list of plant species appears in the *Expert Opinion* column edited by ecologist, Ron Frumkin, 2008 that is summarized in ramotenv.com/rls/matseget.pps (last accessed on February 4, 2013). Recently, Professor Shmida identified a rare iris that had not been spotted in the original survey. Accordingly, the number of plant species went up from 532 to 533!

73. For a review of the existing surveys natural history and biodiversity of the site and wonderful photographs, see Ramot for the Environment, "Flora and Fauna of Mitspeh Naftoah" (2012), www.ramotenv.com/blog (last accessed February 4, 2013).

74. National Park Service, "Rock Creek Park" (2012), www.nps.gov/rocr/index.htm (last accessed February 4, 2013).

75. Amir Ben-David, "The Plan: An Overpass in the Jerusalem Forest," *Yedioth Ahronot,* July 11, 2011.

76. City of Paris, municipal website, Mairie de Paris (2012), www.paris.fr/english (last accessed February 4, 2013).

77. Raanan Ben Zur and Shachar Hai, "From Nature to the Celebration: Traffic Jams on the Way Home," *Ynet,* April 14, 2012 (in Hebrew).

78. Lea Shne'or, "Planning the Forests of the Jewish National Fund for Different Population Groups," *A Matter of Accessibility* (Fall 2005): 14–20 (in Hebrew).

79. Gershon Avni, personal communication, January 26, 2012.

80. Ibid., March 11, 2012.

81. Moshe Eloz, JNF development department, personal communication, March 11, 2012.

82. Jewish National Fund, *Jewish National Fund Budget for 2011* (Jerusalem: JNF Budget and Auditing Department, 2011), 13.

83. Eyal Inbar, "A Million Riders Can't Be Mistaken: Bicycle Trails in the Forests," *Ynet,* November 17, 2010.

84. Oshrat Nagar Lavit and Billy Frankel, "Pedaling the Distance," *NRG / Maariv,* March 30, 2008, www.nrg.co.il/online/1/ART1/715/781.html (last accessed on February 4, 2013).

85. See: Joint Distribution Committee, Israel, "People with Disabilities in Israel: Facts and Figures" (Jerusalem: JDC, 2003), www2.jdc.org.il/category/disability-general (last accessed on February 4, 2013).

86. State of Israel, Law of Equal Rights for People with Disabilities, Second Amendment no. 2 (2005): "Accessibility."

87. Lea Shne'or, *Planning Projects in Keren Kayemeth LeIsrael-Jewish National Fund Forests for Diverse Population Groups* (Jerusalem: JNF, 2004).

88. Edith Bowman and Simon Bell, *Accessibility and Disability: A Guide to the Application of Disabled Access Legislation to Scotland's Woodlands,* prepared for Forestry Commission Scotland (2008), at www.forestry.gov.uk/pdf/OPEN space_Disabled_access_Task1.pdf/$FILE/OPENspace_Disabled_access_Task1.pdf (last accessed February 4, 2013).

89. Shne'or, "Planning the Forests," 16.

90. Ibid., 29.

91. Caron Chess and Kristen Purcell, "Public Participation and the Environment: Do We Know What Works?" *Environmental Science and Technology* 33, no. 16 (1999): 2685–93.

92. Gene Rowe and Lynn J. Frewer, "Evaluating Public-Participation Exercises: A Research Agenda," *Science Technology Human Values* (2004): 512–56.

93. Chanan Erez, Chair, Megido Regional Council, presentation, JNF–U.S. Forest Service Workshop, Kibbutz Ramat Rachel, June 16, 2011.

94. Aliza Fleisher and Moshe Sheller, *Survey of Community Forests,* report submitted to the Jewish National Fund Community and Forest Department (Eshtaol: JNF, 2006).

95. Jewish National Fund, "Community Forest: A Shared Covenant between the Com-

munity, the City Council, and the Jewish National Fund" (2006), posted at www.kkl.org.il/community-forests (last accessed February 4, 2013).

96. Aliza Fleisher, *Survey of Community Forests: Ofakim, Rosh Ha Ayin, Kochav Yair, Meitar, Migdal HaEmeq, and Jerusalem,* submitted to the JNF, Jerusalem, September 2010.

97. Fleisher and Sheller, *Survey of Community Forests,* 5.

98. Fleisher, *Survey of Community Forests: Ofakim,* 4.

99. Idit Sappir-Gildor, Anais Worschewoitz, and Alon Tal, *Public Attitudes towards KKL Forests: An Empirical Study,* report submitted to the Jewish National Fund (Jerusalem: JNF, 2002).

100. Noga Collins-Kreiner, "Public Preferences for Bike Paths in the Forests," *Research Abstracts, JNF 2011 Research Symposium* (Beit Dagan, Israel, May 9, 2012), 18.

101. Sappir-Gildor, Worschewoitz, and Tal, *Public Attitudes.*

Chapter 9. Ecosystem Services and Israel's Forests

1. Debra Satz, *Why Some Things Should Not Be for Sale: The Moral Limits of Markets* (Oxford: Oxford University Press, 2010).

2. M. Shechter et al., "On Valuing Natural Resources Damages," in *Preservation of World in the Wake of Challenge,* ed. Y. Steinberger (Jerusalem: ISSES, 1996), 343–49.

3. Ibid.

4. Shiri Zemah Shamir, S and M. C. van der Heide, and Mordechai Shechter, "Economic Valuation of Ecological Characteristics: The Case of Mediterranean Grove" (2011), unpublished manuscript with author.

5. Nir Becker and Shirra Freeman, "The Economic Value of Old Growth Trees in Israel," *Forest Policy and Economics* 1, no. 1: (2009): 608–15.

6. The report found that of the twenty-four ecosystem services assessed, only four were improving, five were stable and fifteen were declining. "The supply of certain ecosystem services has increased at the expense of others. Significant gains in the provision of food and fiber have been achieved through habitat conversion, increased abstraction and degradation of inland waters, and reduced biodiversity. Millennium Ecosystem Assessment, "Forests and Woodland Systems," in *Ecosystems and Human Well-Being: Current State and Trends* (Washington, DC: Island Press, 2005), 1: 16.

7. Ibid., xvi.

8. For example, Yvonne Baskin, *The Work of Nature: How the Diversity of Life Sustains Us* (Washington, DC: Island Press, 1997); Gretchen C. Daily, ed., *Nature's Services: Societal Dependence on Natural Ecosystems* (Washington, DC: Island Press, 1997); Edward Wilson, *The Diversity of Life* (1992; repr., Cambridge, MA: Harvard University Press, 2010); Geoffrey Heal, *Nature and the Marketplace: Capturing the Value of Ecosystem Services* (Washington, DC: Island Press, 2000); Andrew Beattie and Paul Ehrlich, *Wild Solutions: How Biodiversity Is Money in the Bank* (New Haven, CT: Yale University Press, 2001).

9. Gretchen C. Daily and Katherine Ellison, *The New Economy of Nature: The Quest to Make Conservation Profitable* (Washington, DC: Island Press, 2002), 5.

10. The figures ranged from sixteen to fifty-four trillion dollars. Costanza et al., "The

Value of the World's Ecosystem Services and Natural Capital," *Nature* 387 (1997): 253–60.

11. William F. Hyde, "Timber Economics in the Rockies: Efficiency and Management Options," *Land Economics* 57, no. 4 (1991): 630–38. Also Brock Evans, "Ancient Treasures, Old Growth: History and Heritage," *Environmental Law and Litigation* 2 (1987): 141–52.

12. Millennium Ecosystem Assessment, "Forests and Woodland Systems," vii.

13. "Financial income from forests, whether from wood production or otherwise, will be considered a secondary output of forest maintenance, and shall not constitute a primary objective of afforestation in Israel." "JNF Forestry Policy for Israel" (2006), at www.kkl.org.il/eng/about-kkl-jnf/kkl-jnf-id/sustainable-development (last visited February 4, 2013).

14. Moti Kaplan, *National Outline Plan for Forests and Afforestation,* National Master Plan 22, policy document (Jerusalem: JNF, 2011), 4.

15. Ami Zahavi, "Forestry in Israel: The Ecological Alternative; A Response to Avi Perevolotsky's Article," *Forest* 12 (2011): 80 (in Hebrew).

16. Yossi Riov, personal communication, December 27, 2011.

17. Alon Ben-Gal, "Sustainable Water Supply for Agriculture in Israel," in *Water Wisdom: A New Menu for Palestinian and Israeli Cooperation in Water Management,* ed. Alon Tal and Alfred Abed-Rabbo (New Brunswick, NJ: Rutgers University Press, 2010), 211–23.

18. Daniel Hillel, *Out of the Earth: Civilization and the Life of the Soil* (Berkeley, CA: University of California Press, 1992).

19. G. Konukcu, F. Gowing, and J. W. Rose, "Dry Drainage: A Sustainable Solution to Waterlogging and Salinity Problems in Irrigation Areas?" *Agricultural Water* 83 (2006): 1–12.

20. Liron Amdur and Haim Zaban, "Commercial Afforestation in Israel: Economic Feasibility and Policy Study," *Forest* 8 (2006): 14–21.

21. The primary eucalypt species utilized for lumber are: *E. camaldulensis, E. gomphocephala, E. cladocalyx, E. maculta,* and *E. urograndis*. Beehive operators prefer *E. torquata, E. torwood.*

22. Yehiel Zohar and A. Gafni, *The Eucalyptus: From Its Branches Upward,* ed. Abraham Weinstein, Yehiel Zohar, and David Brand (Beit Dagan: Ministry of Agriculture, 2009) (in Hebrew).

23. Liron Amdur and Haim Zaban, "Commercial Afforestation in Israel: Economic Feasibility and Policy Study," *Forest* 8 (2006): 14–21.

24. Great Britain, Forestry Commission, "Direct Production Average Unit Prices" (January 2005 to October 2011), www.forestry.gov.uk/pdf/MONTHLYGRAPHSDec12.pdf/$file/MONTHLYGRAPHSDec12.pdf (last accessed February 4, 2013).

25. Suohil Zedan, personal communication, February 6, 2012.

26. Amiram Cohen, "Prices of Olives Increases? Olive Growers Demand an Increase in the Price of Raw Materials by Tens of Percent," *TheMarker,* October 6, 2011.

27. Zedan, communication.

28. No'am Seligman, personal communication, February 23, 2012.

29. Zeev Sternick, Michael Eliav, and Dalit Goldfus, *Agriculture in Israel 2000: Area and Livestock* (Jerusalem: Central Bureau of Statistics, 2002).

30. Yan Landau, "Tree Planting for Grazing Contributes to Livestock Health Research Abstracts," *Research Abstracts, JNF 2011 Research Symposium* (Beit Dagan, Israel, May 9, 2012), 9–10 (in Hebrew).

31. David Evlagon et al., "Estimating Normative Grazing Capacity of Planted Mediterranean Forests in a Fire-Prone Environment," *Agriculture, Ecosystems and Environment* 155 (2012): 133–41.

32. Alon Tal, "The Logic and Logistics of Grazing Regulations," *Journal of Land Degradation and Development* 20 (2009): 455–67. In fact, irrigated Israeli pastureland never really existed.

33. No'am Seligman, personal communication, February 23, 2012.

34. David Evlagon et al., "How Much Browse Is Available for Goats That Graze Mediterranean Woodlands?" *Small Ruminant Research* 94, no. 1 (2010): 103–8.

35. Landau, "Tree Planting for Grazing," 9.

36. Evlagon et al., "Estimating Normative Grazing Capacity."

37. Seligman, communication.

38. David Evlagon et al., *The Vegetation in the Forest Understory and Its Value as Forage in the Forest Stands in the Central Region of JNF* (Eshtaol, Israel: JNF, 2011).

39. Mark Claton, "Study: Forests Absorb Much More Greenhouse Gas Than Previously Known," *Christian Science Monitor,* July 15, 2011, reporting on Yude Pan et al., "A Large and Persistent Carbon Sink in the World's Forests," *Science* 333, no. 6045 (2011): 988–93.

40. Stephen Pacala and Robert Socolow, "Stabilization Wedges: Solving the Climate Problem for the Next 50 years with Current Technologies," *Science* 305, no. 5686 (2004): 968–71.

41. Brent Sohngen and Sandra Brown, "Extending Timber Rotations: Carbon and Cost Implications," *Climate Policy* 8 (2008): 433–51.

42. Hannes Bottcher and Macus Lidner, "Managing Forest Plantations for Carbon Sequestration Today and Tomorrow," in *Ecosystem Goods and Services from Plantation Forests,* ed. Jurgen Bauhus, Peter J. van der Meer and Markku Kanninen (London: Earthscan, 2010), 55.

43. Ibid., 46.

44. Renee Johnson, Jonathan Ramseur, and Ross Gourte, *Estimates of Carbon Mitigation Potential from Agricultural and Forestry Activities* (Washington, DC: Congressional Research Service, 2009), www.fas.org/sgp/crs/misc/R40236.pdf (last accessed February 4, 2013).

45. Robert Stavins, "What Role for U.S. Carbon Sequestration?" *Huffington Post,* July 21, 2009. See also Robert Stavins and Richard Kenneth, *The Cost of U.S. Forest-Based Carbon Sequestration* (Pittsburgh: Pew Center on Global Climate Change, 2005).

46. Ross W. Gorte, *U.S. Tree Planting for Carbon Sequestration* (Washington, DC: Congressional Research Service, May 4, 2009).

47. McKinsey and Company, *Greenhouse Gas Abatement Potential in Israel: Israel's GHG Abatement Cost Curve, Hebrew Version* (Herzliyah, Israel: McKinsey, 2009), 120.

48. U.S. Department of Agriculture, "Forest Sequestration and Products Storage,"

chap. 4 of *U.S. Agriculture and Forestry Greenhouse Gas Inventory, 1990–2001* (Washington, DC: USDA, 2004).

49. A. Thuille and A. E. Schulze, "Carbon Dynamics in Successional and Afforested Spruce Stands in Thuringia and the Alps," *Global Change Biology* 12 (2006): 325–42.

50. M. Soroos, *The Endangered Atmosphere: Preserving a Global Commons* (Columbia: University of South Carolina Press, 1997).

51. There are two complimentary components to this reservoir: soil organic carbon (SOC)—and soil inorganic carbon (SIC), the smaller of the two. The soil sequestration that takes place on the forest floor is driven by three interrelated mechanisms: humification (the collection of humus from the remains of plants and animals), aggregation and illuviation (the *accumulation* of mineral and organic matter in the lower soil horizons). Rattan Lal, "Forest Soils and Carbon Sequestration," *Forest Ecology and Management* 220 (2005): 242–58.

52. Bottcher and Lidner, "Managing Forest Plantations," 45.

53. K. I. Paul et al., "Change in Soil Carbon Following Afforestation," *Forest Ecology and Management* 168, nos. 1–3 (2002): 241–57.

54. L. Olsson, A. Warren, and J. Ardo, "The Potential Benefits of Dryland Agricultural Soils as Carbon Sinks," *Arid Lands Newsletter* 49 (2001) http://ag.arizona.edu/oals/ALN/aln49/olsson.html (last accessed on February 4, 2013).

55. H. L. Zhao et al., "Effects of Desertification on Soils and Crop Growth Properties in Horquin Sandy Cropland of Inner Mongolia, North China," *Soil and Tillage Research* 87 (2006): 175–85.

56. R. B. Jackson et al., "Ecosystem Carbon Loss with Woody Plant Invasion Of Grasslands," *Nature* 418 (2002): 623–26. At the same time, tree establishment on native grasslands and forests produced a net drop of 10–13 percent in SOC, although as forests grew older—and critically as precipitation dropped, these changes diminished. See: L. I. Paul et al., "Change in Soil Carbon Following Afforestation," *Forest Ecology Management* 168 (2002): 241–57. The invasion of woody vegetation into deserts and grasslands resulted in reduction of SOC at moist sites (660–1,070 mm precipitation), but increased SOC at the driest sites (below 350 mm).

57. Uriel N. Safriel et al., "Soil Erosion-Deserti?cation and the Middle Eastern Anthroscapes," in *Sustainable Land Management: Learning from the Past for the Future,* ed. Selim Kapur, Winfried Blum, and Hari Eswaran (Heidelberg: Springer, 2010), 99.

58. Rattan Lal, "Sequestering Carbon in Soils of Arid Ecosystems," *Land Degradation and Development* 20, no. 4 (2009): 441.

59. J. M. Grunzweig et al., "Biogeochemical Factors Contributing to Enhanced Carbon Storage Following Afforestation of a Semi-Arid Shrubland," *Biogeosciences* 4 (2007): 891–904.

60. Ibid.

61. The Aleppo pines and other trees growing in Yatir are not seen naturally in landscapes with less than four hundred millimeters of rain. So some researchers classify the trees as "introduced species." Gabriel Schiller and Nir Atzmon, "Performance of Aleppo Pine (*Pinus halepensis*) Provenances Grown at the Edge of the Negev Desert: A Review," *Journal of Arid Environments* 73, no. 12 (2009): 1051–57.

62. Dan Yakir, personal communication, February 2, 2012.

63. Grunzweig et al., "Biogeochemical Factors," 891–904.

64. Irrigation also improves net leaf assimilation along with general tree-ring growth. Tamir Klein et al., "Association between Tree-Ring and Needle d13C and Leaf Gas Exchange in *Pinus halepensis* under Semi-Arid Conditions," *Oecologia* 44 (2005): 45–54.

65. Nir Atzmon, Yitzschak Moshe, and Gabriel Schiller, "Eco-Physiological Response to Severe Drought in *Pinus halepensis* Mill Trees of Two Provenances," *Plant Ecology* 171 (2004): 15–22.

66. Klein et al., "Association between Tree-Ring," 45–54.

67. M. D. Nosetto et al., "Carbon Sequestration in Semi-Arid Rangelands: Comparison of *Pinus ponderosa* Plantations and Grazing Exclusion in NW Patagonia," *Journal of Arid Environments* 67 (2006): 142–56.

68. McKinsey and Company, *Greenhouse Gas Abatement,* 116, 120.

69. Holly K Gibbs et al., "Monitoring and Estimating Tropical Forest Carbon Stocks: Making REDD a Reality," *Environmental Research Letters* 2 (2007): 1–13.

70. Carbon Offset Daily, "Second CDM Forestry Project Approved," March 9, 2009, www.carbonoffsetsdaily.com/news-channels/top-stories/second-cdm-forestry-project-approved-5107.htm (last accessed February 4, 2013).

71. Earth Watch Institute, "UNFCC Approves India's First CDM Forestry Project" (2010), www.earthwatch.org/europe/newsroom/science/news-forest2905.html (last accessed February 4, 2013).

72. UNFCC, "Uganda Registers First CDM Forestry Project in Africa," http://cdm.unfccc.int/Projects/DB/TUEV-SUED1229620290.53/view (last accessed February 4, 2013).

73. Alon Tal and Jessica Gordon, "Carbon Cautious: Israel's Afforestation Experience and Approach to Sequestration," *Small-Scale Forestry* 9, no. 4 (2010): 409–20.

74. Eyal Rotenberg and Dan Yakir, "Contribution of Semi-Arid Forests to the Climate System," *Science* 327 (2010): 451–54.

75. (1) "The outgoing longwave radiation for a clear sky at the top of the atmosphere suggests a maximum of 20 percent. (2) Similarly, because of atmospheric absorption and cloud reflection, the local albedo at the top of the atmosphere is lower than the surface value. By not taking into account this energy redistribution, Rotenberg and Yakir may have substantially overestimated the warming effect of forestation and the cooling effect of desertification," Xuhui Lee, "Forests and Climate: A Warming Paradox," *Science* 328 (2010): 1479; and Stephan Leu, "Forests and Climate: The Search for Specifics," *Science* 328 (2010): 1479.

76. Dan Yakir, personal communication, February 2, 2012.

77. An excellent and comprehensive description of the complexities of pollination ecology is found in Haifa University Professor Amots Dafni's book *Pollination Ecology: A Practical Approach* (Oxford: Oxford University Press, 1992).

78. Eric Lonsdorf et al., "Crop Pollination Services," in *Natural Capital, Theory and Practice of Mapping Ecosystem Services,* ed. Peter Kareiva et al. (Oxford: Oxford University Press, 2011), 168.

79. Alexandra-Maria Klein et al., "Understanding the Role of Species Richness for

Crop Pollination Services," in *Biodiversity, Ecosystem Functioning, and Human Well-Being* (Oxford: Oxford University Press, 2009), 196.

80. Mike H. Allsopp, Willem J. de Lange, and Ruan Veldtman, "Valuing Insect Pollination Services with Cost of Replacement," *PLoS ONE* 3 (2008): 9, www.plosone.org/article/info:doi/10.../journal.pone.0003128 (last accessed February 4, 2013).

81. Alon Tal, "To Make a Desert Bloom: The Israeli Agriculture Adventure and the Quest for Sustainability," *Agricultural History* 81, no. 2 (2007): 228–58.

82. Ingrid Williams, *Insect Pollination and Crop Production: A European Perspective; The Conservation Link between Agriculture and Nature* (Brasilia: Ministry of Environment, 2002), 59–65.

83. A recent review of twenty-nine studies around the world reached the conclusion that "wild pollinators are relevant for crop productivity and stability even when honey bees are abundant." Lucas A. Garibaldi et al., "Stability of Pollination Services Decreases with Isolation from Natural Areas Despite Honey Bee Visits," *Ecology Letters* 14 (2011): 1062–72.

84. Yael Mandelik and Uri Roll, "Diversity Patterns of Wild Bees in Almond Orchards and Their Surrounding Landscape," *Israel Journal of Plant Sciences* 57 (2009): 185–91.

85. U.S. National Research Council, *Status of Pollinators in North America* (Washington, DC: U.S. National Academies Press, 2006).

86. BBC News, "Why Are Europe's Bees Dying?" November 20, 2008, http://news.bbc.co.uk/2/hi/7739798.stm (last accessed February 4, 2013). Also: Dennis vanEngelsdorp, Marina Doris Meixner, "A Historical Review of Managed Honey Bee Populations in Europe and the United States and the Factors That May Affect Them," *Journal of Invertebrate Pathology* 103 (2010): supplement, S80–S95.

87. David Cook et al., "Predicting the Economic Impact of an Invasive Species on an Ecosystem Service," *Ecological Applications* 17 (2007): 1832–40.

88. Gidi Ne'eman, personal communication, February 10, 2012.

89. Stephen J. Martin et al., "Global Honey Bee Viral Landscape Altered by a Parasitic Mite," *Science* 336 (2012): 1304–6.

90. Claire Kremen, Neal M. Williams, and Robbin W. Thorp, "Crop Pollination from Native Bees at Risk from Agricultural Intensification," *Proceedings of the Natural Academy of Sciences* 99, no. 16 (2002): 812–16. Also: Claire Kremen et al., "The Area Requirements on an Ecosystem Service: Crop Pollination by Native Bee Communities in California," *Ecology Letters* 7 (2004): 1109–19.

91. Tamar Keasar and Avi Shmida, "An Evaluation of Israeli Forestry Trees and Shrubs as Potential Forage Plants for Bees," *Israel Journal of Plant Sciences* 57 (2009): 49.

92. Yael Mandelik, personal communication, February 15, 2012.

93. Christopher O'Toole and Anthony Raw, *Bees of the World* (London: Blanford, 1991), 192.

94. Mandelik, personal communication.

95. Amots Dafni and Christopher O'Toole, "Pollination Syndromes in the Mediterranean: Generalizations and Peculiarities," in *Plant-Animal Interaction in Mediterranean-Type Ecosystems*, ed. M. Arianoutsou and R. Groves (Dordrecht, Netherlands: Kluwer, 1994), 125–35.

96. Robert Minckley, "Faunal Composition and Species Richness Differences of Bees (*Hymenoptera: Apiformes*) from Two North American Regions," *Apidologie* 39 (2008): 176–88.

97. Simon G. Potts et al., "Linking Bees and Flowers: How Do Floral Communities Structure Pollinator Communities?" *Ecology* 84, no. 10 (2003): 2628–42.

98. Michal Korech and Yair Asaf-Shapira, "Open Spaces in Israel," in *Vital Signs* (Tel Aviv: Heschel Center for Environmental Learning and Leadership, 2001), 184 (in Hebrew).

99. Efrat Hadas, *Allocating Water and Land to Active Agricultural Operations for the Long Term, Report to the Israel Ministry of Agriculture, the Jewish Agency* (Jerusalem: Jewish Agency,2002) (in Hebrew).

100. Ariella Gotlieb, Yael Hollender, and Yael Mandelik, "Gardening in the Desert Changes Bee Communities and Pollination Network Characteristics," *Basic and Applied Ecology,* 12, no. 4 (2011): 310–20.

101. Ofrit Shavit, Amots Dafni, and Gidi Ne'eman, "Competition between Honeybees (*Apis mellifera*) and Native Solitary Bees in the Mediterranean Region of Israel: Implications for Conservation," *Israel Journal of Plant Sciences* 57 (2009): 171–83.

102. Ibid.

103. Ibid.

104. Mandelik, communication.

105. Har'el Agra and Gidi Ne'eman, "Woody Species as Landscape Modulators: Their Effect on the Herbaceous Plants in a Mediterranean Maquis, on Mount Meron, Israel," *Forest* 11 (2009): 20–27.

106. Yisrael Tauber, "Forestry in Israel, the Ecological Alternative: A Response to Avi Perevolotsky's Article," *Forest* 12 (2011): 78.

107. Mandelik, communication.

108. Ibid.

109. Keasar and Shmida, "Evaluation of Israeli Forestry Trees," 49.

110. Mandelik, communication.

111. Simon G. Potts et al., "Response of Plant-Pollinator Communities to Fire: Changes in Diversity, Abundance, and Floral Reward Structure," *Oikos* 101, no. 1 (2003): 103–12.

112. Ibid., 109.

113. Ne'eman, communication.

114. Arnon Dag and Aryeh Regev, "The Economic Value of Honeybee Crop Pollination in Israel," *Yalkut H'michveret* 42 (1999): 96–105 (in Hebrew).

115. Yisrael Tauber and Gidi Ne'eman, *The State of Nature Report* (Jerusalem: Maarag —LTER Network, 2011) (in Hebrew), www.isees.org.il/images/WebMaster/files/Scientist%20Role/Nature%20Report_Part%201.pdf (last accessed February 4, 2013).

116. Ne'eman, communication.

117. Zvi Mendel, personal communication, January 21, 2012.

118. These include Loess (*Calcic haploxeralf*), the Hamra (*Typic rodoxeralf*), and Rendzina along with a variety of mountain soil types that are all given to erosion. See Yoram Benyamini, "Measuring and Monitoring Soil Erosion for Soil Conservation and Soil Protection in Israel," *Soil Conservation for Europe* (Heiloo, Netherlands: Scape,

2008), 148, http://eusoils.jrc.ec.europa.eu/projects/scape/uploads/108/Benyamini.pdf (last accessed February 15, 2013).

119. Walter Clay Lowdermilk, *Palestine: Land of Promise* (New York: Harper and Brothers, 1944), 66–84.

120. Palestine Government, *A Survey of Palestine: Prepared in December 1945 and January 1946 for the Information of the Anglo-American Committee of Inquiry* (Washington DC, Institute of Palestine Studies, 1946), 426–27.

121. Lowdermilk, *Palestine,* 5.

122. Recent watershed studies of the Alexander and the Besor watershed confirms that the high contribution of nonpoint source loadings of nutrients into the streams. See Alon Tal et al., "Chemical and Biological Monitoring in Ephemeral and Intermittent Streams: Lessons Learned from a Study of Two Transboundary Palestinian-Israeli Watersheds," *Journal of River Basin Management* 8 no. 2 (2010): 185–205. The study reports: "The ongoing erosion continuously exposes and mobilizes nutrients in the flow path, resulting in the initial portion of the runoff hydrograph showing a relatively low concentration of nutrients that steadily increases as water travel time increases. . . . The nutrient loadings during large events appear to be a full two orders of magnitudes higher, highlighting the contributions of non-point sources" (197).

123. See, for example, Gurmel Singh et al., "Soil Erosion Rates in India," *Journal of Soil and Water Conservation* 47, no. 1 (1992): 97–99. K. F. Wiersum, "Effects of Various Vegetation Layers in an *Acacia auriculiformis* Forest Plantation on Surface Erosion in Java, Indonesia," in *Soil Erosion and Conservation* (Ankeny, Iowa: Soil Conservation Society of America, 1985), 79–89; also Bojie Fu, "Soil Erosion and Its Control in the Loess Plateau of China," *Soil Use and Management* 5, no. 2 (1989): 76–82.

124. Anne-Gaelle Ausseil and John Dymond, "Evaluating Ecosystem Services of Afforestation on Erosion-Prone Land: A Case Study in the Manawatu Catchment, New Zealand," International Congress on Environmental Modeling and Software Modeling for Environment's Sake, Ottawa, Canada, 2010 (2011), proceedings at www.iemss.org/iemss2010/proceedings.html (last accessed February 4, 2013).

125. John Dymond et al., "Validation of a Regionwide Model of Landslide Susceptibility in the Manawatu-Wanganui region of New Zealand," *Geomorphology* 74 (2005): 70–79.

126. Fu, "Soil Erosion."

127. Tongqian Zhao, Bosu Yang, and Hua Zheng, "Assessment of the Erosion Control Function of Forest Ecosystems Based on GIS: A Case Study in Zhangjiajie National Forest Park, China," *International Journal of Sustainable Development and World Ecology* 16, no. 5 (2010): 356–461.

128. Paolo Porto, Des E. Walling, and Giovanni Callegari, "Investigating the Effects of Afforestation on Soil Erosion and Sediment Mobilisation in Two Small Catchments in Southern Italy," *Catena* 79, no. 3 (2009): 181–88.

129. Lowdermilk, *Palestine.*

130. Israel Soil Erosion Research Station (2012), http://sers.gov.il (last accessed February 4, 2013).

131. More recently there is a greater focus on conservation-tilling management systems among farmers that combine cover crops or soil mulching in orchards. Israel Soil

Erosion Research Station (2011), website, posted February 11, 2011, http://sers.gov.il/en/soil-water-and-environment-conservation-practices (last accessed February 4, 2013).

132. Shmuel Arbel, director of the Israel Soil Erosion Research Station, personal communication, February 16, 2011.

133. Shmuel Arbel, Moshe Getner, and Adit Arazi, *Flood and Rain Data in Extreme Events: A Summary of the 2002–2003 Year* (Emek Hefer, Israel: Soil Erosion Research Station, 2004), 21 (in Hebrew).

134. This refers to floods for which there is only a 2 percent probability of such large total quantities and of flow intensity of water taking place.

135. Arbel, Getner, and Arazi, *Flood and Rain Data,* 21.

136. Shmuel Arbel et al., *Flood and Rain Data for Extreme Events: Summary of the Hydrological Year 2006–2007* (Emek Hefer, Israel: Soil Erosion Research Station. 2008), 22–23.

137. Moshe Getner and Shmuel Arbel, *The Influence of Forests and Soil Conservation Activities and Rangeland Management in the South of Israel on the Aspects of Flow and Soil Erosion,* report no. 90-9-171-03 (Emek Hefer, Israel: Soil Erosion Research Station, 2006).

138. Moshe Getker and Alon Maor, *Final Report of Research Monitoring the Influence of Planted Trees on the Banks of Streams and the Flow Stability in Watersheds,* research no. 90-9-180-06 (Emek Hefer, Israel: Soil Erosion Research Station, 2008).

139. Lea Wittenberg and Moshe Inbar, "The Role of Fire Disturbance on Runoff and Erosion Processes—a Long-Term Approach, Mt. Carmel Case Study, Israel," *Geographical Research* 47, no. 1 (2009): 51.

140. Hemu Kharel Kafle and Hendrik J. Bruin, "Climatic Trends in Israel, 1970–2002: Warmer and Increasing Aridity Inland," *Climatic Change* 96, nos. 1–2 (2009): 63–77.

141. Arbel, communication.

142. "It is estimated that the total annual cost of erosion from agriculture in the USA is about 44 billion U.S. dollars per year, i.e. about 247 U.S. dollars per ha of cropland and pasture. On a global scale the annual loss of 75 billion tons of soil costs the world about 400 billion U.S. dollars per year, or approximately 70 U.S. dollars per person per year." Hari Eswaran, Rattan Lal, and Paul Reich, "Land Degradation: An Overview," in *Responses to Land Degradation, Proceedings Second International Conference on Land Degradation and Desertification, Khon Kaen, Thailand,* ed. E. M. Bridges et al. (New Delhi: Oxford University Press, 2001).

143. For example a study in Mexico estimates the quantity of soil erosion and where the sediment would reach, but considers the economic benefit still unquantified. W. Neil Adger et al., "Total Economic Value of Forests in Mexico," *Ambio* 24, no. 5 (1995): 291–92.

144. For a description of the modeling for economic valuation of sediment retention see: Marc Conte et al., "Retention of Nutrients and Sediment by Vegetation," in *Natural Capital,* 96–97, 102–3.

145. Gretchen Daily, personal communication, February 13, 2012.

146. Gabriel Schiller, "Reducing Noise through Afforestation," in "Silviculture in Israel: A Country at the Edge of Deserts," book manuscript in preparation.

147. Ibid.

148. Jean-Jacques Rousseau, *Emilius and Sophia; or, A New System of Education,* vol. 2 (London: n.p., 1762), 59–72, at Internet Archive, http://archive.org/details /emiliusandsophio1rousgoog (last accessed February 4, 2013).

149. Gregory Bratman, Paul Hamilton, and Gretchen C. Daily, "The Impacts of Nature Experience on Human Cognitive Function and Mental Health," *Annals of the New York Academy of Sciences* 40 (2012): 1–19.

150. R. S. Ulrich, "Natural versus Urban Scenes," *Environmental Behavior* 13 (1981): 523–56.

151. R. H. Schein, "Belonging through Landscape," *Environment and Planning* 41 (2009): 811–26, also references to Hawaiian indigenous perceptions of nature in Rachelle Gould and Nicole Ardoin, "Malama the 'Aina,' Malama the People on the 'Aina': The Reaction to Avatar in Hawai'i," *Journal for the Study of Religion, Nature, and Culture* (2010): 434, 435.

152. Millennium Ecosystem Assessment, *Ecosystems and Human Well-Being,* 16.

153. Nir Becker and Shirra Freeman, "The Economic Value of Old-Growth Trees in Israel," *Forest Policy and Economics* 11, no. 1 (2009): 608–15; Also, Tzipi Eshet, Dafna DiSegni Eshel, and Mordechai Shechter, "The Value of Israel's Forests," CIOSTA XXXIII CIGR Conference and Workshop, International Union of Forest Research Organizations (IUFRO), Mediterranea, University of Reggio Calabria, Italy, June 2009.

154. Nir Becker and Yael Choresh, "An Economic Evaluation of Recreation in Nature: The Economic Value of the Biriya Forest Using a Travel Cost Method," *Forest* 9 (2007): 33–39 (in Hebrew).

155. Hagai Einav, "More Than Half a Million People Participated in Tree Planting in Honor of Tu Bishvat," *Ynet,* January 29, 2010 (In Hebrew).

156. "There are four new years: On the first of Nisan is a new year for kings and for festivals. On the first of Elul is the new year for the tithe of cattle. R. Eleazar and R. Simeon, however, place this on the first of Tishrei. On the first of Tishrei is new year for release and jubilee years, for plantation and for [tithe of] vegetables. On the first of Shvat is the new year for trees. According to the ruling of Beit Shammai; Beit Hillel, however, place it on the fifteenth of that month." Mishnah, "Rosh Hashana" 2a.

157. After recording a squabble over the precise date between two competing rabbinic schools the fifteenth day of Shevat, was selected and ultimately became the tradition.

158. Sharon Udasin, "JNF-JNF Launches Social Forest for Tu Bishvat," *Jerusalem Post,* February 6, 2012.

159. Joel Cohen, "Human Population Grows Up," *Scientific American*, September 2005, 48–55.

160. Yehiel Leket, as quoted in Irus Braverman, *Planted Flags: Trees, Land, and Law in Israel/Palestine* (Cambridge: Cambridge University Press, 2009), 71.

161. Jeremy Benstein, *The Way into Judaism and the Environment* (Woodstock, VT: Jewish Lights, 2006), 179–80.

162. Zedan, communication.

163. Ibid.

164. Shimon Gapso, presentation at a JNF–U.S. Forest Service workshop, Kibbutz Ramat Rachel, June 16, 2011.

165. Idit Sappir-Gildor, Anais Worschewoitz, and Alon Tal, *Public Attitudes towards*

JNF Forests: An Empirical Study; Report Submitted to the Jewish National Fund (Jerusalem: JNF, 2002) (in Hebrew).

166. Becker and Choresh, "Economic Evaluation," 33–39.

167. Ibid.

168. The fact that JNF foresters are not empowered to issue fines under Israel's litter law (the Protection of Cleanliness Law) constitutes a major legal barrier to interventions that might improve the common sight of garbage scattered across the forest, particularly after the Saturday picnic rush.

169. Schiller, "Recreation and Relaxation in the Forest," in "Silviculture in Israel."

170. Jennifer Spagnolo and Richard de Dear, "A Field Study of Thermal Comfort in Outdoor and Semi-Outdoor Environments in Subtropical Sydney Australia," *Building and Environment* 38, no. 5 (2003): 721–38.

171. Measuring human perspiration rates as well as general climatic conditions, they found that properly designed and sufficiently shaded forests can offer consider relief to visitors during the summer months. Under optimal conditions—with 80 percent cover to allow some ventilation—groves will lower effective temperatures by an average of six degrees Celsius for visitor during most of the day time conditions relative to open spaces. (During the colder winter months, open spaces are often more comfortable.) René Karschon and Gabriel Schiller, "Forest Trees as Ameliorators of Heat Stress of Man in a Mediterranean Environment," in *Trees and Forests for Human Settlements*, ed. J. W. Andersen (Toronto: Centre for Urban Studies, 1976), 382–87.

172. Limor Shashua-Bar, David Pearlmutter, and Evyatar Erell, "The Influence of Trees and Grass on Outdoor Thermal Comfort in a Hot-Arid Environment," *International Journal of Climatology* 31 (2011): 1498–1506.

173. It seems that pine stands provide cooler microclimates than broadleaf trees do because they can provide shade without excessively blocking the breeze. Gabriel Schiller, "The Relation of Microclimate to Thermal Stress of Man in Two Forest Types on Mt. Carmel," *La-Yaaran* 24 (1974): 1–6 (in Hebrew). Recreational areas facing north are more comfortable, with southern facing ones providing superior lighting in the spring. René Karschon, Gabriel Schiller, and A. Weinstein, "Some Effects of Slope Aspect in a Mediterranean Environment," in *Wissenschaftilche Mitteilungen* 35 (University of Munich, Meteorologisches Institut, 1979): 121–25.

174. Avi Dickstein, personal communication, February 15, 2012.

175. Richard Louv, *Last Child in the Woods: Saving Our Children from Nature-Deficit Disorder* (Chapel Hill, NC: Algonquin, 2008).

176. Or Kashti, "Most of the Members of Youth Movements Are Wealthy and Urban," *Haaretz,* October 25, 2009.

177. David Ashkenazi, Education Department, Jewish National Fund, personal communication, February 24, 2012.

178. Zoe Tal, personal communication, Palo Alto California, February 20, 2012.

179. Bratman, Hamilton, and Daily, "Impacts of Nature Experience," 1–19.

180. B. Park et al., "Physiological Effects of Shinrin-yoku (Taking in the Atmosphere of the Forest): Using Salivary Cortisol and Cerebral Activity as Indicators," *Journal of Physiology and Anthropology* 26 (207): 123–28.

181. R. Kaplan and S. Kaplan, *The Experience of Nature: A Psychological Perspective* (Cambridge: Cambridge University Press, 1989).

182. Riov, communication.

Epilogue

1. 1 Samuel 38:8.

2. 2 Samuel 1:21.

3. Joseph Weitz, *Forests and Afforestation in Israel* (Jerusalem: Masada Press, 1974), 181.

4. Azariah Alon, interview with author, February 20, 2012.

5. Azariah Alon, interview with author, September 15, 1997.

6. See details of site at www.skigilboa.co.il (last accessed February 4, 2013).

7. Alon, interview.

8. Avi Perevolotsky, *Ecology, Theory, and Reality in Israel* (Jerusalem: Carta, 2001).

9. Avi Perevolotsky, interview by author, Latrun, Israel, August 3, 2012.

10. Avi Perevolotsky, "Forestry in Israel: The Ecological Alternative," *Forest* 12 (2011): 66–73.

11. Ibid.

12. Ibid.

13. Zvika Mendel, "Alternative Ecological Forestry? With Such an Alternative, There Will Be No Forest," *Forest* 12 (2011): 73–77.

14. Ibid., 75.

15. Yosef Weitz quotes an article from the 1960s printed in the JNF technical journal of the time "LaYaaran," that lists benefits such as water augmentation erosion control, climate amelioration, tourism, landscape aesthetics concluding: "The value of the forest must therefore not be gauged simply the income from its timber." Weitz, *Forests and Afforestation,* 312.

16. Gidi Ne'eman, interview with author, Moshav Amirim, August 10, 2011.

17. Mendel, "Alternative Ecological Forestry?" 77.

18. Perevolotsky, interview.

19. Ayala Lavi, Hai Kigel, and Imanuel Noy-Meir, "The Spread of *Pinus halepensis* from Planted Forests to Natural Vegetation," *Forest* 5–6 (2004): 12–18.

20. Yoni Weitz, "The Dynamics of Dispersion of Pines from Planted Forests to Open Spaces and Implications on the Vegetative Landscape in Israel," JNF Research Symposium, Beit Dagan, Israel, May 9, 2012.

21. Perevolotsky, interview.

22. Perevolotsky, "Forestry in Israel," 71.

23. Yossi Riov, personal communication, December 27, 2011.

24. Ibid.

25. Nir Herr et al., "Factors Affecting Survival and Development of Mature Transplanted Tabor Oaks: *Quercus ithaburensis,*" *Forest* 11 (2009): 32–40.

26. Nir Herr, Yossi Moshe, and Anat Madmony, "Sowing of *Quercus ithaburensis* Acorns as a Method of Forest Establishment in the Ramat Menasheh Biosphere Park," *Forest* 12 (2011): 28–36.

27. Riov, communication.

28. Yagil Osem, Nir Atzmon, and Avi Perevolotsky, "Sustainable Forest Management: What Is It?" *Forest* 7 (2005): 3–11.

29. Jewish National Fund, "The Forestry Policy of JNF," available on the JNF website in Hebrew, www.kkl.org.il/?catid=%7B4D0F7750-804D-4B7D-8842-D8E50A0D26B5%7D (last accessed February 4, 2013).

30. Alon Tal, "Israel's New Bible of Forestry and the Pursuit of Sustainable Dryland Afforestation," *Geography Research Forum* 32 (2013): 81–95.

31. Jewish National Fund, *The Bible of Forestry in Israel: Policies and Directives for Planning and Managing the Forest* (Eshtaol, Israel: JNF, February 13, 2012), 33.

32. Ibid., 54–56.

33. Ibid., 55.

34. Perevolotsky, interview. Perevolotsky is quick to point out that the Nature Reserve Authority suffers from the same sentimentality, only with a weakness for oak trees.

35. Genia Denisyok, personal communication, February 3, 2013.

36. Pinhas Alpert et al., "Climatic Trends to Extremes Employing Regional Modeling and Statistical Interpretation over the E. Mediterranean," *Global and Planetary Change* 63 no. 2–3 (September 2008): 163–70.

37. Yuval Azulay, "Environment Minister: Desertification Intensifying, Rainfall Decreasing," *Globes,* January 10, 2012.

38. JNF, *Forestry Bible* (Eshtaol: JNF, 2012), 40. Fruit trees and more sensitive species such as cedars require more prolonged support.

39. Ira Mor-Hayitin et al., "What Is the Optimal Time for Inoculating Seedlings with Mycorrhiza? Experiments with *Laurus nobilis* and *Eucalyptus camaldulensis,*" *Forest* 12 (2011): 3–12.

40. Zohar Litmanovich et al., "Influence of Autumn Planting on Forest Tree-Seedlings Survival and Growth," *Forest* 9 (2008): 35–37.

41. David Brand, "Effect of Root Zone Temperature on the Development and Physiology of Forest Tree Seedlings and the Implications on the Planting Date in Israel" (Ph.D. diss., Hebrew University of Jerusalem, August 2011).

42. JNF, *Forestry Bible,* 39.

43. "In arid regions wide spacing is recommended for all species: Forest spacing of 1.5 by 1.5 meters which is considered favorable in some temperate regions is much too close for arid regions. Here, 2.5 by 2.5 meters or even 3.5 by 3.5 meters and even wider spacing is normal practice." Amihud Goor, *Tree Planting Practices for Arid Areas* (Rome: FAO, 1955), 70.

44. Gabriel Schiller, "What Is the Desirable Density for Pine Forests in Yatir in Light of Climate Change?" *Research Abstracts, JNF 2011 Research Symposium* (May 9, 2012, Beit Dagan, Israel), 7 (in Hebrew).

45. Yagil Osem et al., "Thinning Treatments in a Mature Jerusalem Forest and Their Influence on the Vitality of the Trees: Natural Regeneration and Biodiversity," *Research Abstracts, JNF 2011 Research Symposium* (May 9, 2012, Beit Dagan, Israel), 6 (in Hebrew).

46. Zion Madar, "Decline of Cedar Trees in Israel," *Forest* 10 (2008): 38–41.

47. Amir Perlberg and Amit Dolev, *Large Mammal Survey in the Moddi'in Area* (Tel Aviv: Society for the Protection of Nature in Israel, 2007).

48. Perevolotsky, "Forestry in Israel,"70.

49. Gilead Friedman, Ido Yizhaki, and Yossi Leshem, "Changes in the Biology of Nesting for Two Avian Raptors as a Result of Human-Induced Ecological Pressures," *Research Abstracts, JNF 2011 Research Symposium* (May 9, 2012, Beit Dagan, Israel), 12 (in Hebrew).

50. Interview with David Brand, Tel Aviv, July 27, 2012.

51. Ne'eman, interview.

52. Many of the most popular nature reserves now require entrance fees. To get to the beach, parking fees are invariably levied. Public swimming pools cost money. Indeed, the city of Raanana wanted to charge visitors in its central municipal park until a legal action by the Israel Union for Environmental Defense led to Supreme Court intervention prohibiting it.

53. Psalm 96:12.

Index

Page numbers in italics refer to photographs. JNF in index refers to Jewish National Fund.